T0229597

APOPTOSIS in NEUROBIOLOGY

Edited by
Yusuf A. Hannun
Rose-Mary Boustany

CRC Press
Boca Raton   London   New York   Washington, D.C.

Library of Congress Cataloging-in-Publication Data

Catalog record is available from the Library of Congress

Dedication

To Raymond D. Adams, MD, a mentor, friend, and guiding light for scores of neurologists, neuroscientists, and many others who are destined to carry the fields of neurobiology and applied neuroscience into the next millenium.

Rose-Mary Boustany

To my father, Awni Hannun, for his unwavering confidence and support.

Yusuf A. Hannun

Preface

In the last few years, the scientific community has synchronously and overwhelmingly come to the realization that the study of cell death is a highly rewarding and important endeavor. Relegated to the sidelines of modern cell biology research for most of the last century, cell death, nonetheless, has received some attention from investigators who noted several forms of morphologic cell death and speculated on the relevance of this process. Indeed, major breakthroughs in cell biology came from the investigation of neurotrophic factors that prevented the otherwise default cell death of neurons. Biologists had also noted the significance of programmed or predetermined cell death in developmental biology, and botanists had labeled the periodic death of leaves as senescence.

Understandably, general interest in cell death was lacking, due to the preconception that cell death is a default process that shows little if any regulation, and therefore, does not lend itself to investigation or interest. Major events and observations in cell biology occurred in the last three decades that slowly began to change this perception and ultimately created the current avalanche of interest in this field of study. First, different forms of cell death were clearly distinguished and defined: necrosis was applied to the usual forms of direct cell death due to (usually harsh) physical conditions, and apoptosis was applied to a more slowly developing process that could be distinguished morphologically from necrosis. This alerted keen observers to perceive that not all forms of cell death are identical, and therefore, by implication, there must be distinct mechanisms that operate during cell death. Second, it became appreciated that apoptosis is accompanied by activation of specific endonucleases that cleave DNA at internucleosomal junctions, whereas necrotic cells showed diffuse and generalized (nonspecific) breakdown of DNA. This singular observation heralded the biochemical approach to apoptosis since it demonstrated, and very clearly, that apoptotic stimuli generate signals that result in specific biochemical effects. This approach eventually led to the discovery of the role of proteases (the caspases) in apoptosis, and to the unraveling of mechanisms in receptro-mediated cell death. Third, evaluation of molecular mechanisms of oncongenesis disclosed that one prominent "anti-oncongene," p53, functioned primarily as a mediator of growth arrest and apoptosis whereas the oncongenic Bcl-2 functioned primarily as an inhibitor of apoptosis. Finally, genetic studies in C. elegans identified several genes specifically involved in apoptosis. Elucidation of the structure of those genes, as homologues of Bcl-2 and caspases, allowed for the convergence of these different approaches in the study of apoptosis. This convergence has catapulted the study of apoptosis to its current heights, and it promises rapid unfolding of many of the remaining mysteries on the significance of apoptosis and its mechanisms.

The field of neurobiology is particularly rich in potential understanding and application of apoptosis study. It appears that disorders of neurodevelopment as well as neurodegenerative disorders are a direct result of activation of apoptotic programs (either due to primary defects in these programs or, more commonly, as a consequence of insults and injuries that activate these programs). Therefore, the study of apoptosis in neurobiology promises significant rewards in understanding such diverse disorders as Alzheimer's disease, Parkinson's disease, and the many neurodegenerative diseases of the central and peripheral nervous systems.

This volume was compiled with the singular purpose of allowing the uninitiated neuroscientist intellectual and practical access to the study of apoptosis, with special consideration to the nervous system. The book is divided into two major sections. The first concentrates on conceptual approaches to the study of apoptosis in neurobiology and its significance in the nervous system. The second part provides for a user-friendly approach to methods and techniques in the study of apoptosis and, where appropriate, as specifically applied to neurobiology.

We would like to take this opportunity to thank our contributors for outstanding and timely contributions. We would also like to thank our many colleagues and students who make these efforts worthwhile.

Yusuf A Hannun and Rose-Mary Boustany

Contents

Section A   Diseases and Concepts

Section B   Methods

The Editors

Yusuf Hannun, M.D., is currently the Ralph Hirschmann Professor of Biomedical Research and Chair, Department of Biochemistry and Molecular Biology and Professor of Medicine at the Medical University of South Carolina in Charleston. Dr. Hannun obtained his M.D. degree from the American University of Beirut in 1981 where he recieved training in internal medicine. He then trained in hematology and medical oncology at Duke University. He also trained in biochemistry with Dr. Robert Bell. He then joined the faculty of Duke University where he spent 15 years, becoming the R. Wayne Rundles Professor of Medical Oncology. His research interests are in the areas of lipid-mediated cell regulation, the chemistry and biochemistry of sphingolipids, mechanisms of cell death, and cancer biology. He has authored or co-authored more than 170 manuscripts, edited 4 books and holds three patents on the use of sphingolipid-derived molecules in the treatment of human diseases. He has been named a Pew Scholar in the biomedical sciences and he is the 13th Mallinckrodt Scholar in biomedical research.

Rose-Mary Boustany, M.D., is tenured Associate Professor in Pediatrics and Neurobiology at Duke University Medical Center. She obtained her M.D. in 1979 at the American University of Beirut where she completed her training in pediatrics. Dr. Boustany spent almost nine years (1980 to 1988) in Boston at Massachusetts General Hospital and the Shriver Center for Mental Retardation. There she trained in pediatric neurology and neurogenetics and later joined the neurology faculty at Massachusetts General Hospital and was associate director of the Lysosomal Storage Diseases Laboratory at the Shriver Center. She moved to Duke University at the end of 1988 where she joined the division of pediatric neurology. She also spent two years in the laboratory of Kuni Suzuki at the University of North Carolina at Chapel Hill. Her fields of interest include neurogentics, the cell and molecular biology of inherited neurodegenerative diseases, and basic mechanisms of neuronal apoptosis.

Contributors

Andrea Amalfitano, Duke University Medical Center, Durham, North Carolina

Tim R. Bilderback, Department of Pharmacology and Toxicology, University of Kansas, Lawrence, Kansas

Rose–Mary Boustany, Department of Pediatrics, Duke Medical Center, Durham, North Carolina

Joseph M. Corless, Duke University, Durham, North Carolina

Rick T. Dobrowsky, Department of Pharmacology and Toxicology, University of Kansas, Lawrence, Kansas

William C. Earnshaw, Institute of Cell and Molecular Biology, University of Edinburgh, Edinburgh, U.K.

Harris A. Gelbard, University of Rochester Medical Center, Rochester, New York

Yussef A. Hannun, Medical University of South Carolina, Charleston, South Carolina

Kam M. Hoffman, Department of Pharmacology and Toxicology, University of Kansas, Lawrence, Kansas

Scott H. Kaufmann, Division of Oncology Research and Department of Pharmacology, Mayo Medical School, Rochester, Minnesota

Timothy J. Kottke, Division of Oncology Research, Mayo Medical School, Rochester, Minnesota

L. Miquel Martins, Institute of Cell and Molecular Biology, University of Edinburgh, Edinburgh, U.K.

Peter W. Mesner, Jr., Division of Oncology Research, Mayo Medical School, Rochester, Minnesota

Vinodh Narayanan, Department of Pediatrics, Neurology, and Neurobiology, The Children's Hospital of Pittsburgh, Pittsburgh, Pennsylvania

Kasturi L. Puranam, Department of Pediatrics, Duke University Medical Center, Durham, North Carolina

Donald E. Schmechel, Division of Neurology, Departments of Medicine and Neurobiology, Duke University Medical Center, Durham, North Carolina

Nina Felice Schor, Departments of Pediatrics, Neurology, and Pharmacology, University of Pittsburgh, and Division of Child Neurology, The Children's Hospital of Pittsburgh, Pittsburgh, Pennsylvania

Sidney A. Simon, Department of Neurobiology, Duke University Medical Center, Durham, North Carolina

Miriam J. Smyth, Department of Medicine, Duke University Medical Center, and Geriatric Educational and Clinical Center, Department of Veterans Affairs Medical Center, Durham, North Carolina

Section A

Diseases and Concepts

1

Introduction: Occurrence, Mechanisms, and Role of Apoptosis in Neurobiology and in Neurologic Disorders

Rose-Mary Boustany and Yusuf A. Hannun

CONTENTS

Apoptosis in the developing nervous system results in naturally occurring cell death (NOCD), a necessary and desirable process. NOCD effectively eliminates neurons that have made faulty synapses or have not reached appropriate targets.[1] In the rest of the organism, apoptosis is essential for organogenesis, sculpts digits and extremities, and plays a role in determining polarity of structures by contributing to directional growth of cell populations.[2]

Failure of carefully orchestrated and effective apoptosis in the developing fetus can have serious and long-lasting effects in the adult. Congenital brain malformations such as heterotopias, schizencephaly, myelomeningocoele, and many others probably represent poorly designed and/or incomplete apoptosis.

An accelerated rate of apoptosis is purposefully induced when cancers are treated with radiation and various chemotherapeutic agents. In fact, cancers are frequently thought of as failure of enactment of apoptosis. Mutations in *p53*, that normally is a suppressor of growth, occur in a large number of human tumors.[3] In addition, there are numerous endogenous factors that protect normal and tumor cells from apoptotic death. Nerve growth factor (NGF) bound to its low affinity P75 or high affinity Trk A receptors is an example.[4] NGF binding to the p75 receptor on neuroblastoma tumor cells explains their resistance to chemotherapy induced apoptosis. Chapter 4 on neurooncology delves into this issue in greater detail.

If cancer is a state of transformation, unbridled cell proliferation, or failure of enactment of apoptosis, neurodegenerative diseases on the other hand represent accelerated apoptosis in the face of fully differentiated nondividing neurons. In fact, the repertoire of most neurons in the adult nervous system is limited to healthy quiescence, senescence, or death. Neurodegenerative disease is the phenotypic expression of undesirable and inappropriate neuronal death occurring in the adult brain. These diseases can be due to autosomal recessive defects in genes involved in the apoptotic pathway. Examples include defects in the antiapoptotic *CLN3* gene in the juvenile form of Batten disease or defects in the survival motor neuron (SMN) or neuronal apoptosis inhibitory protein (NAIP) defective in spinal muscular atrophy.[5-7] Alternatively, neurodegenerative disorders can result from defects in dominant genes, as seen in the expanded triplet repeat diseases. These represent a deleterious gain of function model where the expanded CAG/polyglutamine tract in the mutant protein results in novel toxic protein–protein interaction in part responsible for the death of neurons. Some of these diseases are Huntington disease (*huntingtin*), spinocerebellar ataxia type-1 (*ataxin-1*), Machado-Joseph disease (*ataxin-3*) and dentatorubro-pallipallidoluysian atrophy or DRPLA (*atrophin-1*). There are other neurodegenerative diseases where apoptosis has been implicated as the mechanism of neuronal death. These include a subset of Alzheimer cases, amyotrophic lateral sclerosis, Parkinson's disease, and various forms of retinitis pigmentosa resulting from mutations in rhodopsin or other retinal proteins. A more complete discussion of these disorders is addressed in Chapter 3 on neurodegenerative diseases.[8]

Acquired diseases representing neuronal apoptosis triggered by an infectious agent include HIV-1 encephalitis and prionic encephalopathies. It is thought that the HIV-1 infection initiates an apoptosis-signaling cascade in the central nervous system. The reader is referred to Chapter 5 on HIV-1.[9]

1.1 Molecular Mechanisms of Apoptosis

We are just beginning to unravel the complexities and intricacies of the regulation of apoptosis. Insight has developed rapidly in the last decade from (1) studies on cytokine- and chemotherapeutic agent-induced cell death, (2) genetic regulation of cell death in the nematode C. *elegans*,[11] and (3) studies on proapoptotic tumor suppressor genes such as *p53* and antiapoptotic oncongenes, most notably *bcl-2*.[12]

Control of apoptosis is possible at many levels. This regulation can be expressed as a positive or negative modulating effect (Table 1.1): transcriptional regulation, induction of early intermediate genes; stage of the cell cycle

TABLE 1.1

Positive and Negative Modulators of Apoptosis

Negative	Positive
Bcl-2	Bax
Bcl-x_L	Bcl-x_s
Bag	Bcl-x_β
Baculovirus p35	Bag, Bak, Mcl-1, Bok
Cowpox virus serpin crm A	TNF superfamily (Fas, TNFR-1, Reaper)
NAIP?	Chemotherapeutic agents
SMN?	Radiation
CLN3	ceramide
NGF	p53
IL-6, IL-3, erythropoetin	*c-myc*

and relative levels of cyclins; presence or absence of nerve growth factor and its receptors; TNF-α and related receptors[13]; Fas–Fas-L interactions,[14] ceramide as proapoptotic lipid second messenger and the sphingomyelin cycle[15]; the neuroprotective *bcl-2* oncogene and its homologues,[16] *p53* and retinoblastoma genes as inducers of growth arrest and apoptosis[17]; the early initiator and later executionary caspase cascades and their triggers and inhibitors[18]; the role of the mitochondrion as central processor of incoming messages, and the role of translocation of inner mitochondrial membrane proteins such as cytochrome c, Apaf-1, and other factors to be found.[19] A hypothetical and simplified choreography depicting possible interactions, as best illustrated with apoptosis-inducing cytokines, is outlined in the scheme shown. According to this model, the action of proapoptotic cytokines, such as TNF, Fas-L, or NGF, on their membrane receptors (P75 receptor in the case of NGF) results in recruitment/activation of a number of adapter proteins such as FADD, TRAFs, and TRADs. These proteins, though poorly understood mechanisms, couple the occupied receptors to distinct pathways of signaling and cell regulation. Whereas Fas appears to be a more dedicated proapoptotic receptor, the TNF receptors couple to apoptotic, antiapoptotic, and inflammatory pathways. Thus, TNF can activate the following: (1) NF-kB, which predominantly functions as antiapoptotic transcription factor; (2) the jun kinase (JNK) or stress-activated kinase (SAPK) pathway, which primarily functions in the regulation of stress, at times promoting apoptosis and at other times inhibiting it; and (3) the MACH/Flice protease, a member of the caspase family of proteases, which launches the apoptotic functions of TNF.[20]

It is not yet clear how MACH/Flice turns on the apoptotic program. In the case of Fas, it has been proposed that a cascade of proteases is turned on, and that it is necessary and sufficient to cause apoptosis. This proposed mechanism now appears as an over-simplified explanation, especially in the case of TNF, where many endogenous pathways are activated and regulated in response to TNF and Fas and contribute to the terminal apoptotic outcome. These pathways include the formation of reactive oxygen intermediates and changes in mitochondrial permeability and function.[21] Also implicated are ceramide- and sphingolipid-derived molecules as stress-induced mediators that promote and enhance the apoptotic program.

Noncytokine stresses, such as heat, oxidative damage, and DNA-damaging agents also activate apoptosis by generating poorly understood internal signals. It is not yet determined whether these processes overlap cytokine-induced apoptosis, but in the case of DNA-damaging agents the proapoptotic protein P53 plays an important role in driving the response of the cells either through induction of cell cycle arrest or the induction of apoptosis.[22]

Significant results now implicate cytochrome c as a key mediator of the apoptotic pathways (Figure 1.1). Many, but not all, inducers of apoptosis cause the release of cytochrome c from the mitochondria. Also, it is now assumed that the mitochondrial membrane is the site of action of members of the Bcl-2 family of pro- and antiapoptotic proteins.[22] It is suggested that *bcl-2*, the mammalian homologue of the *ced-9* gene from *C. elegans*, functions primarily by inhibiting the release of cytochrome c, whereas proapoptotic relatives of *bcl-2* may promote this event. The released cytochrome c interacts with Apaf-1, a positive regulator of apoptosis with homology to the *C. elegans ced-4* proapoptotic gene. This collaboration results in activation of downstream caspases such as caspase 3, which are homologues of the *C. elegans ced-3* gene. It is the action of these caspases on their substrates that results in the systematic degradation of key substrates such as nuclear lamins, PARP, fodrin, protein kinases, and other structural or regulatory proteins. This process culminates in the organized collapse of the nucleus, membranes, and cellular organelles. Many neuronal proteins are now recognized as substrates of caspases, including presenilins and huntingtin.[23,24] The orderly breakdown of dying cells through the apoptotic mechanisms results in the packaging of cellular debris into apoptotic bodies which are then cleared by reticuloendothelial cells as well as normal adjacent cells, thus preventing inflammatory reactions to cell fragments.

The study of existing apoptotic developmental and neurodegenerative diseases, be they caused by a genetic defect or a triggering environmental factor, provide us with naturally occurring human models that validate existing hypotheses in neuronal culture systems and provide new information pertinent to basic cell biology.

1.2 Mechanisms of Apoptosis in Neurological Disorders

One theory invoked to explain Alzheimer cases that are apoptosis positive is that the accumulation of amyloidogenic protein results in excess intracellular calcium, a known trigger for the endonuclease responsible for the DNA fragmentation seen during the final stages of apoptosis.[25] Oxidative stress due to defects in energy and/or mitochondrial metabolism contributes to apoptosis in anterior horn cells in amyotrophic lateral sclerosis, in the substantia

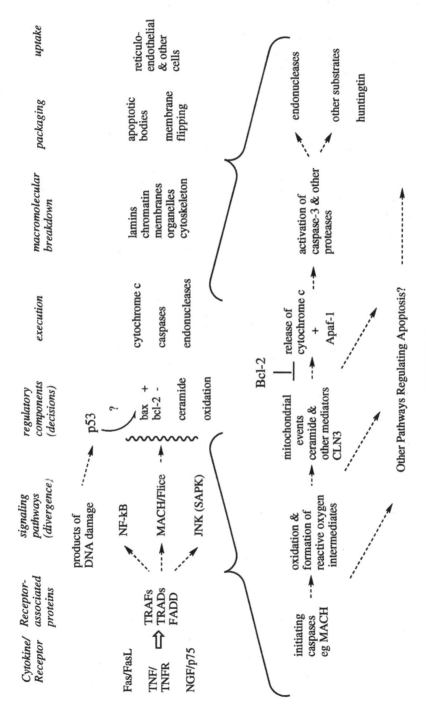

FIGURE 1.1

nigra in Parkinson's disease, and in the penumbral region of infarcts seen in cerebrovascular disease.[26-29] Excitotoxicity induced via activation of NMDA receptors and also that induced by application of kainic acid to the hippocampi has linked apoptosis and hippocampal sclerosis to the occurrence of epilepsy.[30] Defects in antiapoptotic genes such as *CLN3* and *NAIP* and *SMN* in the juvenile form of Batten disease and spinal muscular atrophy type 1, respectively, coexist with massive neuronal loss and have inexorably coupled these inherited neurodegenerative diseases most intimately with the occurrence of apoptosis.[31-33]

The fact that some sites in the body are immunologically privileged was recognized over 100 years ago. The brain, like the eye and the testes and placenta, is an immune-privileged site. Obviously, it is important to protect the central nervous system and the eye from the ravages of invasive immunopathologic injury.[34] There are multiple hypothetical mechanisms surrounding the concept of immune privilege. One such theory is based on the Fas–Fas ligand interaction. The expression of Fas is high in the cornea, in photoreceptors, and in neurons, whereas expression of Fas-L is high in endothelium. The strong Fas–Fas ligand interaction provides a tight "apoptotic" vise curtailing the entry of activated macrophages, lymphokines, and other growth-supporting factors into the sanctuary of the eye or brain. The existence of conditions such as a defect in an antiapoptotic gene, oxidative stress, or the presence of a toxic element that engages the apoptotic process, tips the balance in the direction of neuronal or photoreceptor death in the brain and eye, but remains phenotypically silent in other tissues.

1.3   Methodologies in Apoptosis

A long list of techniques exists that facilitates the study of apoptosis in neurobiology and other disciplines. The second part of this book covers many techniques that the authors have found useful. These include TUNEL staining and other staining techniques that capitalize on the morphologic changes of the nucleus and biochemical changes occurring in the cell and nucleus during apoptosis (see Chapter 8). Also, they include cell viability assays, electron microscopy (Chapter 9), flow cytometry (Chapter 10), measurement of ceramide and sphingolipids (Chapter 12), and the use of viral vectors to introduce genes of interest into cells (Chapter 13). The production of proapoptotic or antiapoptotic gene knockout and/or specific gene overexpressing mice, as well as the creation of mice with tissue-specific gene expression have aided in the elucidation of apoptotic pathways and mechanisms as we know them.[35] Once the role of a gene or the protein it codes for has been ensconced as significant, homologous genes and/or interacting proteins can be fished out using traditional library screening or yeast two- and three-hybrid systems.[36]

Ultimately, the choreography of neuronal apoptotic pathways will become more complex, detailed, and specific. A better understanding of molecular mechanisms in apoptotic pathways will make it possible to design effective drugs targeting defined subsets of neurons at precise points in development or adult life.

References

1. Hamburger V and Levi-Montalcini R. Proliferation differentiation and degeneration of the spinal ganglia of the chick embryo under normal and experimental conditions. *J. Exp. Zool.* 111:457, 1949.
2. Kerr JFR, Wyllie AH, and Currie AR. Apoptosis, a basic biological phenomenon with wide-ranging implications in tissue kinetics. *Br. J. Cancer* 26:239, 1972.
3. El-Deiry WS, Tokino T, Velsulesko VE, Levy DB, Parsons R, Trent JM, Lin D, Mercer WE, Kinzler KW, and Vogelstein B. WAF1, a potential mediator of p53 tumor suppression. *Cell* 75:817, 1993.
4. Ibanez CE, Ebendal T, Barbany G, Murray-Rust J, Blundell TL, and Persson H. Disruption of the low affinity receptor-binding site in NGF allows neuronal survival and differentiation by binding to the trk gene product. *Cell* 69:329, 1992.
5. Lerner TJ, Boustany RM, Anderson JW et al. and the International Batten Disease Consortium. Isolation of a novel gene underlying Batten disease, CLN3. *Cell* 82:949, 1995.
6. Lane SC, Jolly RD, Schmechel DE, Alroy J, and R-M Boustany. Apoptosis as the mechanism of neurodegeneration in Batten disease. *J. Neurochem.* 67:677-683, 1996.
7. Burke JR, Wingfield MS, Lewis KE, et al. The Haw River syndrome: dentatorubropallidoluysian atrophy in an African-American family. *Nat. Genet.* 7:521-524, 1994.
8. Gelbard HA, Boustany R-M, and Schor NF. Apoptosis in development and disease of the central nervous system. *Pediatric Neurol.* 16(2):93-97, 1997.
9. Everall IP, Luthbert PJ, and Lantos PL. Neuronal loss in the frontal cortex in HIV infection. *Lancet* 337:1119, 1991.
10. Gamard CJ, Dhbaibo GS, Lie B, Obeid LM, and Hannun YA. Selective involvement of ceramide in cytokine–induced apoptosis. *J.Biol.Chem.* 272(26):16474, 1997.
11. Ellis RE, Yuan JY, and Horwitz HR. Mechanisms and functions of cell death. *Annu. Rev. Cell. Biol.* 7:663, 1991.
12. Wang XW and Harris CC. p53 tumor-supressor gene: clues to molecular carcinogenesis. *J. Cell. Physiol.* 173(2): 247, 1997.
13. Bayaert R and Fiers W. Molecular mechanisms of TNF-induced cytotoxicity, *FEBS Lett.* 340:9, 1994.
14. Yonehara S., Ischii A and Yonehara M. A cell-killing monoclonal antibody (anti-fas) to a cell surface antigen co-downregulated with the receptor of tumor necrosis factor. *J. Exp. Med.* 169:1747-1756, 1989.
15. Obeid LM, Linardic CM, Karolak LA, and Hannun YA. Programmed cell death induced by ceramide. *Science* 259:1769, 1993.

16. Reed JC. Double identity for proteins of the Bcl-2 family. *Nature* 387:773, 1997.
17. Hochhauser D. Modulation of chemosensitivity through altered expression in cancer. *Anti-Cancer Drugs* 8(10):903, 1997.
18. Nicholson DW, and Thornberry NA. Caspases: killer proteases. *Trends Biochem. Sci.* 22:229, 1997.
19. ZouH, Henzel WJ, Liu X, Lutschg A, and Wang X. Apaf-1, a human protein homologous to C. elegans ced-4, participates in cytochrome c dependent activation of caspase-3. *Cell* 90:405, 1997.
20. Hu S., Vincenz C, Ni J, Gentz R, and Dixit VM:1-FLICE, a novel inhibitor of TNFR-1 and CD-95-induced apoptosis. *J. Biol. Chem.* 272:17255, 1997.
21. Marchetti P, Casteldo M, Susin SA, Zamzami N, Hirsch T, Macho A, Haeffner A, Hirsh F, Geuskins M. and Kroemer G. Mitochondrial permeability transition is a central coordinating event of apoptosis. *J. Exp. Med.* 184:1155, 1996.
22. Yang J, Liu X, Bhalla K, Kim CN, Ibrado AM, Cai J, Peng T-I, Jones DP, and Wang X. Prevention of apoptosis by Bcl-2: Release of cytochrome c from mitochondria blocked. *Science* 275:1129, 1997.
23. Goldberg YP, Nicholson DW, Rasper DM, Kalchman MA, Koide HB, Graham AK, Bromm M, Kazimi-Esfarjani P, Thornberry NA, Vaillancourt JP, and Haydn MA. Cleavage of huntingtin by apopain, a proapoptotic cysteine protease, is modified by the polyglutamine tract. *Nat. Genet.* 13(4):380, 1996.
24. Loetsher H., Deutschle U, Brackhaus M, Reinhardt D, Nelboek P, Mous J, Grunberg J, Haass C, and Jacobson H. Pesenilins are processed by caspase type proteases. *J. Biol. Chem.* 272:20655, 1997.
25. Forloni G, Chiesa R, Smirolda S et al. Apoptosis mediated neurotoxicity induced by application of beta-amyloid fragment 25-35. *Neuroreport* 4:523, 1993.
26. Dipasquale B, Marini AM, and Youle R. Apoptosis induced by 1-methyl-4-phenylpyridinium in neurons. *Biochem. Biophys. Res. Commun.* 1181:1442, 1991.
27. Rabizadeh S, Gralla EB, Borchelt DR, Gwinn R, Valentine JS, Sisdia S, Wong O, Lee M, Hahn H, and Bredeson DE. Mutations associated with ALS convert SOD from an antiapoptotic gene to a proapoptotic gene: studies in yeast and cancer cells. *Proc. Natl. Acad. Sci. U.S.A.* 92:3024, 1995.
28. Mills EM, Gunasekar PG, Pavlakovic G, and Isam GE. Cyanide-induced apoptosis and oxidative stress in differentiated PC-12 cells. *J. Neurochem.* 67:1039, 1996.
29. Okamoto M, Matsumoto M, Ohtsuki T, Taguchi M, Kyanagihara T, Kamada T. Internucleosomal DNA cleavage involved in ischemia-induced neuronal death. *Biochem. Biophys. Res. Commun.* 196:1356, 1993.
30. Pollard H, Cantagrel S, Charriault-Marlangue C, Moreau J, and Yezekiel BA. Apoptosis associated DNA fragmentation in epileptic brain damage. *Neuroreport* 5:1053, 1994.
31. Puranam K, Qian W-H, Nikbakht K, Venable M, Obeid L, Hannun Y, and Boustany R-M. Upregulation of Bcl-2 and elevation of ceramide in Batten Disease. *Neuropediatrics* 28:37, 1997.
32. Puranam K, Qian W-H, Nikbakht K, Guo W-X, and Boustany R. CLN3 defines a novel antiapoptotic pathway operative in neurodegeneration and mediated via ceramide, *Cell Death and Diffeientiation*, in press.
33. Iwahashi H, Eguchi Y, Yasuhara N, Hanafusa T, Matsuzawa Y, and Tsujimoto Y. Synergistic antiapoptotic activity between Bcl-2 and SMN implicated in spinal muscular atrophy. *Nature* 390:413, 1997.

34. Griffith TS, Brunner T, Fletcher SM, Green D, and Ferguson TA. Fas-ligand induced apoptosis as a mechanism of immune privilege. *Science* 270:1189, 1995.
35. Zanjani HS, Vogel MW, Delhaye-Bouchaud N, Martinou JC, and Mariani J. Increased cerebellar Purkinje cell numbers in mice overexpressing a human bcl-2 transgene. *J. Compar. Neurol.* 374(3):332, 1996.
36. Zhou H, and Reed JC. Heterodimerization-independent functions of cell death regulatory proteins Bax and Bcl-2 in yeast and mammalian cells. *J. Biol. Chem.* 272(50): 31482, 1997.

2

Cell Death in the Developing Nervous System

Vinodh Narayanan

CONTENTS

2.1 Introduction

In the context of neural development, programmed cell death refers to the **naturally occurring** cell death seen at various stages of development in almost all neural populations. This term is not entirely synonymous with "apoptosis," which refers to a particular cell death mechanism that is triggered both in developmental cell death and in disease or injury. There are thus many triggers that may initiate the cell death program. Programmed cell death results in the elimination of cells that are not needed, without injury to neighboring cells and without an inflammatory response. This ranges from the removal of extraneous cells that were generated as part of a lineage, abnormal cells, cells that were produced in excess, cells that did not succeed in establishing a proper interaction with other cells, cells that were dependent on a hormone or factor that is not available anymore, or cells that had a role only at a particular developmental stage. The eventual form of the nervous system (morphogenesis) is a result of a balance between the early processes of proliferation and regression, followed by cell growth and maturation.

Apoptotic cell death has been observed in many different cell types, and is relevant to the study of many human diseases. Commonly cited examples of apoptotic cell death that result in the loss of particular tissues include the elimination of the tail from developing vertebrate embryos (frogs and humans) and the elimination of webs between the digits of developing embryos.[1] The dramatic changes that occur during metamorphosis in amphibia and insects are accompanied by apoptotic cell death in tissues that are not required in the adult organism.

Naturally occurring cell death as a phenomenon in neural development has been known for almost a century.[2] However, the systematic and quantitative study of neuronal death began with the work of Viktor Hamburger and Rita Levi-Montalcini. Their work not only quantified cell death in different cell populations, but led to a now generally accepted hypothesis about the role of target tissue and led to the discovery and characterization of neurotrophic factors. We shall review here their original work and that of other neurobiologists, and summarize the current ideas as they pertain to neural histogenesis.

2.2   Target-Independent Cell Death

One form of programmed cell death is an intrinsically programmed genetic cell death that is best exemplified by the developing nematode, *Caenorhabditis elegans*. In this organism, cell identity, cell location, and function are entirely determined by cell lineage. Of the approximately 1090 cells that are generated by cell division during development of the adult hermaphrodite, about 131 undergo programmed cell death.[3] It is known precisely which cells in the developing organism (and at what point in their lineage) are destined to die, this being one of the terminally differentiated states. In *C. elegans*, there are regional differences in the patterns of programmed cell death, and cell death appears to function primarily to generate regional diversity,[1] perhaps by eliminating certain sublineages.[4] Genetic studies have led to the identification of mutations that affect programmed cell death, and to the cloning of genes (*ced-3* and *ced-4*) that are *necessary* for[5-7] and genes (*ced-9*) that *inhibit* cell death.[8,9]

Although the determination of cell fate purely by lineage is not found in more complex organisms, the cellular mechanisms that mediate cell death in *C. elegans* and vertebrates share common features. The *ced-3* gene from *C. elegans* has been cloned and was found to encode a homologue of human and murine interleukin-1β-converting enzyme (ICE).[7] Conversely, expression of the murine ICE gene product in cultured mammalian cells causes apoptotic cell death,[10] indicating a conservation of the molecular mechanisms of programmed cell death through evolution. ICE is a member of a family of proteases which activates its substrate by proteolytic cleavage and initiates a cascade of events leading eventually to apoptotic cell death. The *ced-9* gene product is a structural and functional homologue of the mammalian *bcl-2* protooncogene, which is known to suppress apoptosis in a variety of model systems.[9] Thus, programmed cell death in *C. elegans* provided an example of cell death as a predetermined outcome as a function of lineage, and led to the discovery of the molecular mechanisms mediating apoptosis that are functional in more complex organisms.

A form of target-independent (but not strictly predetermined) cell death has also been observed in vertebrate nervous systems, in the embryonic cerebral cortex, and even earlier in the developing neural tube. There is a fairly widespread and uniform appearance of cells undergoing apoptosis in the embryonic murine cerebral cortex. The peak of this apoptotic cell death occurs around E14–16, and virtually no dying cells are seen at E10 or in the adult.[11] Although many dying cells were observed in regions which contained postmitotic neurons (marginal zone, cortical plate, and intermediate zone), the majority of dying cells were within the proliferative zones.[11] Some large neurons undergoing apoptosis in the border area between the subplate and cortical plate were thought to be subplate neurons.[12] The reason for the observed rate of cell death (average of about 50%) in the proliferative zones of the embryonic cortex is not clear, but it is interesting that the period of maximal cell death (E12–E16) corresponds roughly to the neuronogenetic interval, that time period during which all the terminally postmitotic neurons are generated.[13]

Results similar to the above have also been observed in human fetuses. Apoptotic cells were found in the ventricular zone at the 12th week of gestation, reaching a peak by the 21st week of gestation.[14] In the oldest fetuses studied (23 weeks), apoptotic cells were found primarily in the deep portions of the subplate. As was the case in murine embryos, programmed cell death in the human embryonic cortex was most prominent in the proliferative zones. Whether these represent differentiated cells or undifferentiated cells was not clear. It has been hypothesized that the apoptotic cells represent postmitotic cells that are uncommitted to reach an appropriate position in the cortical plate, and are thus eliminated before migration.[14] However, the precise characteristics that determine whether a cell in the proliferative zone survives or dies remain unknown. Among the postmitotic regions of the developing cortex, most of the observed cell death occurs within the subplate, a transient population of neurons that occupies the layer between the proliferative zone and the cortical plate.[12] Almost all of these cells are eliminated early in postnatal life by apoptosis.[15]

Another example of cell death occurring at a very early stage of neural development, before the stage of neuron-target contact, is that seen in the developing neural tube. In the chick embryo, between the 8- and 12-somite stages, many dead cells are seen, concentrated in the neural folds.[16] In order to determine the function of apoptotic cell death in neural tube closure, chick embryos were cultured at the 8-somite stage, and allowed to develop to the 13-somite stage. In these embryos, the neural tube was completely closed between somites 1 and 8, similar to what is observed *in vivo*. When these embryos were cultured in the presence of specific protease (caspase) inhibitors, programmed cell death and neural tube closure were blocked.[16] These results suggest that programmed cell death is required for neural tube closure, although it is not clear what aspect of neural tube closure depends on apoptosis.

2.3   Target-Dependent Cell Death and the Discovery of Trophic Factors

The studies of Hamburger and Levi-Montalcini on the role of targets in neural development began with the observation that reduction of a peripheral field (limb) resulted in a size reduction of the innervating primary nerve center. Hamburger and Levi-Montalcini studied the mechanisms by which such changes were brought about.[17] They reported on the effect of reduction and augmentation of target size on the development of the spinal ganglia of the chick, examining the rate of proliferation, differentiation, and degeneration. The occurrence of cell degeneration in the spinal ganglia during normal development was noted, and it was observed that there was a distinct topographic pattern of dying cells within the spinal ganglia. It occurred (in normal embryos) most extensively in the cervical and thoracic regions, and was minimal in the brachial and lumbosacral segments. However, limb bud removal caused an extensive cell degeneration within the brachial ganglia (wing bud) or lumbosacral ganglia (hind limb bud). Hamburger proposed that the mechanisms behind cell death in normal and experimental (limb bud extirpation) embryos were the same, and that the enlarged target offered by the developing limb (wing or hind limb) prevented the cell degeneration in the corresponding sensory spinal ganglia. A stated hypothesis was that either synaptic contact with the target or a trophic substance produced in the target area was necessary for cell survival, and that competition for these trophic interactions was what determined whether or not a cell survived or died.

Following these seminal observations, cell death has been noted in many different neuronal populations, and may be a universal developmental phenomenon. Examples include the spinal ganglia and motoneurons, the cranial nerve nuclei, the optic tectum, the retina, and the cerebellum (see Table 1, Reference 18). The phenomenon of cell death is not limited to neurons, but occurs in glial cells as well. About 50% of newly formed oligodendrocytes normally die in the developing rat optic nerve.[19] Using cultured O-2A progenitors and oligodendrocytes, Barres showed that cell death can be prevented by the addition of certain growth factors (PDGFs or IGFs), suggesting that survival of cells that effectively compete for factors that are available in limiting quantity is a feature not limited just to neurons.[19] Overproduction of oligodendrocytes followed by cell death may be the mechanism behind matching the number of oligos required to myelinate the axons within the optic nerve. It may also allow for even spacing of oligodendrocytes along the length of the axon.[19] Peripheral glia, Schwann cells, also undergo programmed cell death during normal development.[20] The rate of Schwann cell death appears to be regulated by axon-derived trophic support. In one study aimed at identifying the phenotype of dying cells in developing (postnatal) murine cortex, up 50% of the pyknotic cells were thought to be glia as they were GFAP positive.[21]

2.4   Cell Death in Sensory and Sympathetic Ganglia

Cell death in the sympathetic nervous system was first noted by Levi-Montalcini while trying to understand the development of regional differences in the spinal motor column.[2] At a time when the preganglionic sympathetic neurons were forming in the thoracic spinal cord of the chick, there was massive cell degeneration in the corresponding region of the cervical spinal cord.[22] In a series of experiments looking at the effect of limb bud extirpation or transplantation, Hamburger and Levi-Montalcini discovered that alteration of target size affected the innervating neuronal population by enhancing or reducing cell degeneration.[17] They also reported normally occurring neuronal death in the spinal ganglia, noting that in the neurons in those regions that had a larger target (the brachial and lumbosacral regions) there was quantitatively less cell death than in those regions (cervical and thoracic) that had a smaller target. In the case of the developing sensory ganglia, the problem of matching the size of the innervating neuronal population with the size of the target was accomplished by initial overproduction of neurons followed by cell death controlled in some manner by the target. The concept of competition for a trophic substance which was produced by the target tissue was introduced at this time. The discovery of the first neurotrophic factor, nerve growth factor (NGF), resulted from studies of the effect of mouse sarcomas implanted in chick embryos. A diffusible agent produced by the sarcomas caused an increase in size of sympathetic and sensory ganglia in the host embryo. This same growth-promoting activity was found in snake venom and in mouse submaxillary glands, leading eventually to the biochemical and molecular characterization of NGF.[23,24] When purified NGF was injected into normal chick embryos or those in which the wing bud was removed, there was a reduction of cell death in the spinal ganglia.[25,26]

2.5   Cell Death in Motoneurons

The analysis of cell death in spinal motoneurons was also started by Hamburger, who demonstrated loss of about 40% of motor neurons in the lateral motor column of the chick embryo during development.[27] He also showed that enlarging the target by transplantation of a supernumerary limb rescued motoneurons that were destined to die.[28] R. Oppenheim and colleagues have continued this investigation, looking at the timing of cell death, the relationship to synapse formation and synaptic activity, and the role of afferent input.[29] Contact between motoneurons and their targets occurred before the period of cell death, even in experimental embryos where the limb bud was removed.[30,31] Thus, it was proposed that motoneurons compete for a target-derived trophic factor, similar to sensory and sympathetic neurons competing for NGF secreted by their targets.[32]

That arrival of motoneuron axons at the target site alone was not sufficient to guarantee survival was demonstrated in studies of the role of synaptic activity on cell death. Neuromuscular blockade during the period of cell death increased motoneuron survival by 50%, suggesting that synaptic activity played a key role in cell death.[33,34] Enhanced neuromuscular activity (by chemical activation of AChR or by electrical stimulation of nerves) increased motoneuron cell death in chick embryo.[35] Blockage of neuromuscular transmission probably resulted in the maintenance of extrajunctional ACh receptors, which could accept additional innervation. In contrast, physiologic activity in the developing synapse (at a certain critical level) may result in only a single synapse being supported, causing death of those neurons that were unsuccessful in establishing a **functional synapse**. Thus, competition at the target would be for secreted trophic factors or for a limited number of functional synaptic sites, and this combination determined the extent of motoneuron survival.

It is still not clear if there is a single target-derived neurotrophic factor for motoneurons. Several neurotrophic factors have now been discovered that promote motoneuron survival *in vitro* and *in vivo*: cholinergic development factor (CDF), ciliary neurotrophic factor (CNTF), brain-derived neurotrophic factor (BDNF), insulin-like growth factors (IGFs), and glial-derived neurotrophic factor (GDNF), to name a few.[36] Some of these are expressed by skeletal muscle while others are not. Whether these act in conjunction with a yet to be discovered muscle derived trophic factor to determine motoneuron survival is not known. The characterization of motoneuron trophic factors have important clinical implications as well. Several of these trophic factors have already been used in experimental protocols for treatment of motoneuron degenerative disorders such as amytrophic lateral sclerosis (ALS).

What has been established already is that target-derived factors are not the only ones that promote motoneuron survival. Afferent input (from sensory ganglia, spinal interneurons, or from other CNS regions) also has a role in the extent of motoneuron cell death. Elimination of afferent input to motoneurons resulted in a decrease in their survival.[37] The increase in cell death as a result of deafferentation is seen in several other neuronal populations. Blockade of afferent synaptic activity also induces the death of some neurons.[38,39] Thus, deafferentation may enhance cell death either because afferents produce a trophic factor, or neurotransmitters released at the synapse may also promote cell survival, or synaptic activity alters the ionic composition of the postsynaptic neuron resulting in enhanced survival.[18]

2.6 Summary

Programmed cell death is a phenomenon that has been observed in almost all parts of the developing nervous system. The reduction of neuronal number by cell death is necessary because neurons are generated in great excess. Early work focused on the peripheral nervous system (sensory and sympathetic ganglia and the motoneurons), leading to the idea that programmed cell death functions to match the size of a given neuronal population with its target. Only those neurons survived that made appropriate and functional connections with their target. This occurred as a result of competition for trophic factors produced (in limiting amounts) by the target and for potential synaptic sites on the target. In addition to the preeminent trophic factor, NGF, numerous other trophic factors have been discovered, and their mechanisms of action characterized. In addition to target-derived and afferent-derived trophic signals, there may be other factors that promote neuronal survival, such as glial-derived signals or circulating hormones.

Other forms of programmed cell death occur well before the neuronal population in question has contacted its target. These include the lineage-dependent cell death seen in *C. elegans*, the cell death seen at the time of neural tube closure, and the cell death in the proliferative zone of the embryonic cortex. The phenotype of the cells that are eliminated by this apoptosis in these two models is not entirely clear, but at least in the developing neural tube, prevention of apoptosis drastically affects neural development.

The programmed cell death which is observed in normal neural development has several possible functions, including: (1) elimination of neurons that fail to make functional synaptic contact with their target (size matching); (2) elimination of cells or separation of cell layers during epithelial sheet fusion and shape change (e.g., neural tube closure); (3) elimination of transient neural populations (e.g., subplate neurons) which serve an important function only during a particular phase of development; and (4) elimination of cells that were necessarily generated in a lineage (e.g.,. *C. elegans*). The cell death program that is seen in normal development is also triggered as a result of injury, and in disease states. The molecular mechanisms behind apoptotic cell death appear to have been conserved through evolution, and are now the subject of intense research. These issues will be discussed in other chapters.

References

1. Purves, D. and Lichtman, J. W., *Principles of Neural Development*, Sinauer Associates, Sunderland, MA, 1985, Chap. 6.
2. Hamburger, V., History of the discovery of neuronal death in embryos, *J. Neurobiol.*, 23, 1116, 1992.
3. Horvitz, H. R., Genetics of cell lineage, in *The Nematode* Caenorhabditis elegans, Wood, W. B., Ed., Cold Spring Harbor, New York, 1988, Chap. 6.
4. Sulston, J., Cell lineage, in *The Nematode* Caenorhabditis elegans, Wood, W. B., Ed., Cold Spring Harbor, New York, 1988, pp. 146-149.
5. Ellis H., M. and Horvitz, H. R., Genetic control of programmed cell death in the nematode *Caenorhabditis elegans, Cell*, 44, 817, 1986.
6. Yuan, J. and Horvitz, H. R., The *Caenorhabditis elegans* genes ced-3 and ced-4 act cell autonomously to cause programmed cell death, *Dev. Biol.*, 138, 33, 1990.
7. Yuan J., Shaham, S., Ledoux, S., Ellis, H. M., and Horvitz, H. R., The *C. elegans* cell death gene ced-3 encodes a protein similar to mammalian interleukin-1β-converting enzyme, *Cell*, 75, 641, 1993.
8. Hengartner, M. O., Ellis, R. E., and Horvitz, H. R., *Caenorhabditis elegans* gene ced-9 protects cells from programmed cell death, *Nature*, 356, 494, 1992.
9. Hengartner, M. O. and Horvitz, H. R., *C. elegans* cell survival gene ced-9 encodes a functional homologue of the mammalian proto-oncogene bcl-2, *Cell*, 76, 665, 1994.
10. Miura, M., Zhu, H., Rotello, R., Hartweig, E. A., and Yuan, J., Induction of apoptosis in fibroblasts by IL-1β-converting enzyme, a mammalian homologue of the *C. elegans* cell death gene ced-3, *Cell*, 75, 653, 1993.
11. Blaschke, A. J., Staley, K., and Chun, J., Widespread programmed cell death in proliferative and postmitotic regions of the fetal cerebral cortex, *Development*, 122, 1165, 1996.
12. Allendoerfer, K. L. and Shatz, C. J., The subplate, a transient neocortical structure: its role in the development of connections between thalamus and cortex, *Annu. Rev. Neurosci.*, 17, 185, 1994.

13. Caviness, Jr., V. S., Takahashi, T., and Nowakowski, R. S., Numbers, time and neocortical neuronogenesis: a general developmental and evolutionary model, *Trends Neurosci.*, 18, 379, 1995.

14. Simonati, A., Rosso, T., and Rizzuto, N., DNA fragmentation in normal development of the human central nervous system: A morphological study during corticogenesis, *Neuropathol. Appl. Neurobiol.*, 23, 203, 1997.

15. Price, D. J., Aslam, S., Tasker, L., and Gillies, K., Fates of the earliest generated cells in the developing murine neocortex, *J. Comp. Neurol.*, 377, 414, 1997.

16. Weil, M., Jacobson, M. D., and Raff, M. C., Is programmed cell death required for neural tube closure?, *Curr. Biol.*, 7, 281, 1997.

17. Hamburger, V. and Levi-Montalcini, R., Proliferation, differentiation, and degeneration in the spinal ganglia of the chick embryo under normal and experimental conditions, *J. Exp. Zool.*, 111, 457, 1949.

18. Burek, M. J. and Oppenheim, R. W., Programmed cell death in the developing nervous system, *Brain Pathol.*, 6, 427, 1996.

19. Barres, B.A., Hart, I.K., Coles, H.S.R., Burne, J.F., Voyvodic, J.T., Richardson, W.D., and Raff, M.C., Cell death in the oligodendrocyte lineage, *J. Neurobiol.*, 23, 1221, 1992.

20. Ciutat, D., Caldero, J., Oppenheim, R. W., and Esquerda, J. E., Schwann cell apoptosis during normal development and after axonal degeneration induced by neurotoxins in the chick embryo, *J. Neurosci.*, 16, 3979, 1996.

21. Soriano, E., del Rio, J.A., and Auladell, C., Characterization of the phenotype and birthdates of pyknotic dead cells in the nervous system by a combination of DNA staining and immunohistochemistry for 5'-bromodeoxyuridine and neural antigens, *J. Histochem. Cytochem.*, 41, 819, 1993.

22. Levi-Montalcini, R., The origin and development of the visceral system in the spinal cord of the chick embryo, *J. Morphol.*, 86, 253, 1950.

23. Levi-Montalcini, R. and Angeletti, P.U., Nerve growth factor, *Physiol. Rev.* 48, 534, 1968.

24. Snider, W.D. and Johnson, Jr., E.M., Neurotrophic molecules, *Ann. Neurol.*, 26, 489, 1989.

25. Hamburger, V., Brunso-Bechtold, J.K., and Yip, J.W., Neuronal death in the spinal ganglia of the chick embryo and its reduction by nerve growth factor, *J. Neurosci.*, 1, 60, 1981.

26. Hamburger, V. and Yip, J.W., Reduction of experimentally induced neuronal death in spinal ganglia of the chick embryo by nerve growth factor, *J. Neurosci.*, 4, 767, 1984.

27. Hamburger, V., Cell death in the development of the lateral motor column of the chick embryo, *J. Comp. Neurol.*, 160, 535, 1975.

28. Hollyday, M. and Hamburger, V., Reduction of the naturally occurring motor neuron loss by enlargment of the periphery, *J. Comp. Neurol.*, 170, 311, 1976.

29. Oppenheim, R.W., Cell death during development of the nervous system, *Annu. Rev. Neurosci.*, 14, 453, 1991.

30. Oppenheim, R.W. and Chu-Wang, I.-W., Spontaneous cell death of spinal motoneurons following peripheral innervation in the chick embryo, *Brain Res.*, 125, 154, 1977.

31. Oppenheim, R.W., Chu-Wang, I.-W., and Maderdrut, J.L., Cell death of motoneurons in the chick embryo spinal cord. III. The differentiation of motoneurons prior to their induced degeneration following limb-bud removal, *J. Comp. Neurol.*, 177, 87, 1978.

32. Oppenheim, R.W., Haverkamp, L.J., Prevette, D., McManaman, J.L., and Appel, S.H., Reduction of naturally occurring motoneuron death *in vivo* by a target-derived neurotrophic factor, *Science,* 240, 919, 1988.

33. Pittman, R.N. and Oppenheim, R.W., Neuromuscular blockade increases motoneurone survival during normal cell death in the chick embryo, *Nature,* 271, 364, 1978.

34. Pittman, R. and Oppenheim, R.W., Cell death of motoneurons in the chick embryo spinal cord. IV. Evidence that a functional neuromuscular interaction is involved in the regulation of naturally occurring cell death and stabilization of synapses, *J. Comp. Neurol.,* 187, 425, 1979.

35. Oppenheim, R.W. and Nunez, R., Electrical stimulation of hindlimb increases neuronal cell death in chick embryo, *Nature,* 295, 57, 1982.

36. Oppenheim, R.W. Neurotrophic survival molecules for motoneurons: An embarassment of riches, *Neuron,* 17, 195, 1996.

37. Okado, N. and Oppenheim, R.W., Cell death of motoneurons in the chick embryo spinal cord: IX. The loss of motoneurons following removal of afferent inputs, *J. Neurosci.,* 4, 1639, 1984.

38. Lipton, S. A., Blockade of electrical activity promotes the death of mammalian retinal ganglion cells in culture, *Proc. Natl. Acad. Sci. U.S.A.,* 83, 9774, 1986.

39. Galli-Resta, L., Ensini, M., Fusco, E., Gravina, A., and Margheritti, B., Afferent spontaneous electrical activity promotes the survival of target cells in the developing retinotectal system of the rat, *J. Neurosci.,* 13, 243, 1993.

3

Apoptosis in Neurodegenerative Disorders

Donald E. Schmechel

CONTENTS

0-8493-3352-0/99/$0.00+$.50
© 1999 by CRC Press LLC

3.1   Introduction

The occurrence of apoptotic cell death during the course of illness has been
clearly demonstrated for a number of human neurodegenerative disorders
and animal models of human disease.[1,2] Programmed cell death of neurons,
glial cells, and other elements of the nervous system is not only a feature of
neurodegeneration, but also a prominent feature of normal prenatal and
postnatal development.[3-6] This is most elegantly displayed in the studies of
development and neurodegeneration in *C. elegans*.[7,8] It is perhaps natural to
expect that the nervous system's ability to conduct graceful and programmed
removal of cells during development might reemerge during abnormal
aging, injury, or neurodegenerative disease. One must ask, however, in each
particular model or neurodegenerative disease whether the timing and number of cells involved in this particular mode of cell death are sufficient to
consider apoptotic cell death a key element of the disease process. In this
chapter, we will review the normal occurrence of apoptotic cell death in the
developing nervous system to use as a standard to judge the potential importance of apoptosis in aging and in neurodegenerative disorders. In particular,
we will consider the "gold standard" of a neurodegenerative disorder, Batten
disease, where the gene defect directly involves apoptotic mechanisms.
Finally, we will consider the case for apoptotic cell death in a number of
other more common neurodegenerative disorders.

3.2   Cell Death in Normal Development and Aging

3.2.1   Spinal Cord

The developing spinal cord and its interactions with inducing structures
such as the notochord and with its targets of innervation, especially striated
muscle, has been extensively studied.[9-12] It is now clear that programmed
cell death resulting in eventual loss of up to 50% of cell classes such as
motoneurons is a normal feature of development.[9] Programmed cell death
occurs at very early stages of spinal cord development and continues
throughout the later stages. Of interest is the finding that early apoptotic
death of spinal motoneurons may be determined by intrinsic programs or
local factors within the spinal cord, whereas later stages of motoneuron
apoptotic death may be dominated by trophic factor influence related to
interactions and innervation of target muscles.[9] Thus, even in this one portion
of the neuraxis and for one discrete set of cells, apoptosis is heterogeneous
in mechanism. Moreover, apoptotic cell death involves not only projection
neurons innervating skeletal muscle and sensory neurons of the dorsal root

ganglia, but also supporting glial cell elements in the spinal cord and peripheral nerves and interneurons intrinsic to the spinal cord.[9-12]

3.2.2 Midbrain, Forebrain, Retina

The occurrence of significant cell death is likewise a feature of development in higher levels of the neuraxis.[3] This has been well documented for cerebellum, whose development can be studied postnatally in rodents, and likewise for the visual pathway which is easily studied in animals with late eye-opening.[3,14-16] An interesting finding in these systems is the occurrence of distinct periods of apoptotic activity and the occurrence of apoptosis in both neuronal and glial populations.[17] These events are likely keyed to the tempo of neurogenesis, establishment of neural connections of postmitotic neurons, and neuronal–glial interactions.[17] With the advent of functionality of these pathways and systems, this transient period of apoptosis ceases. While further gliogenesis, myelination, and alterations in axonal connectivity occur, apoptosis is clearly a major player only during early development of the nervous system under normal circumstances.[15] The occurrence of low background levels of apoptosis in the adult nervous system is of uncertain significance, although low steady rates over time should not be neglected as a significant biological factor in long-lived animals. This may be particularly true for cells which are poised midway in differentiation or which are remnant stem cell populations from the subventricular zone. Likewise, low levels of apoptotic cell death in defined subsets of neurons could signal an eventual significant attrition over time. For example, subclasses of gabaergic inhibitory neurons account for usually 1 to 5% of total neurons as defined by neuropeptide content, and all together usually 20 to 30% of total neurons in most forebrain regions. Significant cell loss through apoptosis could occur in these or similar neuronal subsets without being easily detectable by usual histological methods.

3.2.3 Cerebral Cortex: Example of Cortical Subplate

The cortical subplate is a transient zone in the developing cerebral cortex which integrates a primitive level of organization of the cortical plate.[18] Neurons generated early in the development of the neocortex are situated in this zone beneath the main cortical plate. Subplate neurons interconnect with the neurons in layer I and participate in important cell–cell interactions with migrating neurons and with incoming afferent and efferent projections. This structure is particularly prominent in human and primate brain. Its disappearance during the course of development illustrates another principle of apoptosis, which is the removal of earlier phylogenetic patterns of organization that are developmentally active, but are then superseded by more complex patterns. Thus, the primitive developmental pattern of cerebral cortical organization and activity is changed with subsequent neuro-genesis.[18,19] While some of these neurons are diluted in the marked expansion of the cerebral cortical mantle, there is evidence for neuronal loss. Local circuit neurons in layer I and particularly neurons in the cortical subplate undergo cell death and glial reaction to remove cell and their processes. This timing is also near the remodeling and loss of long axonal projections of other systems. In the visual system, up to 50% of neuronal projections from retina to posterior visual structures are lost. In other regions of cerebral cortex, there is significant loss of projection neurons and the transient sub-plate neurons. During this time period of development, significant microglial activity is observed in the deep white matter underlying the cortex.[20] This raises another issue surrounding the occurrence of cell death and apoptosis during development — glial reaction to the extensive cell death. Microglial recruitment and activation represents an early and significant interaction of immune-competent cells with dying neurons and glial cells. These early interactions might in certain cases alter the subsequent cellular and immune reaction to cell death in the adult nervous system.

3.2.4   Hippocampus

The hippocampus is an important area to consider for apoptotic cell death, given its very ordered geometry and its importance to memory and cogni-tion. A number of studies suggest that the hippocampus may be subject to apoptotic cell death even after development. In particular, the influence of adrenal steroids and hypothalamic–pituitary axis on hippocampal neurons have been demonstrated in a number of studies.[21] The most direct model — adrenalectomy with loss of corticosteroid levels — produces significant cell death by apoptosis in hippocampal granule cell neurons. The organization of the dentate gyrus and its component granule cells is such that it is likely to withstand significant gradual losses of granule cells without much behav-ioral compromise. This suggests two problems: (1) why would these neurons be particularly sensitive to systemic factors and subject to apoptotic cell death?, and (2) would low rates of apoptotic cell death in such nervous system structures eventually serve as a priming event or portal into neuro-degenerative disorders? The important clinical issue is that significant low levels of apoptotic cell death could occur with gradual denervation over time. This process might be clinically relatively silent, and yet be a significant pathological event for disease.

3.2.5   Deafferentation

Deafferentation of adult or mature neurons can clearly be a cause of apoptosis for many classes of neurons.[22] The significant issue for neurodegenerative illness is the response of the nervous system to two common events during aging: denervation of a neuron from its target cells or tissue, and activation of the immune-competent glial cells from systemic factors or acquired illness. In the first case, it is very clear that many neurons are dependent on trophic factors from their innervated targets for their moment-to-moment sustenance and biological program. Thus, during development, one of the factors that may lead to apoptotic cell death is failure to innervate a target tissue (e.g., striated muscle for motoneurons), failure to successfully maintain innervation once achieved (e.g., competition, disuse atrophy), injury or compromise to the integrity of that connection (e.g., axonal injury), and/or loss of trophic factor production by the target. The significance of loss of trophic factors for adult neurons, varies for different classes of neurons depending on their connectivity, redundancy of connections, and age of the nervous system. Glial response and the immune competence of glial cells interacts directly with the above dependence of the neuron on its connections. For example, when an axon suffers reversible crush injury, there is an immediate response and proliferation of satellite glial cells around the motoneuron cell body at the level of the spinal cord. On the positive side, these activated glial cells may assist in removal of afferent connections to such a cell and participate in a program of recovery for the neuron (retrograde cell reaction of the neuron or chromatolysis). On the downside, these cells are immunocompetent and may be part of the substrate for adverse genetic factors to result in neurodegenerative illness. For example, amyotrophic lateral sclerosis (ALS) or motor neuron disease is preceded in some cases by discrete injury to peripheral nerve trunks and ALS may "begin" at this level in the spinal cord before progression to other levels. The general principle is that any injury to neurons or their processes is accompanied by reaction of glial cells at the level of the injury and at the parent neuronal cell body. Thus, consideration of deafferentation and loss of trophic factors or their influence must be connected with the idea that glial cells are also involved in the outcome from the earliest time points of injury.

3.3 Apoptotic Cell Death in Early Onset Neurodegenerative Disorders

3.3.1 Necrosis–Apoptosis Continuum

The role of apoptotic cell death in neurodegenerative disorders has been controversial because of the difficulty of precise identification of apoptosis in human tissues. The methods brought to bear must include anatomical methods employing TUNEL (terminal deoxynucleotidyl dUTP nick end labeling) staining, ultrastructural study, and morphological analysis and biochemical methods with identification of endonucleosomal DNA cleavage and characteristic laddering of DNA fragments. Furthermore, in the nervous system, the timing with regard to disease onset, the tempo of apoptosis during the disease course, and the potential restriction to one cell class in a complex tissue with many cell classes further complicates the identification and evaluation of apoptosis as a disease process in many illnesses. In addition, the potential inciting factors (e.g., ischemia, excitotoxic injury, toxic injury to mitochondria) and the resulting process of active tissue injury are capable of producing both cell necrosis and apoptosis in the same tissue. This brings up the issue of assessing cell death over a potential continuum of necrotic or accidental cell death to apoptotic or programmed cell death.[23,24]

Experimental studies make clear that studying the role of apoptosis in a given disease may be heavily weighted by other secondary factors. Persons dying late in the course of their illness have perimortem morbidity from other illnesses such as sepsis and dehydration. The degree of hypoxia and rate of decline during the dying process, and status of the hypothalamic-pituitary-adrenal axis with the effect of steroid levels can affect the levels of apoptosis detected in postmortem examination.[25] This is a particular problem for human studies where almost all brain tissues are obtained postmortem late in the course of the neurodegenerative illness.

With the knowledge that common environmental assaults on the nervous system as well as the complex process of cell injury and glial response can produce both apoptotic cell death and necrotic cell death, there is a real issue of assessing the role of apoptotic cell death for a given neurodegenerative disease. The most convincing case can be made for those illnesses with a genetic defect in a gene directly related to the apoptotic cell response (e.g., Batten disease). Not surprisingly, such a major genetic influence on apoptosis is associated with onset of disease in early life. Many other neurodegenerative diseases have defects in genes that influence the response of a cell to injury or cellular protective mechanisms such as free radical protective enzymes, or involve genes which can be shown *in vitro* to influence apoptosis. Many of these diseases have catastrophic later onset of illness in midlife. These illnesses can be viewed as having a secondary genetic relationship to apoptosis. However, even environmental events in a genetically neutral or "normal" host can result in apoptosis. Two common examples are ischemia and toxic injury. For example, carbon monoxide poisoning and probably other mitochondrial toxins can result in delayed neuronal death through apoptosis.[26] Such environmental injury may interact with genetic factors to result in neurodegenerative disease in late life. Thus, a further category of neurodegenerative disease can be characterized as having a secondary or environmental relationship to apoptosis. These three categories are represented in Table 1.1. In the following sections, we will study these issues in several selected neurodegenerative diseases where apoptosis is a major and/or clear-cut mechanism of pathogenesis.

TABLE 3.1

Apoptosis and Neurodegenerative Disease

Role of Apoptosis	Gene/Protein	Disease Example
Primary — genetic		
Gene directly related to apoptosis	*NAIP*, ?SMN	Spinal muscular atrophy
Gene directly related to apoptosis	CLN3	Battens disease
Secondary (1) — genetic		
Gene affecting G-protein signaling transduction	Rd gene, rhodopsin	Retinal degeneration
Gene affecting apoptotic pathway	Presenilin 2	Alzheimer's disease
	Amyloid precursor protein	Alzheimer's disease
	?APOE	Alzheimer's disease
Gene affecting cell injury/or cell protection	Cu,Zn-SOD	Motor neuron disease (ALS)
Gene acted on by apoptotic genes	Huntingtin	Huntington's disease
	MJD1	Machado-Joseph disease
	DRPLA-protein	Dentatorubro pallidoluysian atrophy (DRPLA)
	Ataxin-1	Spinocerebellar ataxia
Gene affecting mitochondrial function	mtDNA defect	e.g., Leber's hereditary optic Neuropathy
Secondary (2) — environmental injury		
Excitotoxic injury		?Alzheimer's disease, ALS
Ischemia		Vascular dementia
Mitochondrial injury		?Alzheimer's disease, PD e.g., Minimata disease
Toxin: cadmium, mercury		
Toxin: carbon monoxide		Delayed neuronal degeneration

3.3.2 Batten Disease

Batten disease is an eponymic term for a family of autosomal recessive or apparently sporadic neurodegenerative disorders, also termed neuronal ceroid lipofuscinosis, and refers most often to the juvenile form.[27] The neuropathology of these progressive disorders is characterized by massive neuronal death and, in most subtypes, death of photoreceptors in the retina with resultant blindness. Thus, they represent a devastating syndrome of decline in both cognitive and motor skills with loss of milestones, visual loss, and seizures. Their descriptive name stems from the occurrence of inclusions in involved cells which become autofluorescent and represent lipofuscin deposits. Ultrastructural analysis can demonstrate various inclusions, including fingerprint-like bodies, granular osmiphilic deposits, and curvilinear profiles. These disorders are also ordered by their age of onset: infantile neuronal ceroid lipofuscinosis (INCL), late infantile neuronal ceroid lipofuscinosis (LINCL), Batten disease or juvenile neuronal ceroid lipofuscinosis (JNCL), and the adult variant or Kuf's disease. Interestingly, the accumulation of subunit 9 of the mitochondrial ATPase synthase complex in lysosomes of the last three forms, together with an associated alteration in subunit 9 (nuclear coded) mRNA, and protein expression was not accompanied by defects in the gene encoding this protein.

Recent advances in the understanding of these disorders has resulted from the histological demonstration that cell death in Batten disease is represented by apoptotic cell death.[28] In addition, other neurochemical studies have shown upregulation of Bcl-2 and ceramide levels, implicating a close relationship to the regulation of apoptosis.[29] These findings linking apoptosis with Batten disease or juvenile NCL have been further advanced by the discovery of the responsible gene, CLN3 on chromosome 16, which has been cloned and sequenced.[30] Recent work shows that the 438-amino acid protein product of this gene is operative in a novel antiapoptotic pathway. The *CLN3* peptide modulates both endogenous and vincristine stimulated levels of ceramide suggesting mediation of its antiapoptotic effect by attenuating ceramide levels. The human disease is associated with deletions of the *CLN3* gene, which apparently result in loss of function. The gene defect in LINCL, a lysosomal peptidase encoded on chromosome 11, may also be implicated in this apoptotic pathway. Infantile neuronal ceroid lipofuscinosis is due to mutation in the *CLN1* gene (palmitoyl protein thioesterase) on chromosome 1 with unknown relationship to apoptotic mechanisms.[30]

The lesson from Batten disease and likely LINCL is probably that devastating, multisystem diseases with marked cell death by apoptosis can be produced by genetic defects in proteins involved directly in apoptotic pathways.[30] The hallmark of this family of diseases is early onset, although it is interesting that relatively normal development is achieved prior to disease onset and that the course of illness can be quite prolonged.[27] Thus, Batten disease provides one "gold standard" example of neurodegenerative disorder directly linked to apoptosis by gene defect and by pathobiology (see Table 1.1).

3.3.3 Spinal Muscular Atrophy

The prominent apoptosis of spinal motor neurons that occurs during normal development and that results in the normal complement of motor neurons continues unabated in spinal muscular atrophy (SMA). SMA occurs in several forms, from the most common form SMA I, occurring in young infants with death often before age 2; to a milder form SMA II, with somewhat shortened lifespan; to SMA III, with onset late in the first decade and almost normal lifespan. SMA I is due to deletions in the neuronal apoptosis inhibitor protein (*NAIP*) gene on chromosome 5.[31-33] Other copies of the *NAIP* gene exist and some of these truncated versions of the *NAIP* gene may confer partial biological activity. Thus, the different forms of SMA, differing in age of onset and severity, may represent the ability of residual copies of the *NAIP* gene to compensate for loss of the full-length form. It is interesting that the *NAIP* gene is similar to a viral gene inhibiting apoptosis.

In the same region of chromosome 5, other cases of spinal muscular atrophy have been mapped to a second gene called survival motor neuron gene (*SMN*).[34] Both of these genes imply that the absence of apoptosis in the mature nervous system is also dependent on active genetic mechanisms that are antiapoptotic. The relevant proteins are present in the tissues at risk.[35]

3.3.4 Retinal Degeneration

Retinal degeneration is a well-studied system where genes producing apoptotic cell death and photoreceptor loss have been analyzed in animal models and in human disease.[36-41] The involved genes result in photoreceptor loss with resultant retinitis pigmentosa in early to midlife with devastating effects and loss of vision. Other pro- and antiapoptotic and related genes are activated during this process.[42-45] The directly involved defective genes, such as rhodopsin, are involved in signal transduction. The defective proteins result apparently in altered G-protein-related signaling so as to promote a balance favoring apoptotic cell death. This imbalance, since it involves signaling mechanisms for physiological transduction of vision, can also be influenced by environmental factors such as light flux, ischemia, or retinal detachment.[46-48] Thus, in experimental models, the process of apoptotic cell death can be triggered by altering the light and wavelength input to the retinal tissue at risk.[47,48]

Retinal degeneration repeats the theme of spinal muscular atrophy considered above. The survival of cells in the mature nervous system is not automatically assured, and in fact, hangs on a delicate balance of proper interactions with other cells and elements of a complex biological system. Thus, the same factors that lead to the gracious and programmed death of excess cells and connections during the development of the nervous system can supervene to cause exit of mature elements that are unfortunately no longer in excess. Abnormal signaling and balance in the G-protein system can lead to activation of the apoptotic pathway. Unlike spinal muscular atrophy where mutations in antiapoptotic genes cause loss of function and direct involvement of the apoptotic pathway, retinal degeneration represents disorders one step removed, where alteration in normal signaling pathways is "interpreted" by the cell as a sign of physiological failure and activates the genetic program for cell removal.[49,50]

3.3.5 Triplet Repeat Disorders

Triplet repeat expansion disorders include a number of diseases that result in apoptotic cell death for selected sets of cells in the nervous system.[51] These disorders include spinocerebellar ataxia type I (SCA-1), Huntington's disease,

Machado-Joseph disease, dentatorubral pallidoluysian atrophy (DRPLA), and others.[52-57] The common theme is gain of abnormal function through augmentation of a CAG repeat that results in lengthening of a polyglutamine hinge region for the particular protein. Depending on the particular gene and its tissue and cellular distribution, this results in neurodegeneration in early to midlife in the caudate-putamen and cerebral cortex (Huntington's disease), spinocerebellar tracts (Machado-Joseph disease), and cerebellum, red nucleus, and pallidoluysian nuclei (DRPLA). Genetic animal models and cell culture models for these diseases support the concept that abnormal gain of function is the key mechanism which is common to all of the above disorders.[52,57,58] Specificity of disease may be conferred by the involved gene product, its distribution, and stochastic risk of adverse protein–protein interaction depending on length of repeat and expression level.[59]

The mechanism of injury is the ability of the polyglutamine region to aggregate above a certain length of over 35 amino acids, and thereby to influence intracellular metabolism and function through abnormal protein–protein interactions. One enzyme that is affected is glyceraldehyde-6-phosphate dehydrogenase (GAPDH), a key enzyme in glycolysis. Post-translational modification of GAPDH with presumed impact on intermediate metabolism is also implicated in apoptosis.[60,61] These effects result in an apparent stochastic increase in likelihood for certain sets of neurons to exit by apoptotic cell death, with devastating effects on the nervous system. It is interesting that even though this process is relatively specific anatomically, and results from apoptotic cell death, it is not necessarily easy to demonstrate evidence for apoptotic cell death at every point in the illness. Thus, at some time points early in the disease, apoptotic cell death can be demonstrated easily in the caudate-putamen, whereas at the very latest time points (where few neurons remain), apoptotic figures are relatively rare.

3.3.6 Mitochondrial Disorders

When one considers the delicate balance of proapoptotic and antiapoptotic pathways in cells and the need in a complex organism and tissue for removal of cells that are not functioning properly or that are irreparably injured, one must almost ask whether all neurodegenerative disorders must not involve apoptotic mechanisms of cell death. The key questions remain those posed in the introduction: whether in each particular model or neurodegenerative disease, the timing and number of cells involved are sufficient to consider apoptotic cell death a key element of the disease process. Loss of afferent or efferent connections with associated trophic/growth factors or influence, loss of protective/antiapoptotic genes, imbalance of signaling pathways, and serious compromise to the energy metabolism and membrane integrity of the cell can all be key signals for apoptotic cell death. Clearly, one of the most important signals for apoptotic cell death can be mitochondrial injury, excess free radical production, and/or defective oxidative protection mechanisms.[62] If sufficient, such injury can yield energy failure for the cell and immediate or necrotic/accidental cell death. If less severe impairment of oxidative metabolism occurs, even minor mitochondrial dysfunction with release of cytochrome c and pore transition for a lesser number of mitochondria might be a signal for apoptosis.[63-67]

With the knowledge that mitochondrial dysfunction can result in cytochrome c release, a potent inducer of apoptotic cell death, it is perhaps not surprising that mitochondrial disorders, whether genetic or acquired, can result in apoptosis.[62] This should be weighed against the fact that in most models of cellular apoptosis, mitochondrial morphology is apparently intact in the face of marked nuclear changes.[68] Most mitochondrial disorders are characterized by variable onset of disease. The issue of heteroplasmy of affected mitochondria during vertical maternal transmission of the mitochondrial genome and the subsequent possibility of clonal selection during development and life raise the possibility that cells at risk may contain a variable and not necessarily high percentage of genetically compromised mitochondria. Environmental and genetic compromise of mitochondrial function may be an important mechanism or portal for the development of neurodegenerative disorders.[62,69]

Mitochondrial disorders can result in apoptotic cell death such as displayed in Leber's hereditary optic atrophy. In these disorders, the principle for cell injury and induction of apoptosis may depend on heteroplasmy (i.e., the actual proportion of inherited deficient mitochondria apportioned to a particular cell class or tissue during development) and environmental factors yielding a critical level of oxidative injury to the cells at risk. Thus, many of these disorders result in tissue-specific patterns of injury that involve cells with high metabolic rates and high level of oxidative metabolism (e.g., cardiac muscle, extraocular muscles, proximal renal tubule cells, etc.). Since a natural stress or level of oxidative injury in such a cell class at risk could then be amplified by a genetically defective electron transport chain, the potential for a stochastic (need for environmental event or physiological stressor) and catastrophic (amplification of injury and further mitochondrial dysfunction through injury to the "naked" mitochondrial genome) element to mitochondrial disorders is almost an unavoidable consequence of the state condition of mitochondrial DNA and mitochondrial energy production. Thus, many mitochondrial disorders yield further "nonspecific" deletion events of the mitochondrial genome during the expression of the illness. Such events also occur during normal aging and presumably can reflect exogenous environmental injury to an otherwise normal mitochondrial genome. Thus, apoptotic cell death through mitochondrial cytochrome c release is perhaps a very necessary signal to allow a particular organ or tissue to react to adverse mitochondrial injury prior to more catastrophic failure with necrotic cell death and attendant inflammatory reaction.

3.4 Apoptosis in Alzheimer's, Parkinson's, and Motor Neuron Disease

3.4.1 Modeling Chronic Neurodegenerative Diseases

The issue in chronic neurodegenerative disease of onset in mid to late life and lasting from 5 to 25 years from onset to death is modeling the various stages of disease. Most information on AD and other similar chronic neuro-degenerative disorders is gathered from postmortem data at the very end stages of disease when many other factors have intervened, most notably the process of dying. Thus, much information on apoptosis in AD has been gathered, but mostly from the endpoints of the disease process. It is of crucial importance to model the beginning of AD pathology and, particularly, the probable subclinical pathology that exists in persons at genetic and/or environmental risk even at a very early age. The way to circumvent this problem is the creation of *in vitro* and *in vivo* models of AD pathology in animals or human-derived tissue and, particularly with the discovery of specific genes for AD, to create genetic models of AD.

3.4.2 Evidence for Apoptosis in AD

The evidence for apoptotic cell death in AD brain is quite compelling and stems back to work by Cotman and others to demonstrate the anatomical and biochemical features of apoptotic cell death in AD brain tissue.[70-89] This includes demonstration of TUNEL-positive staining as well as correlation with markers associated with apoptosis such as Bcl-2, Bax, c-Jun, Fos, and others.[72,73,78,81-87] There has been relatively good agreement among different workers, although there is relatively little published ultrastructural evidence for apoptosis.

3.4.3 Role of AD Genes: Presenilins, Amyloid Precursor Polypeptide, Apolipoprotein E

The proposed role for apoptotic cell death in AD has been strengthened by the discovery that many or all of the genes discovered for AD influenced cellular apoptosis.[2] Not surprisingly, the strongest case exists for the three genes involved in early onset, autosomal dominant AD, presenilin 1 (chromosome 12), presenilin 2 (chromosome 1), and amyloid precursor polypeptide (*APP*) on chromosome 21.[2] However, unlike the clear-cut case for Batten disease, it is not clear that these gene products are directly placed in the pathway for apoptosis. Rather, by influencing protein trafficking, membrane events, and free radical production, mutations in these three AD genes alter the outcome of cell injury and repair in such a way as to activate apoptotic pathways.

Proposed mechanisms that would initiate apoptotic programs include potentiating oxidative injury, sensitizing the cell to trophic factor deficiency, altering protein trafficking in the rough endoplasmic reticulum, altering APP metabolism, and influencing signal transduction, particularly G-protein related. A strong case has been made for a common mechanism based on G-protein signaling for production of apoptosis in AD without necessarily involving the extracellular events of AB deposition.[2] However, there is also evidence for increased oxidative injury in AD related to intra- and extracellular protein–protein interactions, particularly with APP and its fragments, presenilin, and oxidative priming events all working together to sensitize the cell for apoptosis.[90-104] Thus, at the moment, there is an abundance of potential mechanisms for presenilin or APP mutations to result in apoptotic cell death. Interestingly, however, there is little evidence to date for apoptotic cell death or even neuronal degeneration in the genetic animal models based on these mutations.[105,106] It is also important to realize that most cases of AD in late life with normal presenilin and *APP* genes. It may be necessary to "humanize" the rodent more extensively with human versions of *APOE*, tau, and other AD-related gene products.[107]

The role of the AD susceptibility gene *APOE* in the most common late-onset cases of AD is played out in the setting of normal presenilin 1 and 2 and APP genes. There is yet no evidence for increased apoptotic cell death with *APOE* allele 4 (high risk for AD), but there is *in vitro* evidence that *APOE* and *APP* can interact to produce increased oxidative damage.[108]

3.4.4 Evidence for Apoptosis in Parkinson's Disease

Parkinson's disease (PD) has been classically considered a predominantly environmentally related disorder with known etiologies, including MPTP toxin exposure, carbon monoxide exposure, postencephalitic related (1919 epidemic), manganese exposure, and other putative chemical exposures involving injury to mitochondrial complex I enzymes. Nevertheless, most cases are idiopathic without demonstrable etiology, and clearly some cases are genetically influenced. The knowledge that PD is related to progressive dopaminergic cell loss in the substantia nigra with symptomatology usually appearing at 80 to 90% cell depletion and the evidence connecting Parkinsonism to environmental exposures has led to an oxidative stress theory for the pathogenesis of PD and an emphasis on apoptosis as a possible mechanism of cell loss.[109,110]

Evidence for apoptosis in Parkinson's disease has been provided from both animal models and human material for the key cell population at risk, the dopaminergic neurons of the substantia nigra.[111-118] In some experimental models of PD, there has been "mixed" evidence for classical cellular apoptosis with support for perhaps a variant of apoptotic cell death.[111] It is important to point out that both in experimental animals and in human disease, cell death of dopaminergic neurons does occur in the setting of some amount of cellular inflammation and microglial response. This may indicate that cell death for some dopaminergic neurons may be relatively abrupt and proceed along the pathway of accidental or necrotic cell death with attendant inflammation. Likely, a continuum between apoptotic and necrotic cell death may occur, with the proportion depending on the magnitude and timing of the environmental toxin or stressor.[23,24]

A further issue in Parkinson's disease is the possible role of both endogenous L-dopa produced by the surviving dopaminergic neurons and exogenous L-dopa used in therapy as neurotoxins. A number of experiments have demonstrated the ability of L-dopa to induce apoptosis in cell culture.[119-124] L-dopa synthesis and release is increased in surviving dopaminergic neurons to compensate for loss of neurons during the illness and may therefore contribute in a secondary manner to further injury. In addition, toxins affecting complex I in mitochondria, calcium stress, aging-related, or inherited mtDNA defects have been suggested as portals into apoptotic loss of dopaminergic neurons. [125-127] Protective therapy may be directed at mitochondrial or apoptotic mechanisms of cell injury and death.[128,129]

3.4.5 Evidence for Apoptosis in Motor Neuron Disease (ALS) and Related Disorders

Motor neuron disease or amyotrophic lateral sclerosis (ALS or Lou Gehrig's disease) represents a neurodegenerative disorder in which apoptotic cell death has been invoked as a key pathogenetic mechanism. In ALS, progressive loss of upper and lower motor neurons results in extreme disability and ultimately death. As in AD and Parkinson's disease, both environmental and genetic factors have been suggested.[131] Apoptotic cell death of motoneurons has been related to immunological attack and to failure of cell defenses against oxidative damage.[132,133]

The discovery of Cu,Zn-superoxide dismutase mutations as a cause for some cases of familial ALS has strongly supported a role for apoptosis since the effect of the mutation is to convert an antioxidant defense enzyme into a proapoptotic gene.[134] Furthermore, other rare cases of familial ALS have been associated with *NAIP* mutations.[135] This suggests that ALS may well be very similar to Parkinson's disease and late-onset AD in that a number of specific gene defects or allelic effects may be identified, accounting for 25 to 50% of familial cases. Environmental factors most likely interact with these genes to determine onset and character of illness. Many of the remaining cases may result from multiple genetic factors in combination with stronger and more adverse environmental or aging-related events. It is important to realize that the identified genes to date only account for a proportion of the familial ALS cases, and that most ALS cases (95%) are sporadic.

The importance of understanding the pathogenesis of ALS for other neu-rodegenerative disorders is that apoptotic cell death may be the major mode of motor neuron loss.[136-139] The issue raised for excitoxic injury to the nervous system is the potential for both necrotic and apoptotic cell death with per-haps all intermediates depending on the timing and amount of injury.[23,24] In ALS, there is considerable evidence for excitotoxicity as a mode of cell injury.[140-144]

Alterations in buffering of extracellular glutamate during cellular reactions to injury may result in neuronal cell injury and triggering of apoptosis.[143,144] For example, alterations in glutamate transporters might result during aging or as a result of previous cellular injury. ALS and motor neuron disorders stand, therefore, as an example of apoptotic cell injury with modulation by both genetic and environmental events. In addition, some of these defects may arise from defective RNA splicing, as has been suggested for spinal muscular atrophy cases with mutations in the *SMN* gene. This gene may code for an RNA-splicing enzyme.[145] Abnormal RNA splicing may be a common mechanism in age-related neurodegenerative diseases.

3.4.6 Role of ALS Genes: Copper, Zinc Superoxide Dismutase Mutations

The mutation in Cu,Zn-SOD that is involved in roughly 15 to 25% of familial ALS cases results in an enzyme that still retains enzymatic activity, and yet represents a proapoptotic gene. One potential mechanism is through aggre-gation, much like the triplet repeat diseases, and perturbation of other intra-cellular processes, or effective sequestration of activity in a nonuseful location. Another possibility is that intracellular copper stores are depleted through aggregation and loss of these proteins, resulting in inadequate cop-per charging of other copper proteins and cuproenzymes.[69]

3.4.7 Evidence for Apoptosis in Toxic Environmental Exposures

Apoptotic cell death is likely the result of a number of toxic environmental exposures that can damage the nervous system.[26,146] Two prominent exam-ples are organic mercury or cadmium poisoning, such as occurred in the Minimata disease outbreak in Japan, and in subacute carbon monoxide poi-soning. Both exposures can result in chronic and prolonged nervous system injury, with syndromes of delayed neuronal degeneration either in actual human cases and/or *in vitro* models.[26,146] Since well-studied experimental models of physical, ischemic, and toxic injury to the retina result in apoptotic cell death of photoreceptor cells, it is likely that many environmental expo-sures, particularly those that are subacute and sublethal with less tissue injury and necrotic cell death, result in apoptotic cell death. Such environ-mental exposures (e.g., head injury, ischemic stroke, lack of estrogen) are strong contenders for contributing to the onset of neurodegenerative disor-ders in persons at increased genetic risk.

3.5 Summary

The key issue in considering the role of apoptosis in neurodegenerative disorders is whether this information will yield practical scientific and therapeutic advances that result in the prevention or cure of these terrible diseases. On one level, certain neurodegenerative disorders are clearly disorders of apoptosis through involvement of key genes or pathways. These disorders include spinal muscular atrophy, Batten disease, retinal degeneration, and most likely triplet repeat disorders and many mitochondrial disorders. They are characterized by genetic inheritance, relative cell class and tissue specific pathology, onset before late life, and progressive and fatal outcome. Success in diagnosis, prevention, and treatment of these disorders may be solved on an individual level by gene transfer therapy or on a more general level by a greater understanding of the key events that can be modulated or prevented in the genetic and posttranslational events in cellular apoptosis. None of these diseases currently can be successfully treated in any true sense with other than symptomatic therapy.

The larger question of the involvement of apoptosis in other neurodegenerative disorders such as motor neuron disease, Parkinson's disease, and Alzheimer's disease is less settled. One can certainly make the case that genetically inherited disease such as Cu,Zn-SOD-ALS, early-onset familial AD (mutations in presenilin genes or amyloid precursor polypeptide), and rare cases of genetically inherited Parkinson's disease, may operate early on in their pathogenesis through apoptosis. The relevant genes (see above) can certainly affect apoptosis in cell systems, apparently indirectly through altering signal transduction, intracellular trafficking of proteins, or by altering cellular oxidative defense. However, the genetic load in the vast majority of cases of motor neuron disease, Parkinson's disease, and Alzheimer's disease may be more modest or involve susceptibility genes. In these cases, environmental factors and/or other complex age-related or genetic "gating" events may play the major role in disease onset and progression. In these neurodegenerative diseases, one must ask whether the timing and number of cells involved in this particular mode of cell death are sufficient to consider apoptotic cell death a key element of the disease process. It is clear that neuronal loss is a key feature of these disorders (motor neurons in ALS, dopaminergic nigral neurons in PD, cortical neurons in AD), and proximate cause of clinical deficit. However, the key issue may well turn out to be failure of other cell classes to undergo apoptosis (e.g., glial cells, persistence of neurons with retrograde reaction and physiological dysfunction) and/or abnormalities in the immune competent cells involved in all these disorders (e.g., astrocytes and microglial cells).[147] Excitoxic injury and its effects on neurons, the various classes of glial cells, and other cell classes such as activated macrophages and endothelial cells may be a major factor in apoptosis.[148]

In summary, knowing the developmental history of the nervous system, the issue is not whether apoptotic cell death occurs in neurodegenerative disorders, but establishing the timing and context of this phenomenon. Even if apoptosis turns out not to be the primary event in some or all of these late-onset disorders, apoptosis may still be viable as a target for modulating or preventing disease activity. Given the complexity of these diseases and the constant involvement of immune mechanisms during the disease course, one should remain open to the possibility that the issue is not preventing apoptosis of neurons, but producing apoptosis and removal of immune-competent cells promoting further tissue reaction. The best proof of principle for a key role for apoptosis in the late-onset, complex neurodegenerative disorders may well turn out to be the success of specific therapies modulating cellular apoptosis. These therapies may well need to be long-term and applied at very early time points in the illness prior to clinical onset, and/or during times of environmental stress or exposure to mitigate against initiation events such as head injury, ischemic events, and surgical or systemic stress. The great need in applying the lessons learned and to be learned about cellular apoptosis and neurodegenerative disorders will be biological markers for measuring disease activity and efficacy of treatment.

References

1. Wyllie, A.H., Kerr, J.F.R., and Currie, A.R., Cell death: the significance of apoptosis, *Int. Rev. Cytol.*, 68, 251, 1968.
2. Nishimoto, I., Okamoto, T., Giamberella, U., and Iwatsubo, T., Apoptosis in neurodegenerative diseases, *Adv. Pharmacol.*, 41, 337, 1997.
3. Narayanan, V., Cell death in the developing nervous system, In: *Apoptosis in Neurobiology: Concepts and Methods*, Hannum, Y.A. and Boustany, R.-M., Eds., CRC Press, Boca Raton, FL, Chapter 2, 1999.
4. Schweichel, J.-U. and Merker, H.-J., The morphology of various types of cell death in prenatal tissues, *Teratology*, 7, 253, 1972.
5. Schweichel, J.-U., Das elektronenmikroskopische Bild des Abbaues der epithelialen Scheitelleiste wahrend der Extremitatenentwicklung bei Rattenfeten, *Z. Anat. Entwicklungs-gesch.*, 136, 192, 1972.
6. Clarke, P.G.H., Developmental cell death: morphological diversity and multiple mechanisms, *Anat. Embryol.*, 181, 195, 1990.
7. Robertson, A.M.G. and Thompson, J.N., Morphology of programmed cell death in the ventral nerve cord of *Caenorhabditis elegans* larvae, *J. Embryol. Exp. Morphol.*, 67, 89, 1982.
8. Driscoll, M., Cell death in *C. elegans*: molecular insights into mechanisms conserved between nematodes and mammals, *Brain Path.*, 6, 411, 1996.
9. Burek, M.J. and Oppenheim, R.W., Programmed cell death in the developing nervous system, *Brain Pathol.*, 6, 427, 1996.

10. Yaginuma, H., Tomita, M., Takashita, N., McKay, S.E., Cardwell, C., Yin, Q.W., and Oppenheim, R.W., A novel type of programmed neuronal death in the cervical spinal cord of the chick embryo, 16, 3685, 1996.

11. Ciutat, D., Caldero, J., Oppenheim, R.W., and Esquerda, J.E., Schwann cell apoptosis during normal development and after axonal degeneration induced by neurotoxins in the chick embryo, *J. Neurosci.*, 16, 3979, 1996.

12. Lawson, S.J., Davies, H.J., Bennett, J.P., and Lowrie, M.B., Evidence that spinal interneurons undergo programmed cell death postnatally in the rat, *Eur. J. Neurosci.*, 9, 794, 1997.

13. Krueger, B.K., Burne, J.F., and Raff, M.C., Evidence for large-scale astrocyte death in the developing cerebellum, *J. Neurosci.*, 15, 3366, 1995.

14. Ferrer, I., Pozas, E., Marti, M., Blanco, R., and Planas, A.M., Methylazoxymethanol acetate-induced apoptosis in the external granule cell layer of the developing cerebellum of the rat is associated with strong c-Jun expression and formation of high molecular weight c-Jun complexes, *J. Neuropathol. Exp. Neurol.*, 56, 1, 1997.

15. Wood, K.A., Dipasquale, B., and Youle, R.J., *In situ* labeling of granule cells for apoptosis-associated DNA fragmentation reveals different mechanisms of cell loss in developing cerebellum, *Neuron*, 11, 621, 1993.

16. Lowin, B., French, L., Martinou, J.C., and Tschopp, J., Expression of the CTL-associated protein TIA-1 during murine embryogenesis, *J. Immunol.*, 157, 1448, 1996.

17. Nakao, J., Shinoda, J., Nakai,Y., Murase, S., and Uyemura, K., Apoptosis regulates the number of Schwann cells at the premyelinating stage, *J. Neurochem.*, 68, 1853, 1997.

18. Honig, L.S., Herrmann, K., and Shatz, C.J., Developmental changes revealed by immunohistochemical markers in human cerebral cortex, *Cerebral Cortex*, 6, 794, 1996.

19. Allendoerfer, K.L. and Shatz, C.L., The subplate, a transient neocortical structure: its role in the development of connections between thalamus and cortex, *Annu. Rev. Neurosci.*, 17, 185, 1994.

20. Ferrer, I., Bernet, E., Soriano, E., del Rio, T., and Fonseca, M., Naturally occurring cell death in the cerebral cortex of rat and removal of dead cells by transitory phagocytes, *Neuroscience*, 39, 451, 1990.

21. Sloviter, R.S., Dean, E., and Neubort, S., Electron microscopic analysis of adrenalectomy-induced hippoocampal granule cell degeneration in the rat:apoptosis in the adult central nervous system, *J. Comp. Neurol.*, 330, 337, 1993.

22. Capurso, S.A., Calhoun, M.E., Sukhov, R.R., Mouton, P.R., Price, D.L., and Koliatsos, V.E., Deafferentation causes apoptosis in cortical sensory neurons in the adult rat, *J. Neurosci.*, 17, 7372, 1997.

23. Portera-Cailliau, C., Price D. L., and Martin, L. J., Excitotoxic neuronal death in the immature brain is an apoptosis-necrosis morphological continuum, *J. Comp. Neurol.*, 378, 70, 1997.

24. Portera-Cailliau, C., Price, D. L., and Martin L. J., Non-NMDA and NMDA receptor-mediated excitotoxic neuronal deaths in adult brain are morphologically distinct: further evidence for an apoptosis-necrosis continuum, *J. Comp. Neurol.*, 378, 88, 1997.

25. Middleton, G., Reid, L.E., and Harmon, B.V., Apoptosis in the human thymus in sudden and delayed death, *Pathology*, 26, 81, 1994.

26. Piantadosi, C.A., Zhang, J., Levin, E.D., and Schmechel, D.E., Apoptosis and delayed neuronal damage after carbon monoxide poisoning in the rat, *Exp. Neurol.*, 147, 103, 1997.

27. Boustany, R.-M., Batten disease or neuronal ceroid lipofuscinosis, In: *Handbook of Clinical Neurology*, Vol. 22 (66): *Neurodystrophies and Neurolipidoses*, H.W. Moser, Ed., Elsevier Sciences, 1996, Chapter 12.

28. Lane, S.C., Jolly, R.D., Schmechel, D.E., Alroy, J., and Boustany, R.-M., Apoptosis is the mechanism of neurodegeneration in Batten disease, *J. Neurochem.*, 67, 677, 1996.

29. Puranam, K., Qian, W.-H., Nikbakht, K., Venable, M., Obeid, L., Hannum, Y., and Boustany, R.-M., Upregulation of Bcl-2 and elevation of ceramide in Batten disease, *Neuropediatrics*, 28, 37, 1997.

30. Puranam, K.L., Qian, W.-H., Nikbaht, K., Guo, W.-X., and Boustany, R.-M., CLN3 defines a novel antiapoptotic pathway operative in neurodegeneration and mediated by ceramide, 1998, in press.

31. Lefebvre, S., Burglen, L., Reboullet, S. et al., Identification and characterization of a spinal muscular atrophy-determining gene, *Cell*, 80, 155, 1995.

32. Roy, N., Mahadevan,M.S., McLean, M., Shutler, G., Yaraghi, Z., Farahani, R., Baird, S., Besner-Johnston, A., Lefebvre,C., Kang, X. et al., The gene for neuronal apoptosis inhibitory protein is partially deleted in individuals with spinal muscular atrophy, *Cell*, 80, 167, 1995.

33. Liston, P., Roy, N., Tamai, K., Lefebvre, C., Baird, S., Cherton-Horvat, G., Farahani, R., McLean, M., Ikeda, J.E., MacKenzie, A., and Korneluk, R.G., Suppression of apoptosis in mammalian cells by NAIP and a related family of IAP genes, *Nature*, 379, 349, 1996.

34. Iwahasi, H., Eguchi, Y., Yasuhara, N., Hanafusa, T., Matsuzawa, Y., and Tsujimoto, Y., Synergistic antiapoptotic activity between Bcl-2 and SMN implicated in spinal muscular atrophy, *Nature*, 390, 413, 1997.

35. Tews, D.S. and Goebel, H.H., Apoptosis-related proteins in skeletal muscles of spinal muscular atrophy, *J. Neuropath. Exp. Neurol.*, 56, 150, 1997.

36. Chang, G. Q., Hao Y., and Wong F., Apoptosis: final common pathway of photoreceptor death in rd, rds, and rhodopsin mutant mice, *Neuron*, 11, 595, 1993.

37. Lolley, R. N., The rd gene defect triggers programmed rod cell death, *Invest. Ophthalmol. Vis. Sci.*, 35, 4182, 1994.

38. Portera-Cailliau, C., Sung, C. H., Nathans J., and Adler R., Apoptotic photoreceptor cell death in mouse models of retinitis pigmentosa, *Proc. Natl. Acad. Sci. U.S.A.*, 91, 974, 1994.

39. Lolley, R. N., Rong, H., and Craft, C. M., Linkage of photoreceptor degeneration by apoptosis with inherited defect in phototransduction, *Invest. Ophthalmol. Vis. Sci.*, 35, 358, 1994.

40. Tso, M. O., Zhang, C., Abler, A. S., Chang, C. J., Wong, F., Chang, G. Q. and Lam, T. T., Apoptosis leads to photoreceptor degeneration in inherited retinal dystrophy of RCS rats, *Invest. Ophthalmol. Vis. Sci.*, 35, 2693, 1994.

41. Slack, R.S., Skerjanc, I.S., Lach, B., Craig, J., Jardine, K., and McBurney, M.W., Cells differentiating into neuroectoderm undergo apoptosis in the absence of functional retinoblastoma family proteins, *J. Cell. Biol.*, 129, 779, 1995.

42. Jomary, C., Ahir, A., Agarwal, N., Neal, M. J. and Jones, S. E., Spatio-temporal pattern of ocular clusterin mRNA expression in the rd mouse, *Brain Res. Mol. Brain Res.*, 29, 172, 1995.

43. Joseph, R. M. and Li, T., Overexpression of Bcl-2 or Bcl-XL transgenes and photoreceptor degeneration, *Invest. Ophthalmol. Vis. Sci.*, 37, 2434, 1996.

44. Rich, K. A., Zhan, Y. and Blanks, J. C., Aberrant expression of c-Fos accompanies photoreceptor cell death in the rd mouse, *J. Neurobiol.*, 32, 593, 1997.

45. Isenmann, S., Wahl C., Krajewski, S., Reed, J. C. and Bahr, M., Up-regulation of Bax protein in degenerating retinal ganglion cells precedes apoptotic cell death after optic nerve lesion in the rat, *Eur. J. Neurosci.*, 9, 1763, 1997.

46. Cook, B., Lewis, G. P., Fisher, S. K. and Adler, R., Apoptotic photoreceptor degeneration in experimental retinal detachment, *Invest. Ophthalmol. Vis. Sci.*, 36, 990, 1995.

47. Nash, M. L., Peachey, N. S., Li, Z. Y., Gryczan, C. C., Goto, Y., Blanks, J., Milam, A. H., and Ripps, H., Light-induced acceleration of photoreceptor degeneration in transgenic mice expressing mutant rhodopsin, *Invest. Ophthalmol. Vis. Sci.*, 37, 775, 1996.

48. Abler, A. S., Chang, C. J., Ful, J., Tso, M. O. and Lam, T. T., Photic injury triggers apoptosis of photoreceptor cells, *Res. Commun. Mol. Pathol. Pharmacol.*, 92, 177, 1996.

49. Rosenbaum, P. S., Gupta, H., Savitz, S. I. and Rosenbaum, D. M., Apoptosis in the retina, *Clin. Neurosci.*, 4, 224, 1997.

50. Travis, G.H., Mechanisms of cell death in the inherited retinal degenerations, *Am. J. Hum. Genet.*, 62, 503, 1998.

51. Ross, C.A., When more is less: pathogenesis of glutamine repeat neurodegenerative diseases, *Neuron*, 15, 493, 1995.

52. Zeitlin, S., Liu, J.P., and Chapman, D.L., Arginis, P., and Arginis, E., Increased apoptosis and early embryonic lethality in mice nullizygous for the Huntington's gene homologue, *Nat. Genet.*, 11, 155, 1995.

53. Zoghbi, H.Y., Spinocerebellar ataxia type I, *Clin. Neurosci.*, 3, 5, 1995.

54. Koide et al., Unstable expansion of CAG repeat in hereditary dentatorubral-pallidoluysian atrophy (DRPLA), *Nat. Genet.*, 6, 9, 1994.

55. Nagafuchi, S., Yanisawa, M., Sata, K., Shirayama, T., Ohsaki, E., Bundo, M., Takeda, T., Tadokoro, K., Kondo, I., and Murayama, N., Dentatorubral and pallidoluysian atrophy: expansion of an unstable CAG trinucleotide repeat on chromosome 12p, *Nat. Genet.*, 6, 14, 1994.

56. Burke, J.R., Wingfield, M.S., Lewis, K.E., Roses, A.D., Lee, J.E., Hulette, C., Pericak-Vance, M.A., and Vance, J.M., The Haw River syndrome: dentatorubropallidoluysianatrophy in an African-American family, *Nat. Genet.*, 7, 721, 1994.

57. Portera-Cailliau, C., Hedreen, J.C., Price, D.L., Koliatsos, V.E., Evidence for apoptotic cell death in Huntington's disease and excitotoxic animal models, *J. Neurosci.*, 15, 3775, 1995.

58. Ikeda, H., Yamaguchi, M., Sugai, S., Aze, Y., Narumiya, S., and Kakizuka, A., Expanded polyglutamine in the Machado-Joseph disease protein induces cell death *in vitro* and *in vivo*, *Nat. Genet.*, 13, 196, 1996.

59. Chuang, D.M. and Ishitani, R., A role for GAPDH in apoptosis and neurodegeneration, *Nat. Med.*, 2, 609, 1996.

60. Ishitani, R., Sunaga, K., Tanaka, M., Aishita, H., and Chuang, D.M., Overexpression of glyceraldehyde-3-phosphate dehydrogenase is involved in low K+-induced apoptosis but not necrosis of cultured cerebellar granule cells, *Mol. Pharm.*, 51, 542, 1997.

61. Cooper, A.J.L., Sheu, K.R., Burke, J.R., Onodera, O., Strittmatter, W.J., Roses, A.D., and Blass, J.P., Transglutaminase-catalyzed inactivation of glyceralde-hyde-3-phosphate andalphaketoglutarate dehydrogenase complex by poly-glutamine domains of pathological length, *Proc. Nat. Acad. Sci. U.S.A.*, 94,12604, 1997.

62. Schapira, A.H., Mitochondrial disorders, *Curr. Opin. Neurol.*, 10, 43, 1997.

63. Marchetti, P., Casteldo, M., Susin, S.A., Zamzami, N, Hirsch, T., Macho, A., Haeffner, A., Hirsch, F., Geuskins, M., and Kroemer, G., Mitochondrial perme-ability transition is a central coordinating event of apoptosis, *J. Exp. Med.*, 184, 1155, 1996.

64. Yang, J., Liu, X., Bhalla, K., Kim, C.N., Ibrado, A.M., Cai, J., Peng, T.-I., Jones, D.P., and Wang, X., Prevention of apoptosis by Bcl-2: release of cytochrome c from mitochondria blocked, *Science*, 275, 1129, 1997.

65. Antonsson, B., Conti, F., Ciavatta, A., Montessuit, S., Lewis, S., Martinou, I., Bernasconi, L., Bernard, A., Mermod, J.-J., Mazzei, G., Maundrell, K., Gambale, F., Sadoul, R., and Martinou, J.-C., Inhibition of Bax channel-forming activity by Bcl-2, *Science*, 277, 370, 1997.

66. Okazaki, M., Ishibashi, Y., Asoh, S., and Ohta, S., Overexpressed mitochondrial hinge protein, a cytochrome c-binding protein, accelerates apoptosis by en-hancing the release of cytochrome c from mitochondria, *Biochem. Biophys. Res. Commun.*, 243, 131, 1998.

67. Shimizu, S., Eguchi, Y., Kamiike, W., Funahashi, Y., Mignon, A.,Lacronique, V., Matsuda, H., and Tsujimoto, Y., Bcl-2 prevents apoptotic mitochondrial dys-function by regulating proton flux, *Proc. Nat. Acad. Sci. U.S.A.*, 95, 1455, 1998.

68. Schmechel, D., Assessment of ultrastructural changes associated with apopto-sis, in *Aptosis in Neurobiology: Concepts and Methods*, Hannun, Y.A. and Boustany, R.-M. Eds., CRC Press, Boca Baton, FL, Chap. 3, in press.

69. Schmechel, D.E., Burkhart, D.S., Ange, R., and Izard, M.K., Cholinergic axonal dystrophy and mitochondrial pathology in prosimian primates, *Exp. Neurol.*, 142, 111, 1996.

70. Su, J. H., Anderson, A. J., Cummings, B. J., and Cotman, C. W., Immunohis-tochemical evidence for apoptosis in Alzheimer's disease, *Neuroreport*, 5, 2529, 1994.

71. Cotman, C. W., Whittemore, E. R., Watt, J. A., Anderson, A. J., and Loo, D. T., Possible role of apoptosis in Alzheimer's disease, *Ann. N. Y. Acad. Sci.*, 747, 36, 1994.

72. Satou, T., Cummings, B. J. and Cotman, C. W., Immunoreactivity for Bcl-2 protein within neurons in the Alzheimer's disease brain increases with disease severity, *Brain Res.*, 697, 35, 1995.

73. Nishimura, T., Akiyama, H., Yonehara, S., Kondo, H., Ikeda, K., Kato, M., Iseki E., and Kosaka, K., Fas antigen expression in brains of patients with Alzheimer-type dementia, *Brain Res.*, 695, 137, 1995.

74. Lassmann, H., Bancher, C., Breitschopf, H., Wegiel, J., Bobinski, M., Jellinger, K. and Wisniewski, H. M., Cell death in Alzheimer's disease evaluated by DNA fragmentation *in situ*, *Acta Neuropathol. (Berl.)*, 89, 35, 1995.

75. Cotman, C. W. and Anderson, A. J., A potential role for apoptosis in neurode-generation and Alzheimer's disease, *Mol. Neurobiol.*, 10, 19, 1995.

76. Smale, G., Nichols, N. R., Brady, D. R., Finch, C. E. and Horton, W. E., Jr., Evidence for apoptotic cell death in Alzheimer's disease, *Exp. Neurol.*, 133, 225, 1995.

77. Dragunow, M., Faull, R. L., Lawlor, P., Beilharz, E. J., Singleton, K., Walker, E. B. and Mee, E., *In situ* evidence for DNA fragmentation in Huntington's disease striatum and Alzheimer's disease temporal lobes, *Neuroreport*, 6, 1053, 1995.

78. Anderson, A. J., Su, J. H., and Cotman, C. W., DNA damage and apoptosis in Alzheimer's disease: colocalization with c-Jun immunoreactivity, relationship to brain area, and effect of postmortem delay, *J. Neurosci.*, 16, 1710, 1996.

79. Cotman, C. W. and Su, J. H., Mechanisms of neuronal death in Alzheimer's disease, *Brain Pathol.*, 6, 493, 1996.

80. Sugaya, K., Reeves, M., and McKinney, M., Topographic associations between DNA fragmentation and Alzheimer's disease neuropathology in the hippocampus, *Neurochem. Int.*, 31, 275, 1997.

81. Nagy, Z., Esiri, M. M., and Smith, A. D., Expression of cell division markers in the hippocampus in Alzheimer's disease and other neurodegenerative conditions, *Acta Neuropathol. (Berl)*, 93, 294, 1997.

82. MacGibbon, G. A., Lawlor, P. A., Walton, M., Sirimanne, E., Faull, R. L., Synek, B., Mee, E., Connor, B., and Dragunow, M., Expression of Fos, Jun, and Krox family proteins in Alzheimer's disease, *Exp. Neurol.*, 147, 316, 1997.

83. MacGibbon, G. A., Lawlor, P. A., Sirimanne, E. S., Walton, M. R., Connor, B., Young, D., Williams, C., Gluckman, P., Faull, R. L., Hughes, P. and Dragunow, M., Bax expression in mammalian neurons undergoing apoptosis, and in Alzheimer's disease hippocampus, *Brain Res.*, 750, 223, 1997.

84. Su, J. H., Deng, G., and Cotman, C. W., Bax protein expression is increased in Alzheimer's brain: correlations with DNA damage, Bcl-2 expression, and brain pathology, *J. Neuropathol. Exp. Neurol.*, 56, 86, 1997.

85. Vyas, S., Javoy-Agid, F., Herrero, M. T., Strada, O., Boissiere, F., Hibner, U., and Agid, Y., Expression of Bcl-2 in adult human brain regions with special reference to neurodegenerative disorders, *J. Neurochem.*, 69, 223, 1997.

86. Drache, B., Diehl, G. E., Beyreuther, K., Perlmutter, L. S. and Konig, G., Bcl-xl-specific antibody labels activated microglia associated with Alzheimer's disease and other pathological states, *J. Neurosci. Res.*, 47, 98, 1997.

87. de la Monte, S. M., Sohn, Y. K. and Wands, J. R., Correlates of p53- and Fas (CD95)-mediated apoptosis in Alzheimer's disease, *J. Neurol. Sci.*, 152, 73, 1997.

88. Lucassen, P. J., Chung, W. C., Kamphorst, W., and Swaab, D. F., DNA damage distribution in the human brain as shown by *in situ* end labeling; area-specific differences in aging and Alzheimer disease in the absence of apoptotic morphology, *J. Neuropathol. Exp. Neurol.*, 56, 887, 1997.

89. Li, W. P., Chan, W. Y., Lai, H. W., and Yew, D. T., Terminal dUTP nick end labeling (TUNEL) positive cells in the different regions of the brain in normal aging and Alzheimer patients, *J. Mol. Neurosci.*, 8, 75, 1997.

90. LaFerla, F. M., Tinkle, B. T., Bieberich, C. J., Haudenschild, C. C., and Jay, G., The Alzheimer's A beta peptide induces neurodegeneration and apoptotic cell death in transgenic mice, *Nat. Genet.*, 9, 21, 1995.

91. Yamatsuji, T., Matsui, T., Okamoto, T., Komatsuzaki, K., Takeda, S., Fukumoto, H., Iwatsubo, T., Suzuki, N., Asami-Odaka, A., Ireland, S., Kinane, T. B., Giambarella, U. and Nishimoto, I., G protein-mediated neuronal DNA fragmentation induced by familial Alzheimer's disease-associated mutants of APP, *Science*, 272, 1349, 1996.

92. Forloni, G., Bugiani, O., Tagliavini, F. and Salmona, M., Apoptosis-mediated neurotoxicity induced by beta-amyloid and PrP fragments, *Mol. Chem. Neuropathol.*, 28, 163, 1996.

93. Wolozin, B., Iwasaki, K., Vito, P., Ganjei, J.K., Lacana, E., Sunderland, T., Zhao, B., Kusiak, J.W., Wasco, W., and D'Adamio, L., Participation of presenilin 2 in apoptosis: enhanced basal activity conferred by an Alzheimer mutation, *Science,* 274, 1710, 1996.

94. Guo, Q., Furukawa, K., Sopher, B. L., Pham, D. G., Xie, J., Robinson, N., Martin, G. M. and Mattson, M. P., Alzheimer's PS-1 mutation perturbs calcium homeostasis and sensitizes PC12 cells to death induced by amyloid beta-peptide, *Neuroreport,* 8, 3790, 1996.

95. Paradis, E., Douillard, H., Koutroumanis, M., Goodyer, C., and LeBlanc, A., Amyloid beta peptide of Alzheimer's disease downregulates Bcl-2 and upregulates bax expression in human neurons, *J. Neurosci.,* 16, 7533, 1996.

96. Deng, G., Pike, C. J., and Cotman, C. W., Alzheimer-associated presenilin-2 confers increased sensitivity to apoptosis in PC12 cells, *FEBS Lett.,* 397, 50, 1996.

97. Vito, P., Wolozin, B., Ganjei, J. K., Iwasaki, K., Lacana, E., and D'Adamio, L., Requirement of the familial Alzheimer's disease gene PS2 for apoptosis. Opposing effect of ALG-3, *J. Biol. Chem.,* 271, 31025, 1996.

98. Yamatsuji, T., Okamoto, T., Takeda, S., Murayama, Y., Tanaka, N., and Nishimoto, I., Expression of V642 APP mutant causes cellular apoptosis as Alzheimer trait-linked phenotype, *Embo. J.,* 15, 498, 1996.

99. Janicki, S. and Monteiro, M. J., Increased apoptosis arising from increased expression of the Alzheimer's disease-associated presenilin-2 mutation (N141I), *J. Cell. Biol.,* 139, 485, 1997.

100. Zhang, L., Zhao, B., Yew, D. T., Kusiak, J. W., and Roth, G. S., Processing of Alzheimer's amyloid precursor protein during H_2O_2-induced apoptosis in human neuronal cells, *Biochem. Biophys. Res. Commun.,* 235, 845, 1997.

101. Giambarella, U., Yamatsuji, T., Okamoto, T., Matsui, T., Ikezu, T., Murayama, Y., Levine, M. A., Katz, A., Gautam, N., and Nishimoto, I., G protein betagamma complex-mediated apoptosis by familial Alzheimer's disease mutant of APP, *Embo. J.,* 16, 4897, 1997.

102. Mattson, M. P. and Guo, Q., Cell and molecular neurobiology of presenilins: a role for the endoplasmic reticulum in the pathogenesis of Alzheimer's disease?, *J. Neurosci. Res.,* 50, 505, 1997.

103. Guo, Q., Sopher, B. L., Furukawa, K., Pham, D. G., Robinson, N., Martin, G. M., and Mattson, M. P., Alzheimer's presenilin mutation sensitizes neural cells to apoptosis induced by trophic factor withdrawal and amyloid beta-peptide: involvement of calcium and oxyradicals, *J. Neurosci.,* 17, 4212, 1997.

104. Yaar, M., Zhai, S., Pilch, P. F., Doyle, S. M., Eisenhauer, P. B., Fine, R. E., and Gilchrest, B. A., Binding of beta-amyloid to the p75 neurotrophin receptor induces apoptosis. A possible mechanism for Alzheimer's disease, *J. Clin. Invest.,* 100, 2333, 1997.

105. Masliah, E., Sisk, A., Mallory, M., Mucke, L., Schenk, D., and Games, D., Comparison of neurodegenerative pathology in transgenic mice overexpressing V717F beta-amyloid precursor protein and Alzheimer's disease, *J. Neurosci.,* 16, 5795, 1996

106. Irizarry, M. C., Soriano, F., McNamara, M., Page, K. J., Schenk, D., Games, D., and Hyman, B. T., A-beta deposition is associated with neuropil changes, but not with overt neuronal loss in the human amyloid precursor protein V717F (PDAPP) transgenic mouse, *J. Neurosci.,* 17, 7053, 1997.

107. Schmechel, D.E., Xu, P.-T., Gilbert, J.D., and Roses, A.D., Model of genetic susceptibility to late-onset Alzheimer's Disease: mice transgenic for human apolipoprotein E alleles In: *Mouse Models of Genetic Neurological Disease*, Popko, B., Ed., Plenum Press, New York, 1998.
108. Miyata, M. and Smith, J.D., Apolipoprotein E allele-specific antioxidant activity and effects on cytotoxicity by oxidative insults and beta-amyloid peptides, *Nat. Genet.*, 14, 55, 1996.
109. Simonian, N. A. and Coyle, J. T., Oxidative stress in neurodegenerative diseases, *Annu. Rev. Pharmacol. Toxicol.*, 36, 83, 1996.
110. Stern, G., Parkinson's disease. The apoptosis hypothesis, *Adv. Neurol.*, 69, 101, 1996.
111. Jeon, B. S., Jackson-Lewis, V., and Burke, R. E., 6-Hydroxydopamine lesion of the rat substantia nigra: time course and morphology of cell death, *Neurodegeneration*, 4, 131, 1995.
112. Mochizuki, H., Goto, K., Mori, H., and Mizuno, Y., Histochemical detection of apoptosis in Parkinson's disease, *J. Neurol. Sci.*, 137, 120, 1996.
113. Mogi, M., Harada, M., Kondo, T., Mizuno, Y., Narabayashi, H., Riederer, P., and Nagatsu, T., The soluble form of Fas molecule is elevated in parkinsonian brain tissues, *Neurosci. Lett.*, 220, 195, 1996.
114. Vyas, S., Javoy-Agid, F., Herrero, M. T., Strada, O., Boissiere, F., Hibner, U., and Agid, Y., Expression of Bcl-2 in adult human brain regions with special reference to neurodegenerative disorders, *J. Neurochem.*, 69, 223, 1997.
115. Anglade, P., Vyas, S., Javoy-Agid, F., Herrero, M. T., Michel, P. P., Marquez, J., Mouatt-Prigent, A., Ruberg, M., Hirsch, E. C., and Agid, Y., Apoptosis and autophagy in nigral neurons of patients with Parkinson's disease, *Histol. Histopathol.*, 12, 25, 1997.
116. Hunot, S., Brugg, B., Ricard, D., Michel, P. P., Muriel, M. P., Ruberg, M., Faucheux, B. A., Agid, Y., and Hirsch, E. C., Nuclear translocation of NF-kappaB is increased in dopaminergic neurons of patients with Parkinson disease, *Proc. Natl. Acad. Sci. U.S.A.*, 94, 7531, 1997.
117. Marshall, K. A., Daniel, S. E., Cairns, N., Jenner, P. and Halliwell, B., Upregulation of the antiapoptotic protein Bcl-2 may be an early event in neurodegeneration: studies on Parkinson's and incidental Lewy body disease, *Biochem. Biophys. Res. Commun.*, 240, 84, 1997.
118. Drache, B., Diehl, G. E., Beyreuther, K., Perlmutter, L. S., and Konig, G., Bcl-xl-specific antibody labels activated microglia associated with Alzheimer's disease and other pathological states, *J. Neurosci. Res.*, 47, 98, 1997.
119. Ziv, I., Melamed, E., Nardi, N., Luria, D., Achiron, A., Offen, D. and Barzilai, A., Dopamine induces apoptosis-like cell death in cultured chick sympathetic neurons — a possible novel pathogenetic mechanism in Parkinson's disease, *Neurosci. Lett.*, 170, 136, 1994.
120. Walkinshaw, G. and Waters, C. M., Induction of apoptosis in catecholaminergic PC12 cells by L-DOPA. Implications for the treatment of Parkinson's disease, *J. Clin. Invest.*, 95, 2458, 1995.
121. Ziv, I., Barzilai, A., Offen, D., Stein, R., Achiron, A. and Melamed, E., Dopamine-induced, genotoxic activation of programmed cell death. A role in nigrostriatal neuronal degeneration in Parkinson's disease?, *Adv. Neurol.*, 69, 229, 1996.

122. Maruyama, W., Naoi, M., Kasamatsu, T., Hashizume, Y., Takahashi, T., Kohda, K., and Dostert, P., An endogenous dopaminergic neurotoxin, N-methyl-(R)-salsolinol, induces DNA damage in human dopaminergic neuroblastoma SH-SY5Y cells, *J. Neurochem.*, 69, 322, 1997.

123. Offen, D., Ziv, I., Barzilai, A., Gorodin, S., Glater, E., Hochman, A., and Melamed, E., Dopamine-melanin induces apoptosis in PC12 cells; possible implications for the etiology of Parkinson's disease, *Neurochem. Int.*, 31, 207, 1997.

124. Ziv, I., Offen, D., Barzilai, A., Haviv, R., Stein, R., Zilkha-Falb, R., Shirvan, A., and Melamed, E., Modulation of control mechanisms of dopamine-induced apoptosis — a future approach to the treatment of Parkinson's disease?, *J. Neural. Transm. Suppl.*, 49, 195, 1997.

125. Hartley, A., Stone, J. M., Heron, C., Cooper, J. M., and Schapira, A. H., Complex I inhibitors induce dose-dependent apoptosis in PC12 cells: relevance to Parkinson's disease, *J. Neurochem.*, 63, 1987, 1994.

126. Michel, P. P., Vyas, S., Anglade, P., Ruberg, M., and Agid, Y., Morphological and molecular characterization of the response of differentiated PC12 cells to calcium stress, *Eur. J. Neurosci.*, 6, 577, 1994.

127. Ozawa, T., Hayakawa, M., Katsumata, K., Yoneda, M., Ikebe, S. and Mizuno, Y., Fragile mitochondrial DNA: the missing link in the apoptotic neuronal cell death in Parkinson's disease, *Biochem. Biophys. Res. Commun.*, 235, 158, 1997.

128. France-Lanord, V., Brugg, B., Michel, P. P., Agid, Y., and Ruberg, M., Mitochondrial free radical signal in ceramide-dependent apoptosis: a putative mechanism for neuronal death in Parkinson's disease, *J. Neurochem.*, 69, 1612, 1997.

129. Offen, D., Ziv, I., Sternin, H., Melamed, E., and Hochman, A., Prevention of dopamine-induced cell death by thiol antioxidants: possible implications for treatment of Parkinson's disease, *Exp. Neurol.*, 141, 32, 1996.

130. Tatton, W. G., Wadia, J. S., Ju, W. Y., Chalmers-Redman, R. M., and Tatton N. A., (–)-Deprenyl reduces neuronal apoptosis and facilitates neuronal outgrowth by altering protein synthesis without inhibiting monoamine oxidase, *J. Neural. Transm. Suppl.*, 48, 45, 1996.

131. Eisen, A. and Krieger, C., Pathogenic mechanisms in sporadic amyotrophic lateral sclerosis, *Can. J. Neurol. Sci.*, 20, 286, 1993.

132. Rothstein, J. D., Bristol, L. A., Hosler, B., Brown, R. H., Jr., and Kuncl, R. W., Chronic inhibition of superoxide dismutase produces apoptotic death of spinal neurons, *Proc. Natl. Acad. Sci. U.S.A.*, 91, 4155, 1994.

133. Alexianu, M. E., Mohamed, A. H., Smith, R. G., Colom, L. V., and Appel, S. H., Apoptotic cell death of a hybrid motoneuron cell line induced by immunoglobulins from patients with amyotrophic lateral sclerosis, *J. Neurochem.*, 63, 2365, 1994.

134. Rabizadeh, S., Gralla, E. B., Borchelt, D. R., Gwinn, R., Valentine, J. S., Sisodia, S., Wong, P., Lee, M., Hahn, H., and Bredesen, D. E., Mutations associated with amyotrophic lateral sclerosis convert superoxide dismutase from an antiapoptotic gene to a proapoptotic gene: studies in yeast and neural cells, *Proc. Natl. Acad. Sci. U.S.A.*, 92, 3024, 1995.

135. Jackson, M., Morrison, K. E., Al-Chalabi, A., Bakker, M., and Leigh, P. N., Analysis of chromosome 5q13 genes in amyotrophic lateral sclerosis: homozygous NAIP deletion in a sporadic case, *Ann. Neurol.*, 39, 796, 1996.

136. Mu, X., He, J., Anderson, D. W., Trojanowski, J. Q., and Springer, J. E., Altered expression of bcl-2 and bax mRNA in amyotrophic lateral sclerosis spinal cord motor neurons, *Ann. Neurol.*, 40, 379, 1996.

137. Siklos, L., Engelhardt, J., Harati, Y., Smith, R. G., Joo, F. and Appel, S. H., Ultrastructural evidence for altered calcium in motor nerve terminals in amyotropic lateral sclerosis, *Ann. Neurol.*, 39, 203, 1996.

138. Ghadge, G. D., Lee, J. P., Bindokas, V. P., Jordan, J., Ma L., Miller, R. J., and Roos, R. P., Mutant superoxide dismutase-1-linked familial amyotrophic lateral sclerosis: molecular mechanisms of neuronal death and protection, *J. Neurosci.*, 17, 8756, 1997.

139. Durham, H. D., Roy, J., Dong, L. and Figlewicz, D. A., Aggregation of mutant Cu/Zn superoxide dismutase proteins in a culture model of ALS, *J. Neuropathol. Exp. Neurol.*, 56, 523, 1997.

140. Rothstein, J.D., Van Kammen, M., Levey, A.I., Martin, L., and Kuncl, R.W., Selective loss of glial glutamate transporter GLT-1 in amyotrophic lateral sclerosis, *Ann. Neurol.*, 38, 73, 1995.

141. Ginsburg, S., Martin, L., and Rothstein, J.D., Regional deafferentation down regulates subtypes of gluamate tranporters, *J. Neurochem.*, 65, 2800, 1995.

142. Rothstein, J.D., Dykes-Hoberg, M., Pardo, C.A., Bristol, L.A., Jin, L., Kuncl, R.W., Kanai, Y., Hediger, M., Wang, Y., Shielke, J., and Welty, D.F., Knockout of glutamate transporters reveals a major role for astroglial transport in excitotoxicity and clearance of glutamate, *Neuron*, 16, 675, 1996.

143. Davis, K.E., Straff, D.J., Weinstein, E.A., Bannerman, P.G., Correale, D.M., Rothstein, J.D., and Robinson, M.B., Multiple signaling pathways regulate cell surface expression and activity of the excitatory amino acid carrier 1 subtype of Glu transporter in C6 glioma, *J. Neurosci.*, 18, 2475, 1998.

144. Lin, C.L., Bristol, L.A., Jin, L., Dykes-Hoberg, M., Crawford, T., Clawson, L. and Rothstein, J.D., Aberrant RNA processing in a neurodegenerative disease: the cause for absent EAAT2, a glutamate transporter, in amyotrophic lateral sclerosis, *Neuron*, 20, 589, 1998.

145. Hahnen, E., Schonling, J., Rudnik-Schoneborn, S., Raschke, H., Zerres, K, and Wirth, B., Missense mutations in exon 6 of the survival motor neuron gene in patients with spinal muscular atrophy (SMA), *Hum. Mol. Genet.*, 6, 821, 1997.

146. Nagashima, K., A review of organic methylmercury toxicity in rats: neuropathology and evidence for apoptosis, *Toxicol. Pathol.*, 25, 624, 1997.

147. Schmied, M., Breitschopf, H., Gold, R., Zischler, H., Rothe, G., Wekerle, H., and Lassmann, H., Apoptosis of T lymphocytes in experimental autoimmune encephalomyelitis: evidence for programmed cell death as a mechanism to control inflammation in the brain, *Am. J. Pathol.*, 143, 446, 1993.

148. Leist, M. and Nicotera, P., Apoptosis, excitotoxicity and neuropathology, *Exp. Cell Res.*, 239, 183, 1998.

4

Neurooncology and Cancer Therapy

Nina Felice Schor

CONTENTS

The role of apoptosis in cancer biology and therapy in general, and neurooncology in particular, has taken many forms. Aberrancy of the forces that regulate apoptosis initiation and enactment has been blamed for tumorigenesis in many systems. Molecular species that act to protect cells from undergoing apoptosis are being increasingly recognized as mediators of resistance to currently available cancer chemo- and radiotherapy. Most recently, and as a direct outgrowth of this, novel targets for cancer therapy and prevention include most prominently those species that play a role in inducing, enacting, and/or preventing programmed cellular death. This chapter will deal in turn with each of these aspects of apoptosis in neurooncology.

Both inside and outside of the nervous system, it has become clear that apoptosis results from cells embarking on a final common pathway, and there are many roads that lead to this distal approach.[1-3] Which of the many proximal feeders a cell traverses, and whether or not a particular stimulus or condition ultimately results in apoptosis, depend in part on the particular agent used to induce apoptosis, and in part on the intrinsic characteristics of that cell.[4-6] As a consequence, it becomes impossible to discuss in a generic way apoptosis of nervous system neoplastic cells. Rather, one must distinguish findings and conclusions in this area from one another on the basis of the particular cell type (even at the molecular level) and inducer of apoptosis in question. Much of the published work on apoptosis in nervous system neoplasia has involved gliomas and glioblastomas in the central nervous system, and neuroblastomas in the peripheral nervous system. The emphasis

0-8493-3352-0/99/$0.00+$.50
© 1999 by CRC Press LLC

on these particular tumors is perhaps the result of their relative frequency in the clinical arena and the ready availability of cell lines derived from them. This emphasis is reflected in the discussion that follows. In this discussion, the names of genes related to apoptosis are indicated by italicized, lower case letters (e.g., *p53*); the names of the proteins encoded by these genes are indicated by plain type with a capitalized first letter (e.g., P53).

4.1 Apoptosis Regulation and the Pathogenesis of Nervous System Neoplasia

The control of apoptosis induction is a major arbiter, not only of individual cell fate, but of the sculpting and maintenance of organs and organisms throughout embryogenesis, postnatal development, and adult life. It is a determinant of cell number and the complement of cell types that comprise each organ and organ system. It is therefore not surprising that many of the regulators of apoptosis have been either proposed and investigated or fully implicated in the pathogenesis of nervous system neoplasia.

Mutation of the tumor suppressor gene, *p53*, is the most commonly identified genetic aberration in glioma and glioblastoma cells. It is thought that mechanistic information obtained from glioma cell lines on the role of *p53* mutations in prevention of apoptosis is a valid mirror of the pathogenesis of *in vivo* tumors because the frequency of *p53* mutations in such lines is the same as that found in primary tumors.[7]

In general, mutational inactivation of P53 leads to prevention of apoptosis and induction of tumor formation.[2,3] The role of wild-type P53 in inducing the synthesis of proteins important for apoptosis induction was first determined in a glioblastoma multiforme cell line in which *p53* was of the mutant type. While this native cell line did not express the protein WAF-1/CIP-1, induction of wild-type *p53* in these cells resulted in production of this protein, a key element in one pathway for apoptosis induction.[8] Inactivation of P53 has since been shown to be an early event in tumorigenesis, and induction (by transfection) of wild-type *p53* expression in mutant *p53*-expressing glioma cells results in apoptosis.[7] In this regard, it is interesting that those astrocytomas that overexpress wild-type *p53* also express higher levels of the anti-apoptosis gene, *bcl-2*. It has been proposed that the increased expression of *bcl-2* is a mechanism by which these cells evade P53-mediated apoptosis and thereby establish themselves as tumorigenic.[9]

A role has also been proposed for protein kinase C in the pathogenesis (by prevention of apoptosis) and invasiveness of gliomas.[10,11] This hypothesis grows out of the observations that glioma cells frequently overexpress protein kinase C, and pharmacologic inhibition of this enzyme results in apoptosis induction in these cells.[11] Furthermore, fibroblast and epidermal growth factors enhance glioma growth rate and invasiveness in animal model studies, and both are known to initiate signal transduction pathways that involve activation of protein kinase C.[10]

It has long been known that nuclear condensation during mitosis is accompanied by phosphorylation of histone H3 and deubiquination of histone H2A. Studies in glioma cells demonstrating similar postsynthetic modifications of histones have raised the question of whether defects in the ability to effect these modifications might underlie the genesis of gliomas.[12] However, because these chemical changes are unique to chromatin condensation *per se* and not to apoptosis, it is doubtful that they represent primary events in the initiation of apoptosis.

Bcl-2 is one of a family of anti-apoptosis proteins that have been implicated in both tumorigenesis and tumor resistance to therapy in a host of cell types.[2,3] In the case of central nervous system neoplasia, however, expression of *bcl-2* is generally fairly low.[13] While it is true that the *bcl-2* gene is expressed in some cells within medulloblastomas and gliomas, studies of both tumor and normal astrocytes suggest that the variation in *bcl-2* expression seen in cells within these brain tumors reflects the developmental regulation of Bcl-2 production seen in normal maturing and reactive astrocytes.[14]

In contrast to this, overproduction of Bcl-2 and its structural relatives has been implicated in both the etiology and chemoresistance of neuroblastomas.[15-17] The findings that some neuroblastomas are present at birth, and that autopsies of infants that die of other causes reveal neuroblastic rests in a fraction that is larger than that which would be predicted by the incidence of clinically apparent neuroblastoma, have led to the notion that neuroblastoma is a developmental anomaly that results from aberrations in the regulation of apoptosis. The observation that neuroblastomas in particular metastatic sites in young children often disappear without treatment has led some to hypothesize that an alteration in the timing of developmentally governed apoptosis is etiologic in this instance.[18] Consistent with a role for Bcl-2 in the genesis of neuroblastomas, Bcl-2 is prevalent in these tumors, but not in their more "mature," benign counterparts, ganglioneuromas. Accordingly, TUNEL staining for apoptosis reveals abundant apoptotic cells in ganglioneuromas and not in neuroblastomas.[15] In addition, artificially induced overexpression of *bcl-2* in neuroblastomas leads to resistance to apoptosis induction,[16] and many neuroblastomas obtained from patients overproduce either Bcl-2 or the Bcl-2–like protein, Bcl-x_L.[15,17] Unlike the case for gliomas, mutations of the *p53* gene are rare in neuroblastomas.

4.2   Resistance to Therapy: The Role of Antiapoptosis Factors

Many of the chemo- and radiotherapeutic agents used in the treatment of nervous system neoplasia work by inducing apoptosis in the cells of these tumors. As such, the same anti-apoptosis factors that give rise to the tumors in the first place are also responsible for their resistance to therapy. The conventional notion of resistance to a single class of chemotherapeutic agents evolving out of a selection pressure exerted by the use of agents in that class is clearly incomplete. Clinically encountered resistance is now often the result of mechanisms that act well downstream from the proximate action of the drug or radiation itself. These mechanisms are therefore protective against apoptosis induced by a whole host of mechanistically distinct agents.

For example, the prevalence of mutant P53 in cells of glioblastoma multiforme specimens predicts the number of mechanistically different chemotherapeutic agents to which those cells are resistant.[18] Transfection with a constitutive expression construct for Bcl-2 predictably confers chemoresistance upon glioma cells *in vitro*.[19] In an analogous fashion, higher stage neuroblastomas generally have more Bcl-2 and are more chemoresistant than lower stage neuroblastomas.[20] Again, predictably, transfection with *bcl-2* confers chemotherapeutic resistance upon neuroblastoma cells in culture.[16]

Three novel strategies for overcoming Bcl-2–mediated chemotherapeutic resistance in neuroblastoma have been proposed. The first is predicated on the observation that *bcl-2* expression by neuroblastomas is differentially influenced by different differentiation-inducing agents. While most such agents induce increased Bcl-2 levels, differentiation induced by interferon-γ is accompanied by decreased Bcl-2 production.[6] This has led to the suggestion that interferon-γ be used in the therapy of neuroblastoma, not only as a differentiation-inducing agent *per se*, but also as an adjunct to the use of more conventional chemotherapeutic agents to which these cells might otherwise be resistant.

The second strategy exploits the finding that overexpression of *bcl-2* in neuroblastoma cell lines is accompanied by an increase in the cellular concentration of reduced glutathione.[21] This approach uses, as chemotherapy, agents that are prodrugs that require reaction with reduced glutathione for activation. (This is unlike other chemotherapeutic drugs, most of which are detoxified by conjugation with glutathione.) Trials of these prodrugs in cultured neural crest tumor cells have demonstrated "proof-of-concept" in that *bcl-2* overexpression actually potentiates the apoptosis-inducing activity of these drugs.[22] These studies further suggest that this potentiation is the result of increased activation of the prodrugs in cells with increased reducing potential. However, given the proposed mechanism of Bcl-2 in preventing the release of cytochrome c from the mitochondrion,[23,24] it is curious that apoptosis is not still blocked in this system distal to the prodrug activation step. Preliminary studies aimed at this issue suggest that while Bcl-2 offers protection to cells from the oxidation by other drugs of membrane species important in the initiation of apoptosis,[25] it does not protect such species from oxidation by these activated prodrugs (Tyurina, Kagan, and Schor, preliminary results).

Finally, the observation has been made that agents that increase the activity of protein kinase A produce down-regulation of *bcl-2*.[26] It has been proposed that the development of such agents might result in altering for the better the chemosensitivity of tumor cells that, in their native state, overexpress this gene.

Recently, protection from apoptosis of both glioma and neuroblastoma cells by endogenously produced growth factors has been hypothesized, and has also given rise to the proposal of novel targets and strategies for overcoming chemotherapeutic resistance in these tumors. One example of this grows out of the realization that nerve growth factor (NGF) is both an anti-apoptosis factor for and present in the environment of neuroblastoma cells. For at least some neuroblastoma cells, this effect of NGF is mediated by the binding of NGF to its low-affinity, or *p75*, receptor.[27] In these cells, the naked *p75* receptor appears to facilitate apoptosis, while its occupation by NGF prevents apoptosis induced by a variety of conditions,[28] including treatment with chemotherapeutic agents.[27] This raises the possibility that prevention of chemotherapeutic drug-induced apoptosis by endogenous NGF might be responsible for the chemoresistance of some neuroblastomas. The recognition that specific and different regions of the NGF molecule bind to the low-affinity *p75* and high-affinity *trkA* receptors, respectively, has led to the development of ligands for one or the other, but not both, receptors,[29,30] and to the proposal that selective competitive inhibitors of *p75* alone might be effective and nontoxic adjunctive agents for overcoming the chemoresistance of neuroblastomas.[27]

Downstream approaches to the trophic effects of platelet-derived growth factor and epidermal growth factor on glioma cells have also been proposed in an effort to overcome chemotherapeutic resistance. These approaches include inhibition of protein kinase C and tyrosine kinase. Both of these proteins are key components of the signal transduction pathways triggered by binding of platelet-derived and epidermal growth factors to their respective receptors and enzymes, the activities of which are markedly increased in glioma cells.[31] Agents that decrease both protein kinase C and tyrosine kinase activities, including staurosporine and hypericin, have been proposed as apoptosis-inducing chemotherapeutic agents for gliomas.[11] Recently, the protein kinase C-selective inhibitor, calphostin C, has been shown by itself to induce apoptosis and to decrease *bcl-2* expression and Bcl-2 levels in glioma cells.[31]

4.3 Induction of Apoptosis as an Antineoplastic Strategy

Many conventional antineoplastic agents have been demonstrated to induce apoptosis in a host of normal and tumor cell types.[32] The exploitation of apoptosis induction as a useful therapeutic strategy depends critically upon the specific targeting of this effect for the neoplastic cells in question. Novel methods by which to target nervous system neoplasia for apoptosis induction are discussed in this section.

The hydroxymethylglutaryl- (HMG-) CoA reductase inhibitor, lovastatin, has been reported to have a differential apoptosis-inducing effect upon glioma cells, as opposed to normal glial cells in culture.[33,34] In glioma cells, this drug induces growth arrest followed by apoptosis.[35] In normal glial cells, only a reversible decrease in the rate of DNA synthesis occurs.[34] It has been proposed that this differential effect is the result of the requirement for isoprenylation in the course of several pathways, the up-regulation of which is associated with the neoplastic state. Lovastatin inhibits isoprenoid synthesis, and thereby blocks this step. In support of this hypothesis, the addition of mevalonate to lovastatin-treated cells, a means of by-passing the HMG-CoA reductase reaction blocked by lovastatin, prevents lovastatin-induced apoptosis.[35] Lovastatin has therefore been proposed as an antineoplastic agent in glioma therapy.

For reasons that are not mechanistically clear, the estradiol–nitrogen mustard fusion molecule, estramustine, appears also to effect glioma-specific G_2/M arrest followed by apoptosis in a rat model of the disease.[36] Characteristic DNA end-labeled staining and DNA fragmentation are seen in tumor sections from estramustine-treated, but not untreated, rats. Normal brain tissue from estramustine-treated rats did not demonstrate these apoptosis-associated findings. Similarly, treatment with tumor necrosis factor-α induces apoptosis and results in up-regulation of wild-type, but not mutant, *p53* in cultured rat glioma cells, but not in cultured normal newborn rat astroglia or neurons.[37]

One targeted apoptosis strategy currently in clinical trial in childhood and adult central nervous system tumors involves the insertion of the gene for herpes virus thymidine kinase into a retroviral vector. The stereotactically introduced viral gene-containing construct confers sensitivity to the cytotoxic nucleotide analogue ganciclovir upon the cells by facilitating the phosphorylation of ganciclovir and its consequent incorporation into host DNA. This leads to cessation of DNA synthesis and apoptotic cell death inhibitable by overexpression of *bcl-2*. In addition, this approach has been shown to lead to a "bystander" effect on tumor cells that did not incorporate the viral gene, presumably because of the efflux of the toxic ganciclovir phosphate from the cells in which it was produced.[38-40] This is of particular interest because it is virtually impossible to get a foreign gene into 100% of the cells of a tumor. It, of course, leads as well to concern about the toxicity of released drug to surrounding normal nonneural bystander cells.[41]

Targeting of apoptosis induction for neuroblastoma cells has similarly involved identifying those biochemical and cytological characteristics that distinguish these cells from normal neural crest and from normal rapidly dividing nonneural tissues. In addition, it has been determined that, in neural crest-derived tumor cell lines, whether a cell undergoes apoptosis, necrosis, or differentiation in response to a particular drug is a function of the intrinsic characteristics of that particular cell.[4,42,43] This latter point is important, because most neuroblastomas are comprised of a mixture of neural- and epithelial-type cells, and each of these phenotypes is associated with a different response to such stimuli as antimitotic agents[4,43] or differentiation-inducing agents.[44]

One targeted strategy for the induction of apoptosis in neuroblastoma is the development of cytotoxic neurotransmitter analogues such as 6-hydroxydopamine. This dopamine analogue is a generator of reactive oxygen species,[45] and has been demonstrated to induce apoptosis in neural crest tumor cells *in vitro*.[46] It gains selective access to cells, like some neuroblastoma cells, with a dopamine uptake system.[47,48] It is excluded from the CNS and therefore does not produce CNS toxicity, because it is too polar to cross the blood–brain barrier.[49] However, because it is predictably toxic to the normal sympathetic nervous system,[50] its clinical use has been limited to the *ex vivo* purging of bone marrow. Recently, there is some preclinical evidence[51] that 6-hydroxydopamine can be rendered safe and selective for tumor tissue by adjunctive administration with the free radical scavenger and superoxide dismutase mimic, Tempol.[52] Because Tempol is selectively active in normal cells, it protects only them, leaving neoplastic cells susceptible to attack. Studies in a murine allograft model suggest that this adjunctive regimen is both relatively nontoxic and effective.[51] Studies are currently underway to define in neuroblastoma-bearing mice the effectiveness of this regimen against human neuroblastoma and the limits of dose and dosing schedule for this drug combination.

A closely related experimental approach involves the use of a prodrug that can be selectively activated inside neuroblastoma cells by a neurotransmitter analogue specifically designed to chemically react with the prodrug. For example, neocarzinostatin is one of a class of antimitotic prodrugs that requires sulfhydryl activation,[52] and that, once activated, induces apoptosis in neural-type and differentiation with mitotic arrest in epithelial-type neuroblastoma cells.[4,53] By co-administration with 6-mercaptodopamine, a dopamine analogue that has been modified by addition of a sulfhydryl group at the 6-position on the catecholamine ring, neuroblastoma cells can be selectively loaded with reactive sulfhydryl groups, and the concentration–response curve of cultured neuroblastoma cells to neocarzinostatin can be synergistically shifted 2- to 3-fold in the direction of greater efficacy.[54] Theoretically, the activation of neocarzinostatin to its antimitotic metabolite in normal neurons with a dopamine uptake system would not result in adverse effects, since most of these cells are postmitotic. The *in vivo* usefulness of this approach is under examination.

4.4 Apoptosis and Nervous System Cancer

Apoptosis and alterations of its regulation clearly play a role in the origin, resistance to therapy, and effectiveness of therapy of nervous system neoplasia. An understanding of apoptosis and the forces that facilitate and

TABLE 4.1

Proposed Mediators of Tumorigenesis and/or Therapeutic Resistance in Nervous System Neoplasia

Gliomas, Glioblastomas	Mutant P53 protein
	Basic fibroblast growth factor
	Platelet-derived growth factor
	Epidermal growth factor
	Protein kinase C
Neuroblastomas	Bcl-x_L
	Nerve growth factor
	P-glycoprotein (multidrug resistance-associated protein)
	n-*myc*

Note: Both the origins of tumors and their resistance to chemotherapy have been hypothesized to be related to resistance to or blocking of apoptosis. The proximate mediators of these phenomena differ from cell type to cell type. This fact makes imperative the individualization of therapeutic approaches aimed at apoptosis-related targets.

impair its progress is critical to an understanding of the pathogenesis of neoplasia in the nervous system and elsewhere. Novel targets for overcoming therapeutic resistance and inducing selective apoptosis in nervous system tumors are already being identified as a result of advances in our delineation of the pathways that lead ultimately to apoptotic cell death.

Although it is easy philosophically and useful intellectually to refer to drugs as apoptosis inducers and to cells as targets for apoptosis induction, it has become clear that what that means in particular is dependent both upon the drug and the cell of reference. For example, Table 4.1 lists the molecular species proposed to be involved in blocking glioma or neuroblastoma cell apoptosis in response to natural and/or pharmacologic stimuli. The differences between these tumors in the proximal pathway that leads downstream to apoptosis demonstrate the probability that selective pharmacologic induction of apoptosis must rely on delineation and targeting of pathways that may be unique to each tumor type. As such, even within a tumor, different cells have different, "hard-wired" propensities and respond to different signals to undergo apoptosis. This heterogeneity at all levels represents a challenge for the future in cancer therapy.

References

1. Lindenboim, L., Haviv, R., Stein, R., Inhibition of drug-induced apoptosis by survival factors in PC12 cells, *J Neurochem*, 64, 1054, 1995.
2. Savill, J., Apoptosis in disease, *Eur J Clin Invest*, 24, 715, 1994.
3. Ameisen, J. C., Programmed cell death (apoptosis) and cell survival regulation: relevance to AIDS and cancer, *AIDS*, 8, 1197, 1994.

4. Hartsell, T. L., Hinman, L. M., Hamann, P. R., Schor, N. F., Determinants of the response of neuroblastoma cells to DNA damage: the roles of pre-treatment cell morphology and chemical nature of the damage, *J Pharmacol Exp Therap*, 277, 1158, 1996.

5. Marushige, K., Marushige, Y., Induction of apoptosis by transforming growth factor beta 1 in glioma and trigeminal neurinoma cells, *Anticancer Res*, 14, 2419, 1994.

6. Kim, C. J., Kim, H. O., Choe, Y. J., Lee, Y. A., Kim, C. W., Bcl-2 protein expression in neuroblastoma is differentially regulated by differentiation inducers, *Anticancer Res*, 15, 1997, 1995.

7. Gomez-Manzano, C., Fueyo, J., Kyritsis, A. P., Steck, P. A., Roth, J. A., McDonnell, T. J., Steck, K. D., Levin, V. A., Yung, W. K., Adenovirus-mediated transfer of the *p53* gene produces rapid and generalized death of human glioma cells via apoptosis, *Cancer Res*, 56, 694, 1996.

8. El-Deiry, W. S., Tokino, T., Velculescu, V. E., Levy, D. B., Parsons, R., Trent, J. M., Lin, D., Mercer, W. E., Kinzler, K. W., Vogelstein, B, WAF1, a potential mediator of *p53* tumor suppression, *Cell*, 75, 817, 1993.

9. Alderson, L. M., Castleberg, R. L., Harsh, G. R., 4th, Louis, D. N., Henson, J. W., Human gliomas with wild-type *p53* express *bcl-2*, *Cancer Res*, 55, 999, 1995.

10. Couldwell, W. T., Hinton, D. R., Law, R. E., Protein kinase C and growth regulation in malignant gliomas, *Neurosurgery*, 35, 1184, 1994.

11. Couldwell, W. T., Gopalakrishna, R., Hinton, D. R., He, S., Weiss, M. H., Law, R. E., Apuzzo, M. L., Hypericin: a potential antiglioma therapy, *Neurosurgery*, 35, 705, 1994.

12. Marushige, Y., Marushige, K., Disappearance of ubiquinated histone H2A during chromatin condensation in TGF beta 1-induced apoptosis, *Anticancer Res*, 15, 267, 1995.

13. Reed, J. C., Meister, L., Tanaka, S., Cuddy, M., Yum, S., Geyer, C., Pleasure, D., Differential expression of *bcl2* protooncogene in neuroblastoma and other human tumor cell lines of neural origin, *Cancer Res*, 51, 6529, 1991.

14. Krishna, M., Smith, T. W., Recht, L. D., Expression of *bcl-2* in reactive and neoplastic astrocytes: lack of correlation with presence or degree of malignancy, *J Neurosurg*, 83, 1017, 1995.

15. Ikeda, H., Hirato, J., Akami, M., Matsuyama, S., Suzuki, N., Takahashi, A., Kuroiwa, M., Bcl-2 oncoprotein expression and apoptosis in neuroblastoma, *J Pediatr Surg*, 30, 805, 1995.

16. Dole, M., Nunez, G., Merchant, A. K., Maybaum, J., Rode, C. K., Bloch, C. A., Castle, V. P. 1994 Bcl-2 inhibits chemotherapy-induced apoptosis in neuroblastoma, *Cancer Res*, 54, 3253, 1994.

17. Dole, M. G., Jasty, R., Cooper, M. J., Thompson, C. B., Nunez, G., Castle, V. P., Bcl-X$_L$ is expressed in neuroblastoma cells and modulates chemotherapy-induced apoptosis, *Cancer Res*, 55, 2576, 1995.

18. Iwadate, Y., Fujimoto, S., Tagawa, M., Namba, H., Sueyoshi, K., Hirose, M., Sakiyama, S., Association of *p53* gene mutation with decreased chemosensitivity in human malignant gliomas, *Int J Cancer*, 69, 236, 1996.

19. Weller, M., Malipiero, U., Aguzzi, A., Reed, J. C., Fontana, A., Protooncogene bcl-2 gene transfer abrogates Fas/APO-1 antibody-mediated apoptosis of human malignant glioma cells and confers resistance to chemotherapeutic drugs and therapeutic irradiation, *J Clin Invest*, 95, 2633, 1995.

20. Hoehner, J. C., Hedborg, F., Wiklund, H. J., Olsen, L., Pahlman, S., Cellular death in neuroblastoma: *in situ* correlation of apoptosis and *bcl-2* expression, *Int J Cancer*, 62, 19, 1995.
21. Kane, D. J., Sarafian, T. A., Anton, R., Hahn, H., Gralla, E. B., Valentine, J. S., Ord, T., Bredesen, D. E., Bcl-2 inhibition of neural death: decreased generation of reactive oxygen species, *Science*, 262, 1274, 1993.
22. Cortazzo, M., Schor, N. F., Potentiation of enediyne-induced apoptosis and differentiation by Bcl-2, *Cancer Res*, 56, 1199, 1996.
23. Kluck, R. M., Bossy-Wetzel, E., Green, D. R., Newmeyer, D. D., The release of cytochrome c from mitochondria: a primary site for Bcl-2 regulation of apoptosis, *Science*, 275, 1132, 1997.
24. Yang, J., Liu, X., Bhalla, K., Kim, C. N., Ibrado, A. M., Cai, J., Peng, T.-I., Jones, D. P., Wang, X., Prevention of apoptosis by Bcl-2: release of cytochrome c from mitochondria blocked, *Science*, 275, 1129, 1997.
25. Tyurina, Y. Y., Tyurin, V. A., Carta, G., Quinn, P. J., Schor, N. F., Kagan, V. E., Direct evidence for antioxidant effect of Bcl-2 in PC12 rat pheochromocytoma cells, *Arch Biochem Biophys*, 344, 413, 1997.
26. Itano, Y., Ito, A., Uehara, T., Nomura, Y., Regulation of Bcl-2 protein expression in human neuroblastoma SH-SY5Y cells: positive and negative effects of protein kinases C and A, respectively, *J Neurochem*, 67, 131, 1996.
27. Cortazzo, M. H., Kassis, E. S., Sproul, K. A., Schor, N. F., Nerve growth factor (NGF)-mediated protection of neural crest cells from antimitotic agent-induced apoptosis: the role of the low-affinity NGF receptor, *J Neurosci*, 16, 3895, 1996.
28. Rabizadeh, S., Oh, J., Zhong, L.-T., Yang, J., Bitler, C. M., Butcher, L. L., Bredesen, D. E., Induction of apoptosis by the low-affinity NGF receptor, *Science*, 261, 345, 1993.
29. Shih, A., Laramee, G. R., Schmelzer, C. H., Burton, L. E., Winslow, J. W., Mutagenesis identifies amino-terminal residues of nerve growth factor necessary for trk receptor binding and biological activity, *J Biol Chem*, 269, 27679, 1994.
30. Ibanez, C. F., Ebendal, T., Barbany, G., Murray-Rust, J., Blundell, T. L., Persson, H., Disruption of the low affinity receptor-binding site in NGF allows neuronal survival and differentiation by binding to the trk gene product, *Cell*, 69, 329, 1992.
31. Ikemoto, H., Tani, E., Matsumoto, T., Nakano, A., Furuyama, J, Apoptosis of human glioma cells in response to calphostin C, a specific protein kinase C inhibitor, *J Neurosurg*, 83, 1008, 1995.
32. Hickman, J. A., Apoptosis induced by anticancer drugs, *Cancer Metastasis Rev*, 11, 121, 1992.
33. Langan, T. J., Volpe, J. J., Obligatory relationship between the sterol biosynthetic pathway and DNA synthesis and cellular proliferation in glial primary cultures, *J Neurochem*, 46, 1283, 1986.
34. Langan, T. J., Slater, M. C., Isoprenoids and astroglial cell cycling: diminished mevalonate availability and inhibition of dolichol-linked glycoprotein synthesis arrest cycling through distinct mechanisms, *J Cell Physiol*, 149, 284, 1991.
35. Jones, K. D., Couldwell, W. T., Hinton, D. R., Su, Y., He, S., Anker, L., Law, R. E., Lovastatin induces growth inhibition and apoptosis in human malignant glioma cells, *Biochem Biophys Res Commun*, 205, 1681, 1994.
36. Valbo, C., Bergenheim, T., Bergh, A., Grankvist, K., Henriksson, R., DNA fragmentation induced by the antimitotic drug estramustine in malignant rat glioma but not in normal brain — suggesting an apoptotic cell death, *Br J Cancer*, 71, 717, 1995.

37. Yin, D., Kondo, S., Barnett, G. H., Morimura, T., Takeuchi, J., Tumor necrosis factor-alpha induces *p53*-dependent apoptosis in rat glioma cells, *Neurosurgery,* 37, 758, 1995.
38. Short, M. P., Choi, B. C., Lee, J. K., Malick, A., Breakfield, X. O., Martuza, R. L, Gene delivery to glioma cells in rat brain by grafting of a retrovirus packaging cell line, *J Neurosci Res*, 27, 427, 1990.
39. Maria, B. L., Medina, C. D., Gene therapy: the beginning of a new era in treating neurological disease, *Internat Pediat*, 11, 248, 1996.
40. Hamel, W., Magnelli, L., Chiarugi, V. P., Israel, M. A., Herpes simplex virus thymidine kinase/gancyclovir-mediated apoptotic death of bystander cells, *Cancer Res*, 56, 2697, 1996.
41. Ram, Z., Culver, K. W., Walbridge, S., Frank, J. A., Blaese, R. M., Oldfield, E. H., Toxicity studies of retroviral-mediated gene transfer for the treatment of brain tumors, *J Neurosurg*, 79, 400, 1993.
42. Gshwind, M., Huber, G., Apoptotic cell death induced by beta-amyloid 1-42 peptide is cell type dependent, *J Neurochem*, 65, 292, 1995.
43. Smith, T. K., Nylander, K. D., Schor, N. F., The roles of mitotic arrest and protein synthesis in induction of apoptosis and differentiation in neuroblastoma cells in culture, *Dev Brain Res*, 105, 175, 1998.
44. Melino, G., Annicchiarico-Petruzzelli, M., Piredda, L., Candi, E., Gentile, V., Davies, P. J. A., Piacentini, M., Tissue transglutaminase and apoptosis: sense and antisense transfection studies with human neuroblastoma cells, *Mol Cell Biol*, 14, 6584, 1994.
45. Tiffany-Castiglioni, E., Saneto, R. P., Proctor, P. H., Perez-Polo, J. R., Participation of active oxygen species in 6-hydroxydopamine toxicity to a human neuroblastoma cell line, *Biochem Pharmacol*, 31, 181, 1982.
46. Waters, C. M., Walkinshaw, G., Distinct pathways for induction of apoptosis in PC12 cells, *Soc Neurosci Abst*, 21, 2, 1995.
47. Thoenen, H., Tranzer, J. P., Chemical sympathectomy by selective destruction of adrenergic nerve endings with 6-hydroxydopamine, *Naunyn-Schmiedbergs Arch Exp Pathol Pharmakol*, 261, 271, 1968.
48. Uretsky, N. J., Iversen, L. L., Effects of 6-hydroxydopamine on noradrenaline-containing neurons in the rat brain, *Nature (Lond)*, 221, 557, 1969.
49. Malmfors, T., Sachs, C., Degeneration of adrenergic nerves produced by 6-hydroxydopamine, *Eur J Pharmacol*, 3, 89, 1968.
50. Schor, N. F. T., Adjunctive use of ethiofos (WR-2721) with free radical-generating chemotherapeutic agents in mice: new caveats for therapy, *Cancer Res*, 47, 5411, 1987.
51. Purpura, P., Westman, L., Will, P., Eidelman, A., Kagan, V. E., Osipov, A. N., Schor, N. F., Adjunctive treatment of murine neuroblastoma with 6-hydroxydopamine and Tempol, *Cancer Res*, 56, 2336, 1996.
52. Hahn, S. M., Tochner, Z., Murali Krishna, C., Glass, J., Wilson, L., Samuni, A., Sprague, M., Venzon, D., Glatstein, E., Mitchell, J. B., Russo, A., Tempol, a stable free radical, is a novel murine radiation protector, *Cancer Res*, 52, 1750, 1992.
53. Hartsell, T. L., Yalowich, J. C., Ritke, M. K., Martinez, A. J., Schor, N. F., Induction of apoptosis in murine and human neuroblastoma cell lines by the enediyne natural product neocarzinostatin, *J Pharmacol Exp Therap*, 275, 479, 1995.
54. Schor, N. F., Targeted enhancement of the biological activity of the antineoplastic agent, neocarzinostatin: studies in murine neuroblastoma cells, *J Clin Invest*, 89, 774, 1992.

5

Human Immunodeficiency Virus Type 1 Infection: Chronic Inflammation and Programmed Cell Death in the Central Nervous System

Harris A. Gelbard

Human immunodeficiency virus type 1 (HIV-1) is a member of the lentiviral family that includes simian immunodeficiency virus, visna virus, and feline immunodeficiency virus. All lentivirus infections are characterized by multiorgan disease, including infection and inflammation of brain tissue to varying degrees. Current estimates suggest there are approximately 21.8 million individuals worldwide infected with HIV-1, and of this group, at least one million are children. Furthermore, HIV-1–associated dementia (HIV-D) is the most frequent cause of neurologic disease in young adults in the U.S.[1] HIV-D is estimated to occur at an annual rate of 7% in people with AIDS.[2] HIV-1 replicates continuously through the asymptomatic and symptomatic phases of infection. This poses a special problem for the central nervous system (CNS), because new antiretroviral therapies and treatment for opportunistic infections have lengthened the survival time, but have not eradicated the virus from the CNS. This is largely due to impaired immune surveillance and poor penetration of antiretroviral agents through the blood–brain barrier (BBB) and the blood-cerebrospinal fluid (CSF) barrier.

HIV-1 infection of the CNS occurs early in the course of disease. Neurologic disease, including dementia, occurs relatively late in the course of infection and is usually associated with immunosuppression. HIV-D may be associated with the neuropathologic correlates of HIV-1 encephalitis (HIVE).[3] HIV-1 productively infects brain-resident macrophages and microglia.[4,5] In the absence of opportunistic infections of brain, features of HIVE include microglial activation, multinucleated giant cells, astrocytosis, and myelin pallor (decreased staining for myelin). HIV-1–infected children (usually through vertical transmission) have the most striking evidence of this disease complex.[5] Using polymerase chain reaction/*in situ* hydridization techniques, no productive or latent HIV-1 infection of endothelial cells or oligodendrocytes has been demonstrated in the CNS.[6]

However, HIV-1 can "restrictively" infect astrocytes in children and adults; that is, only regulatory gene products such as Tat and nef are made without production of progeny virus.[7,8,9] The actual percentage of restrictively infected astrocytes in brains of patients with HIVE is unknown, but thought to be relatively small. The pathophysiologic significance of restricted infection in astrocytes remains unknown.

Productive infection of neurons with HIV-1 has never been demonstrated. Nuovo et al. demonstrated latent proviral HIV-1 infection of neurons in brain tissue from a single patient with severe HIV-D of 5 years' duration using *in situ* polymerase chain reaction (ISPCR).[10] However, using similar techniques, Takahashi et al. failed to demonstrate HIV-1 infection in neurons from patients with HIV-1–associated dementia.[6] Nevertheless, it is accepted that there is focal neuronal loss in retina, neocortex, and in subcortical brain regions, including putamen, substantia nigra, and cerebellum. There is also a decrease in synaptic density and vacuolation of dendritic spines in affected brain regions.[11,12,13,14,15] However, quantitative analysis of neuronal loss from postmortem studies of patients with HIV-1 and neurologic dysfunction has not demonstrated a clear correlation between the magnitude of neuronal loss and neurologic disease.[16]

As previously noted, cytolytic infection of neurons by HIV-1 is highly unlikely. Thus, changes in neuronal architecture, signal transduction, and number that underlie neurologic disease are likely due to indirect mechanisms; that is, soluble, diffusible factors, including HIV-1 gene products and cellular metabolites released from productively infected macrophages or microglia, present in focal inflammatory infiltrates, mediate dysfunction and death in vulnerable neurons. Numerous reports have demonstrated that HIV-1–infected macrophages and microglia produce soluble neurotoxic factors. Levels of some of these neurotoxins can markedly increase after antigenic stimulation.[17,18] Neurotoxic factors include the HIV-1 coat protein gp120, gp41, and Tat,[19,20,21] as well as cellular metabolites, including eicosanoids (i.e., arachidonic acid and its metabolites), the phospholipid mediator platelet activating factor (PAF), the proinflammatory cytokine tumor necrosis factor alpha (TNF-α), and an as yet unidentified NMDA receptor agonist, NTox, that has been tentatively identified as a phenolic amine with lipophilic properties that lacks the carboxyl groups of quinolinic acid.[22,23,17,24] With the exception of NTox, most of these neurotoxins act by indirect mechanisms that involve dysregulation of a normal cellular process, such as arachidonic acid- and TNF-α–mediated decreases in glutamate uptake in affected astrocytes or gp120-mediated increases in glutamate efflux from affected astrocytes. All of these effects may then result in excess levels of glutamate and overstimulation of NMDA and non-NMDA receptors on vulnerable neurons.[25,26] Dysregulation of high-affinity glutamate uptake in astrocytes could be further amplified by PAF because microglial PAF receptor activation can result in arachidonic acid release.[27] Other indirect mechanisms of neurotoxicity include: activation of autocrine loops in adjacent uninfected macrophages or microglia, such as gp41- and gp120-mediated release of arachidonic acid, cytokines, and nitric oxide;[28,29] and immune activation of uninfected monocytes by proinflammatory cytokines or antigenic stimulation to produce the quinolinic acid, which in turn can directly activate NMDA receptors.[18] Furthermore, glucocorticoids, adrenal hormones released during stress, can potentiate gp120-induced neurotoxicity.[30]

Tat, TNF-α, and PAF may also have direct effects on vulnerable neurons. Tat has been shown to induce hippocampal CA1 neuron depolarization, probably via a non-NMDA receptor mechanism,[31] and induce neuronal apoptosis.[32] TNF-α can induce neuronal cell death, in part by activation of non-NMDA (i.e., AMPA) receptors.[33] This TNF-α–induced neuronal death has the biochemical and morphologic features of apoptosis, involves oxidative stress, and is independent of nuclear factor kappa B (NFκB) activation.[34] PAF can also induce neuronal cell death via a mechanism that involves activation of NMDA receptor channels,[17] and results in neuronal apoptosis.[35] Because these three neurotoxins appear to work in part through activation of non-NMDA and NMDA receptors, it is important to note that NMDA receptor activation and production of nitric oxide (potentially mediated via gp120) can also induce neuronal apoptosis.[36,37]

The hypothesis that a lentiviral infection such as HIV-1 causes cell loss and subsequent tissue atrophy in the brain and the immune system by inappropriate expression of genes involved in programmed cell death (PCD) was first promulgated in 1990.[38] In particular, the authors speculated that HIV-1 is able to modify normal inter- and intracellular signaling to induce pathologic programmed cell death of CD4+ lymphocytes. This hypothesis was confirmed in two subsequent studies.[49,40] A later study demonstrated that CD4+ and CD8+ lymphocytes from patients with both asymptomatic and symptomatic HIV-1 infection have increased levels of reactive oxygen species and a decreased mitochondrial transmembrane potential, which appears to be an early, irreversible step in activation of programmed cell death.[41]

As new techniques became available to identify free 3'-OH ends of newly cleaved DNA *in situ*,[42] coupled with the light microscopic identification of some of the morphologic hallmarks of apoptosis (i.e., chromatin condensation), determination of whether apoptosis of neural cells occurred in postmortem archival formalin-fixed brain tissue from patients with HIV-1 infection became feasible. Using the TUNEL (Terminal deoxynucleotidyl dUTP Nick End Labeling) technique, we demonstrated that apoptotic neurons were present in the cerebral cortex and basal ganglia of children that had HIVE and progressive encephalopathy (HIVE/PE).[43] Double-labeling immunocytochemistry for the TUNEL reagent and the HIV-1 P24 antigen in brain tissue from children with HIVE/PE revealed a spatial association between apoptotic neurons and perivascular inflammatory cell infiltrates containing HIV-1–infected macrophages and multinucleated giant cells (Figure 5.1A–C).[43] Quantitative morphometric analysis of apoptotic neurons

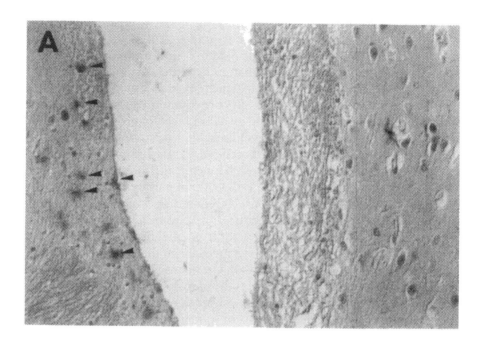

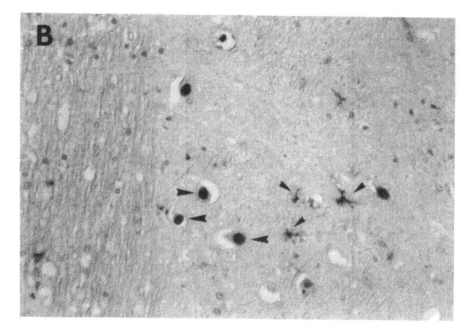

FIGURE 5.1
Montage of cerebral cortex from a pediatric patient with HIVE/PE stained with antisera to p24 (new fuchsin, red immunostain) and ApopTag (i.e., TUNEL) reagent (DAB, brown immunostain). Panel A shows migration of P24-positive perivascular, TUNEL-negative macrophages into brain tissue, Panel B shows an adjacent field of TUNEL-positive neurons (large arrows) and TUNEL-positive microglia (small arrows). Panel C is a higher power magnification of Panel B.

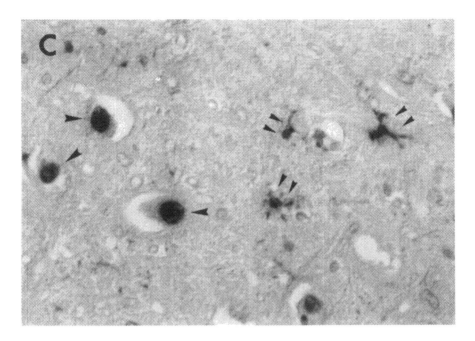

FIGURE 5.1 (continued)

present in basal ganglia from children with HIVE/PE revealed a 12-fold increase in apoptotic neurons relative to children that were seronegative for HIV-1 infection, and a 3-fold increase in apoptotic neurons relative to children that had HIV-1 infection without the neuropathologic features of HIVE, or a pre-mortem diagnosis of PE.[44]

However, apoptotic neurons were infrequently identified in a third of the cases from pediatric HIV-1 seronegative controls.[43] HIV-1 negative control tissue ranged in age between 3 weeks and 16.5 years. Taken together, these studies suggested that neuronal apoptosis was unlikely to be associated with postnatal development, but instead may be the end result of a disease process such as HIVE. Importantly, neuronal apoptosis in the CNS of adult patients with HIV-1 infection was confirmed in four separate reports.[45,46,47,48] Two reports demonstrate that neuronal apoptosis occurs in a small number of cases prior to the onset of clinical AIDS or dementia.[5,38]

Four studies noted the presence of apoptotic macrophages and microglia[45,18,38,45] in the brains of patients with HIVE, but only one report[45] (Petito et al., 1995) noted the presence of apoptotic astrocytes in 2/7 brains of patients with HIVE. This finding was only observed with a DNA polymerase technique, not the TUNEL technique. Interestingly, only one *in vitro* study has demonstrated apoptosis of astrocytes after HIV-1 infection of primary human brain cultures.[25]

In an attempt to reconcile these discrepancies and to further investigate the fate of glial cells in the brains of pediatric patients with HIVE, we have recently analyzed the *in situ* expression of pro- (*Bax*) and anti-apoptosis (*Bcl-*

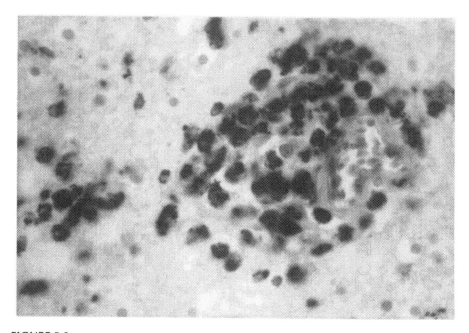

FIGURE 5.2

Human *Bax* expression in white matter containing inflammatory infiltrate from a pediatric patient with HIVE/PE. The chromagen is VIP (purple immunostain). Note that virtually all perivascular macrophages (also identified by CD68 immunostaining, data not shown) are strongly immunoreactive for cytoplasmic *Bax*.

2) gene products in cerebral cortex and basal ganglia.[29] Markedly elevated numbers of microglia, and macrophages immunoreactive for bax were present in the pons, basal ganglia and cerebral cortex of children with HIVE/PE (Figure 5.2), in comparison to HIV-1 infected children and HIV-1 seronegative children. Marked increased numbers of *Bax*-positive microglia were also observed in an adult brain with HIVE.[29] In contrast, astrocytes in brain tissue from patients with HIVE or HIV-1 had little or no *Bax* immunostaining.

Additional findings from Krajewski et al.[29] demonstrated that patients with HIVE/PE, but not HIV-1 or seronegative controls, had increased expression of *Bcl-2* in reactive astrocytes in cortex and basal ganglia. Thus, the lack of TUNEL staining and the presence of increased expression of *Bcl-2* suggest that astrocytes may be resistant to apoptosis in brain tissue that has the neuropathologic hallmarks of HIVE. In contrast, the findings of TUNEL staining, increased *Bax* expression, and decreased or absent *Bcl-2* expression in brain-resident macrophages and microglia suggest that these cells are more prone to undergo apoptosis in patients with HIVE. These findings are consonant with a cellular defense mechanism to limit microglial activation, thereby decreasing the spread of productive HIV-1 infection in brain-resident macrophages and microglia in the CNS of children with HIVE. In support of this hypothesis, a recent *in vitro* study demonstrated that caspase inhibitors

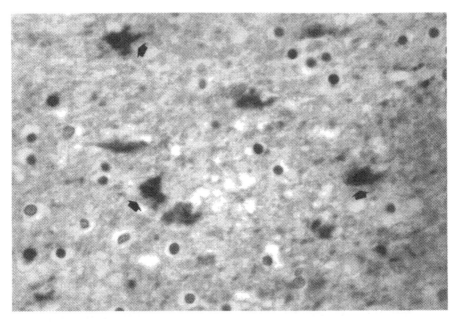

FIGURE 5.3

Human *Bcl-2* expression in white matter containing inflammatory infiltrate from the same pediatric patient with HIVE/PE (see Figure 5.2). The chromagen is VIP (purple immunostain). Note astrocytes (arrows; also identified by glial fibrillary acidic protein immunostaining, data not shown) are strongly immunoreactive for cytoplasmic *Bcl-2*.

actually stimulated HIV-1 production in activated peripheral blood mononuclear cells obtained from HIV-1–infected asymptomatic individuals.[10]

Curiously, TUNEL-positive neurons in basal ganglia and cerebral cortex of children with HIVE did not have cytoplasmic expression of *Bax* with double label immunocytochemistry. Neurons in tissue sections from basal ganglia and cerebral cortex of children with HIVE were not immunoreactive for *Bax* alone. Lack of neuronal *Bax* immunostaining in brain tissue from patients with HIVE may be due to fixation artifacts and suboptimal processing conditions secondary to delays in postmortem fixation. Because microglia and brain-resident macrophages were intensely immunoreactive for *Bax* in brains of patients with HIVE, a more likely explanation would be that neurons in the brains of patients with HIVE may undergo apoptosis secondary to dysregulation of other proapoptotic genes including *Bak, Bcl-x_s* and *Bad*.[43]

Several studies have focused on neuronal changes that occur during SIV infection. One report demonstrated that hippocampal neuronal atrophy occurs in rhesus macaques as early as 3 months following SIV inoculation.[31] In younger monkeys there was a positive association between a reduction in neuronal density and duration of infection. A more recent study by Adamson et al.,[1] using a neurovirulent strain of SIV, demonstrated a spatial association between apoptotic neurons and perivascular inflammatory cell infiltrates containing SIV-infected macrophages and multinucleated giant cells. These findings are consonant with previous reports in patients with HIVE.[3,18] However, this report also noted glial cell apoptosis, in agreement with Petito et al.[45]

Neuronal apoptosis has also been demonstrated in a severe combined immunodeficiency (SCID) mouse model of HIV-1 encephalitis. Here HIV-1–infected monocytes were exogenously infected with a neurovirulent macrophage-tropic strain of HIV-1 (ADA) and stereotactically injected into brain parenchyma, resulting in microglial activation, astrocyte proliferation, and TUNEL-stained neurons.[44] These findings were specific for HIV-1–infected monocytes, since stereotactically injected uninfected monocytes did not elicit the same inflammatory response or induce neuronal apoptosis.

In summary, the available data from *in vitro, in vivo,* and postmortem studies suggest that neurons in CNS with a lentiviral infection such as HIV-1 and SIV-1 undergo apoptosis as opposed to necrosis. Loss of vulnerable neurons by apoptosis is likely to be a gradual process, mediated by a complex, multifactorial network of proinflammatory mediators and excitotoxins that result in impaired glial-neuronal signaling, and ultimately neuronal death. With the development of *in vivo* models of HIVE such as the SCID mouse model,[44] the relative pathophysiologic significance of the many HIV-1–induced neurotoxins in mediating neuronal dysfunction and death can be ascertained. This may in turn help us to design more rational therapeutic strategies to ameliorate the neurologic disease associated with HIV-1 infection of the CNS.

References

1. Adamson, D.C., Dawson, T.M., Zinc, M.C., Clements, J.E., and Dawson, V.L. 1996a. Neurovirulent simian immunodeficiency virus infection induces neuronal, endothelial, and glial apoptosis. *Mol. Med.* 2(4):417-428.
2. Adamson, D.C., Wildemann, B., Sasaki, M., Glass, J.D., McArthur, J.C., Christov, V.I., Dawson, T.M., and Dawson, V.L. 1996b. Immunologic NO synthase: elevation in severe AIDS dementia and induction by HIV-1 gp41. *Science.* 274:1917-1921.
3. Adle-Biassette, H., Levy, Y., Colombel, M., Poron, F., Natchev, S., Keohane, C., and Gray, F. 1995. Neuronal apoptosis in HIV infection in adults. *Neuropathol. Appl. Neurobiol.,* 21:218-227.
4. Ameisen, J.C. and Capron, A. 1990. Cell dysfunction and depletion in AIDS: the programmed cell death hypothesis. *Immunol. Today.* 12:102-104.
5. An, S.F., Giometto, B., Scaravilli, T., Tavolato, B., Gray, F., and Scaravilli, F. 1996. Programmed cell death in brains of HIV-1 positive AIDS and pre-AIDS individuals. *Acta Neuropathol.* 91:169-173.

6. Benos, D.J., Hahn, B.H., Bubien, J.K., Ghosh, S.K., Mashburn, N.A., Chaikin, M.A., Shaw, G.M., and Benveniste, E.N. 1994. Envelope glycoprotein gp120 of human immunodeficiency virus type 1 alter ion transport in astroctyes: implications for AIDS dementia complex. *Proc. Natl. Acad. Sci. U.S.A.* 91:494-498.

7. Bonfoco, E., Krainc, D., Ankarcrona, M., Nicotera, P., and Lipton, S.A. 1995. Apoptosis and necrosis: two distinct events induced, respectively, by mild and intense insults with N-methyl-D-aspartate or nitric oxide / superoxide in cortical cell cultures. *Proc. Natl. Acad. Sci. U.S.A.* 92:7162-7166.

8. Brooke, S., Chan, R., Howard, S., and Sapolsky, R. 1997. Endocrine modulation of the neurotoxicity of gp120: implications for AIDS-related dementia complex. *Proc. Natl. Acad. Sci. U.S.A.* 94:9457-9462.

9. Chao, C.C. and Hu, S. 1994. Tumor necrosis factor-alpha potentiates glutamate neurotoxicity in human fetal brain cell cultures. *Dev. Neurosci.* 16:172-179.

10. Chinnaiyan, A.M., Woffendin, C., Dixit, V.M., and Nabel, G.J. 1997. The inhibition of pro-apoptotic ICE-like proteases enhances HIV replication. *Nature Med.* 3:333-337.

11. Dzenko K.A., Perry S.W., James H., Angel R.A., Epstein L.G., Whittaker S., Dewhurst S., and Gelbard, H.A. 1996. Tumor necrosis factor alpha and ceramide activate platelet activating factor receptors to induce neuronal apoptosis. *Soc. Neurosci. Abstr.* 22(3): 1917.

12. Everall, I.P., P.J. Luthbert, and Lantos, P.L. 1991. Neuronal loss in the frontal cortex in HIV infection. *Lancet* 337:1119-1121.

13. Everall, I.P., Glass, J.D., McArthur, J., Spargo, E., and Lantos, P. 1994. Neuronal density in the superior frontal and temporal gyri does not correlate with the degree of human immunodeficiency virus-associated dementia. *Acta Neuropathol.* 88:538-544.

14. Fine, S.M., Angel, R.A., Perry, S.W., Epstein, L.G., Dewhurst, S., and Gelbard., H.A. 1996. Tumor necrosis factor α inhibits glutamate uptake by primary human astrocytes: implications for pathogenesis of HIV-1 dementia. *J. Biol. Chem.,* 271(26):15303-15306.

15. Gavrieli, Y., Sherman, Y., and Ben-Sasson, S.A. 1992. Identification of programmed cell death *in situ* via specific labeling of nuclear DNA fragmentation. *J. Cell Biol.* 119:493-501.

16. Gelbard H.A., Dzenko K., DiLoreto D., del Cerro C., del Cerro M., and Epstein, L.G. 1993. Neurotoxic effects of tumor necrosis factor in primary human neuronal cultures are mediated by activation of the glutamate AMPA receptor subtype: implications for AIDS neuropathogenesis. *Dev. Neurosci.* 15:417-422.

17. Gelbard H.A., Nottet H.S.L.M., Swindells S., Jett M., Dzenko K.A., Genis P., White R., Wang L., Choi Y.-B., Zhang D., Lipton S.A., Tourtellotte W.W., Epstein L.G., and Gendelman, H.E. 1994. Platelet activating factor: a candidate HIV-1-Induced neurotoxin. *J. Virol.* 68:4628-4635.

18. Gelbard H.A., James H., Sharer L., Perry S.W., Saito, Y., Kazee, A.M., Blumberg B.M., and Epstein, L.G. 1995. Identification of apoptotic neurons in postmortem brain tissue with HIV-1 encephalitis and progressive encephalopathy. *Neuropathol. Appl. Neurobiol.,* 21:208-217.

19. Gelbard, H.A., Boustany, R-M., and Schor, N.F. 1997. Apoptosis in development and disease of the nervous system. 2: Apoptosis in childhood neurologic disease. *Ped. Neurol.* 16:93-97.

20. Genis, P., Jett, M., Bernton, E.W., Gelbard, H.A., Dzenko, K., Keane, R., Resnick, L., Volsky, D.J., Epstein, L.G., and Gendelman, H.E. 1992. Cytokines and arachidonic acid metabolites produced during HIV-infected macrophage-astroglial interactions: implications for the neuropathogenesis of HIV disease. *J. Exp. Med.* 176:1703-1718.

21. Giulian, D., Wendt, E., Vaca, K., and Noonan, C.A. 1993a. The envelope glycoprotein of human immunodeficiency virus type 1 stimulates release of neurotoxins from monocytes. *Proc. Natl. Acad. Sci. U.S.A.* 90: 2769-2773.

22. Giulian, D., Vaca, K., and Corpuz, M. 1993b. Brain glia release factors with opposing actions upon neuronal survival. *J. Neurosci.* 13:29-37.

23. Giulian, D., Yu, J.H., Li, X., Tom, D., Wendt, E., Lin, S.N., Schwarcz, R., and Noonan, C. 1996. Study of receptor-mediated neurotoxins released by HIV-1 infected mononuclear phagocytes found in human brain. *J. Neurosci.* 16:3139-3153.

24. Groux, H., Monte, D., Bourez, J.M., Capron, A., and Ameisen, J.C. 1992. Activation-induced death by apoptosis in CD4+ T cells from HIV-infected asymptomatic individuals. *J. Exp. Med.* 175:331-340.

25. He, J., deCastro, C.M., Vandenbark, G.R., Busciglio, J., and Gabuzda, D. 1997. Astrocyte apoptosis induced by HIV-1 transactivation of the c-kit protooncogene. *Proc. Natl. Acad. Sci. U.S.A.* 94:3954-3959.

26. Janssen, R.S., Nwanyanwu, O.C., Selik, R.M. and Stehr-Green, J.K. 1992. Epidemiology of human immunodeficiency virus encephalopathy in the United States. *Neurology* 42:1472-1476.

27. Ketzler, S., Weis, S. Haug, H. and Budka, H. 1990. Loss of neurons in the frontal cortex in AIDS brains. *Acta Neuropathol.* 80:92-94.

28. Koka, P., He, K., Zack, J.A., Kitchen, S., Peacock, W., Fried, I., Tran, T., Yashar, S.S., and Merrill, J.E. 1995. Human immunodeficiency virus 1 envelope proteins induce interleukin 1, tumor necrosis factor alpha, nitric oxide in glial cultures derived from fetal, neonatal and adult human brain. *J. Exp. Med.* 182:941-952.

29. Krajewski, S., James, H.J., Ross, J., Blumberg, B.M., Epstein L.G., Gendelman, H.E., Gummuluru, S., Dewhurst, S., Sharer, L.R., Reed, J.C., and Gelbard, H.A. 1997. Expression of pro- and anti-apoptosis gene products in brains from pediatric patients with HIV-1 encephalitis. *Neuropathol. Appl. Neurobiol.* 23:242-253.

30. Lipton, S.A. 1993. Human immunodeficiency virus-infected macrophages, gp120, and N-methyl-D-aspartate receptor-mediated neurotoxicity. *Ann. Neurol.* 33:227-228.

31. Luthert, P.J., Montgomery, M.M., Dean, A.F., Cook, R.W., Baskerville, A., and Lantos, P.L. 1995. Hippocampal neuronal atrophy occurs in rhesus macaques following infection with simian immunodeficiency virus. *Neuropathol. Appl. Neurobiol.* 21:529-534.

32. Macho, A., Castedo, M., Marchetti, P., Aguilar, J.J., Decaudin, D., Zamzami, N., Girard, P.M., Uriel, J., and Kroemer, G. 1995. Mitochondrial dysfunctions in circulating T lymphocytes from human immunodeficiency virus-1 carriers. *Blood.* 7:2481-2487.

33. Magnuson, D.S.K., Knudsen, B.E., Geiger, J.D., Brownstone, R.M., and Nath., A. 1995. Human immunodeficiency virus type 1 Tat activates non-N-methy-D-aspartate excitatory amino acid receptors and causes neurotoxicity. *Ann. Neurol.* 37:373-380.

34. McArthur, J.C., Hoover, D.R., Bacellar, H., Miller, E.N., Cohen, B.A., Becker, J.T., Graham, N.M.H., McArthur, J.H., Selnes, O.A., Jacobson, L.P., Visscher, B.R., Concha, M., and Saah, A. 1993. Dementia in AIDS patients: incidence and risk factors. *Neurology* 43:2245-2252.

35. Mori, M., Aihara, M., Kume, K., Hamanoue, M., Kohsaka, S., and Shimizu, T. 1996. Predominant expression of platelet-activating factor receptor in the rat brain microglia. *J. Neurosci.* 16:3590-3600.

36. Muller, W.E.G., Schroder, H.C., Ushijima, H., Dapper, J., and Bormann, J. 1992. gp120 of HIV-1 induces apoptosis in rat cortical cell cultures: prevention by memantine. *Eur. J. Pharmacol.* 226:209-214.

37. Nath, A., Psooy, K., Martin, C., Knudsen, B., Magnuson, D.S.K., Haughey, D., and Geiger, J.D. 1996. Identification of a human immunodeficiency virus type 1 Tat epitope that is neuroexcitatory and neurotoxic. *J. Virol.* 70:1475-1480.

38. Neuen-Jacob, E., Arendt, G., von Giesen, H.-J., and Wechsler, W. 1996. Neuronal cell apoptosis in the basal ganglia occurs early in the course of HIV encephalitis and may precede the clinical signs of HIV-1 associated dementia. *Neuropathol. Appl. Neurobiol.* 22:Suppl. 1:16-17.

39. New, D.R., Ma, M., Epstein, L.G., Nath, A. and Gelbard, H.A. 1997. Human immunodeficiency virus type 1 Tat protein induces death by apoptosis in primary human neuron cultures. *J. Neurovirol.* 3:168-173.

40. Nottet, H.S.L.M., Jett, M., Flanagan, C.R., Zhai, Q.-H., Persidsky, Y., Rizzino, A., Bernton, E.W., Genis, P., Baldwin, T., Schwartz, J., LaBenz, C.J., and Gendelman, H.E. 1995. A regulatory role for astrocytes in HIV-1 encephalitis. *J. Immunol.* 154:3567-3581.

41. Nottet, H.S.L.M., Flanagan, E.M., Flanagan, C.R., Gelbard, H.A., Gendelman, H.E., and Reinhard, Jr. J.F. 1996. The regulation of quinolinic acid in human immunodeficiency virus-infected monocytes. *J. Neurovirol.* 2:111-117.

42. Nuovo, G.J., Galtery, F., MacConnell, P., and Braun, A. 1994. *In situ* detection of polymerase chain reaction-amplified HIV-1 nucleic acids and tumor necrosis factor-α RNA in the central nervous system. *Am. J. Pathol.* 144:659-666.

43. Oltvai, Z.N. and Korsmeyer, S.J. 1994. Checkpoints of dueling dimers foil death wishes. *Cell.* 79:189-192.

44. Persidsky, Y., Limoges, J., McComb, R., Bock, P., Baldwin, T., Tyor, W., Patil, A., Nottet, H.S.L.M., Epstein, L., Gelbard, H., Flanagan, E., Reinhard, S.J., and Gendelman, H.E. 1996. A quantitative analysis of brain immunopathology in SCID mice with HIV-1 encephalitis. *Am. J. Pathol.* 149:1027-1053.

45. Petito, C.K. and Roberts, B. 1995. Evidence of apoptotic cell death in HIV encephalitis. *Am. J. Pathol.* 146:1121-1130.

46. Ranki, A., Nyberg, M., Ovod, V., Matti, H., Elovaara, I., Raininko, R., Haapasalo, H., and Krohn, K. 1995. Abundant expression of HIV Nef and Rev proteins in brain astrocytes *in vivo* is associated with dementia. *AIDS.* 9:1001-1008.

47. Saito, Y., Sharer, L.R., Epstein, L.G., Michaels, J., Mintz, M., Louder, M., Golding, K., Cvetkovich, T.A., and Blumberg, B.M. 1994. Overexpression of *nef* as a marker for restricted HIV-1 infection of astrocytes in postmortem pediatric central nervous tissues. *Neurology.* 44:474-480.

48. Sharer, L.R., Epstein, L.G., Cho, E.-S., Joshi, V.V., Meyenhofer, M.F., Rankin, L.F., and Petito, C.K. 1986. Pathologic features of AIDS encephalopathy in children: Evidence for LAV/HTLV-III infection of brain. *Hum. Pathol.* 17:271-284.

49. Sharer, L.R. 1992. Pathology of HIV-1 infection of the central nervous system (Review). *J. Neuropath. Exp. Neurol.* 51:3-11.

50. Takahashi, K., Wesselingh, S.L., Griffin, D.E., McArthur, J.C., Johnson, R.T., and Glass, J.D. 1996. Localization of HIV-1 in human brain using polymerase chain reaction/*in situ* hybridization and immunohistochemistry. *Ann. Neurol.* 39:705-711.

51. Talley A., Dewhurst S., Perry S., Gummuluru S., Dollard, S.C., Fine, S.M., New, D., Epstein L.G., Gendelman, H.E., and Gelbard, H.A. 1995. Tumor necrosis factor alpha induces apoptosis in human neuronal cells: protection by the antioxidant N-acetylcysteine and the genes *bcl-2* and *crmA*. *Mol. Cell Biol.* 15:2359-2366.

52. Tenhula, W.N., Xu, S.Z., Madigan, M.C., Heller, K., Freeman, W.F., and Sadun, A.A. 1992. Morphometric comparisons of optic nerve axon loss in acquired immunodeficiency syndrome. *Am. J. Ophthalmol.* 15:14-20.

53. Terai, C., Kornbluth, R.S. Pauza, C.D., Richman, D.D., and Carson, D.A. 1991. Apoptosis as a mechanism of cell death in cultured T lymphocytes acutely infected with HIV-1. *J. Clin. Invest.* 87:1710-1715.

54. Tornatore, C., Chandra, R., Berger, J.R., and Major, E.O. 1994. HIV-1 infection of subcortical astrocytes in the pediatric central nervous system. *Neurology.* 44:481-487.

55. Wiley, C.A., Schrier, R.D., Nelson, J.A., Lampert, P.W., and Oldstone, M.B.A. 1986. Cellular localization of human immunodeficiency virus infection within the brains of acquired immune deficiency syndrome patients. *Proc. Natl. Acad. Sci. U.S.A.* 83:7089-7093.

56. Wiley, C.A., Masliah, E., Morey, M., Lemere, C., DeTeresa, R., Grafe, M., Hansen, L., and Terry, R. 1991. Neocortical damage during HIV infection. *Ann. Neurol.* 29:651-657.

57. Wiley, C.A. and Achim, C. 1994. Human immunodeficiency virus encephalitis is the pathological correlate of dementia in acquired immunodeficiency syndrome. *Ann. Neurol.* 36:673-676.

58. Yeung, M.C., Pulliam, L., and Lau, A.S. 1995. The HIV envelope protein gp120 is toxic to human brain-cell cultures through the induction of interleukin-6 and tumor necrosis factor-α. *AIDS,* 9:137-143.

6

The Role of Proteases in Neuronal Apoptosis

Peter W. Mesner, Jr. and Scott H. Kaufmann*

CONTENTS

6.1 Background

Experiments performed over the past 5 years have implicated proteases in the initiation and execution phases of programmed cell death (PCD).

* Supported in part by a grant from the NIH (CA 69008) and a Leukemia Society of America Scholar Award to S.H.K.

Although these studies have outlined an evolutionarily conserved cell death machinery that appears to be similar in various types of cells, important differences between cell types are also emerging. Methods used to study the role of proteases in PCD are summarized in Chapter 11 by Kaufmann et al. In this chapter, we review the current understanding of apoptotic proteases, examine the evidence that they play a role in various types of neuronal cell death, and identify unanswered questions pertinent to the field of neurobiology. We focus primarily on information that has become available since this subject was last reviewed in depth.[1]

6.2   A Primer on Caspases

6.2.1   Caspases: An Evolutionarily Conserved Family of Cysteine Proteases

Apoptosis is a morphologically distinct form of eukaryotic cell death that occurs in a wide range of physiological and pathological settings.[2] The nematode *Caenorhabditis elegans* has provided an elegant model system for the study of this process.[3] During development of this organism, the fate of each cell can be reproducibly defined. Of the 1090 cells produced during ontogeny, precisely 131 undergo PCD. Genetic analysis has defined lineage-specific genes that are thought to be involved in the tissue-specific signaling required for PCD. In addition, a series of genes that appear to be involved in PCD in all cells have been identified. These include several genes whose products are involved in phagocytosis of the dead cells as well as a nuclease required for degradation of genomic DNA in the engulfed cells. Three genes involved further upstream in the cell death process were also genetically identified: *ced-3*, *ced-4* and *ced-9*.[3] Mutation of *ced-3* or *ced-4* abolishes the developmentally regulated cell deaths, permitting the cells to survive and differentiate into mature cells, often neurons.[3] Conversely, mutation of *ced-9* increases the number of cells undergoing PCD. Subsequent analysis has revealed that *ced-9* encodes a homologue of the mammalian protein Bcl-2, a polypeptide that inhibits PCD in a variety of cell types, including neurons;[4-9] *ced-4* encodes a homologue of the recently identified mammalian protein Apaf-1 (apoptotic protease activating factor-1),[10] a docking protein whose function is described below; and *ced-3* encodes cysteine protease related to members of the mammalian interleukin-1β–converting enzyme (ICE) family.[11,12]

At the present time, there are 10 known mammalian members of the ICE/ced-3 family (Table 6.1). These proteases, which are now called caspases (**c**ysteine-dependent **asp**artate-directed prot**ases**),[13] all appear to share several common features (reviewed in References 14 to 17a). First, they are cysteine-dependent proteases that cleave polypeptides on the carboxyl side of aspartate residues, a cleavage site that is distinctly unusual for proteases

TABLE 6.1

Selected Properties of ICE Family Proteases

New Name	Old Name	Identity with		Preferred Small Substrates[a]		Known Polypeptide Substrates
		ced-3	ICE			
Caspase-1	ICE	28%	—	YEVD↓X > YVAD↓X	WEHD↓X	**Cytoplasmic:** pro-IL-1β Not PARP
Caspase-2	Ich-1$_L$/NEDD2	28%	27%	VDVAD↓X	DEHD↓X	
Caspase-3	CPP32/YAMA/apopain	35%	30%	DMQD↓X > DEVD↓X	DEVD↓X	**Cytoplasmic:** Protein kinase Cγ, Protein kinase Cθ, gelsolin, p21-activated kinase 2 (PAK2), D4-GDP dissociation inhibitor (GDI), procaspases -6 and -9, huntingtin, presenilin-1 and -2, ICAD **Nuclear:** PARP, DNA-PK$_{cs}$, 70 kDa U1 SnRNP polypeptide, PITSLRE kinase, Rb, MDM2, replication factor C, C1 and C2 HnRNP
Caspase-4	Tx/ICE$_{rel}$-II	30%	50%	LEVD↓X > YEVD↓X	(W/L)EHD↓X	**Cytoplasmic:** Pro-ICE
Caspase-5	TY/ICE$_{rel}$-III	25%	52%	Unknown	(W/L)EHD↓X	**Nuclear:** Not pro-IL-1β
Caspase-6	Mch2	35%	29%	VEID↓X	VEHD↓X	**Nuclear:** Lamins A/C and B$_1$
Caspase-7	Mch3/CMH-1	33%	<30%	DEVD↓X > DMQD↓X	DEVD↓X	**Nuclear:** PARP, C1 and C2 HnRNP
Caspase-8	Mch5/FLICE	32%		IETD↓X	LETD↓X	**Cytoplasmic:** Procaspases -3, -4, -7, and -9
Caspase-9	Mch6/ICE-LAP6			Unknown	LEDH↓X	**Cytoplasmic:** Procaspases -3 and -6
Caspase-10	Mch4/FLICE2	32%		IEAD↓X	Not reported	**Cytoplasmic:** Procaspases -3, -7, and -8

[a] The left column indicates the preferred substrate specificity reported by Talanian et al.,[35] whereas the right column indicates that reported by Thornberry et al.[34] Additional details are found in Ref. 17a.

in higher eukaryotic organisms. Second, they have a unique quaternary structure compared to other cysteine proteases.[18,19] It appears that each active caspase is an $\alpha_2\beta_2$ tetramer composed of two 17- to 20-kDa and two 10- to 12-kDa subunits, with each active site involving one α and one β subunit.[18-20] Third, the caspases are all synthesized as zymogens, with each precursor polypeptide giving rise to one large subunit and one small subunit during an activation process that involves proteolytic cleavage at Asp-X cleavage sites.[18,20-22] The fact that these precursors are activated by cleavage at the same type of bond they themselves cleave raises the possibility that the various caspases activate each other [23-25] and are also capable of autoactivation.[26-29] This concept will be elaborated upon in subsequent Sections.

   Despite these similarities, there are also differences among the various caspases.[30-32] One of the most prominent differences appears to be their varied substrate specificities.[30,33-35] The caspases all require an asparate in the P1 position (i.e., immediately to the amino-terminal side of the scissile bond) and are tolerant of a variety of amino acids in the P2 position because this side chain projects away from the active site of the enzyme.[18-21] In contrast, the various family members have distinct preferences for the P3–P5 positions. Some of these cleavage specificities are indicated in Table 6.1. Because of hydrophobic residues lining the pocket that binds the P4 residue,[20] caspase-1 prefers tyrosine in the P4 position and is particularly effective at cleaving the YVHD⬇G sequence, thereby converting pro-interleukin-1β to the mature cytokine.[21,36] Conversely, the pocket that binds the P4 residue in caspase-3 is lined with basic residues,[19] resulting in a preference for substrates such as DEVD⬇G, the site at which the repair enzymes poly(ADP-ribose) polymerase and DNA-PK$_{cs}$ are cleaved during apoptosis.[37-39] These differences in substrate specificity are carried to the extreme in the case of caspase-2, which appears to recognize five amino acids on the amino side of the scissile bond compared to the four amino acids recognized by the other caspases.[35]

6.2.2   Two Different Routes to Caspase Activation

In addition to differences in substrate specificity, the caspase precursors differ in the lengths of their amino-terminal precursor domains. It is thought that these variations reflect differences in the mechanism of regulation and/or function of the enzymes.[17] In particular, caspases -1, -2, -4, -5, -8, and -10 have long prodomains. Although the functions of the prodomains in caspases -1, -4, and -5 are unknown, the prodomains of caspases -2, -8, and -10 appear to be involved in targeting these precursors to membrane-bound signaling complexes, where their activation subsequently takes place (see Figure 6.1). The prodomains of caspases -8 and -10 contain two 60-amino acid domains (so-called "death effector domains") that are homologous to protein interaction domains on the adaptor protein FADD.[40-42] Upon ligation of the cell surface receptor CD95, FADD binds to an intracellular domain of this receptor.[43] Protein–protein interactions involving the prodomains of caspase-8 and caspase-10, then recruit these molecules to this complex, resulting in their proteolytic cleavage and activation.[44,45] The mechanism of this activation is not completely understood but most likely involves low level constitutive activity of one or both proenzymes, resulting in autoactivation by cleavage at Asp-X bonds.[17a] In an analogous manner, caspase-2 has a protein interaction motif in its long prodomain that is homologous to the interaction domain on the adaptor protein RAIDD or CRADD.[46,47] Ligation of the type 1 tumor necrosis factor (TNF) receptor results in recruitment of adapter proteins RIP and TRADD, which in turn interact with RAIDD/CRADD, recruiting caspase-2 to this complex. These observations suggest that caspases -2, -8, and -10 might be involved in initiating intracellular proteolytic cascades that constitute the execution phase of cell deaths initiated by signaling at cell death receptors such as CD95 and the type 1 TNF receptor.

In contrast to the caspases with long prodomains, caspases -3, -6, -7, and -9 have short prodomains (reviewed in Reference 17). It has been proposed that these are the workhorse proteases whose catalytic activity actually results in cleavages that disassemble the cell during apoptosis.[17,23,48] These proteases appear to be activated by one of two processes. First, active caspase-8 appears to be capable of directly catalyzing the cleavage of multiple procaspases, including precursors of caspases -3, -4, -7, and -9, *in vitro*.[49-51] Likewise, caspase-10 has been shown to cleave the precursors for caspases -3, -7, and -8 *in vitro*.[40] These observations provide the outline of a direct path from ligation of cell surface death receptors to activation of workhorse caspases.[23,52] On the other hand, a variety of stimuli appear to induce apoptosis through pathways that do not involve ligation of cell surface death receptors.[53,54] Recent studies have provided the outline of an alternative pathway for the activation of these workhorse caspases. Many — perhaps most — apoptosis-inducing stimuli appear to generate intracellular signals whose net result is the loss of cytochrome c from mitochondria.[55-58] The nature of the signals triggering this cytochrome c release is currently unknown. Nonetheless, the release of cytochrome c to the cytosol has three consequences. First, it interrupts electron transport, resulting in collapse of the proton gradient across the inner mitochondrial membrane (measured as mitochondrial transmembrane potential Ψ_m).[58,59] Second, it causes transfer of electrons to nonphysiological acceptors, resulting in generation of reactive oxygen species.[55,58] Third, it directly contributes to caspase activation in the cytosol.

The critical role of cytochrome c in activation of workhorse caspases has been suggested by a series of biochemical fractionation studies from the laboratory of Xiaodong Wang.[10,29,56,60] Based on their earlier observation that caspase-3 can be activated *in vitro* upon prolonged incubation of cytosol at 4°C,[61] this group has utilized column chromatography and FPLC to identify three different polypeptides that are all required to accomplish this activation process. One of these polypeptides, Apaf-1, is a docking protein that shares

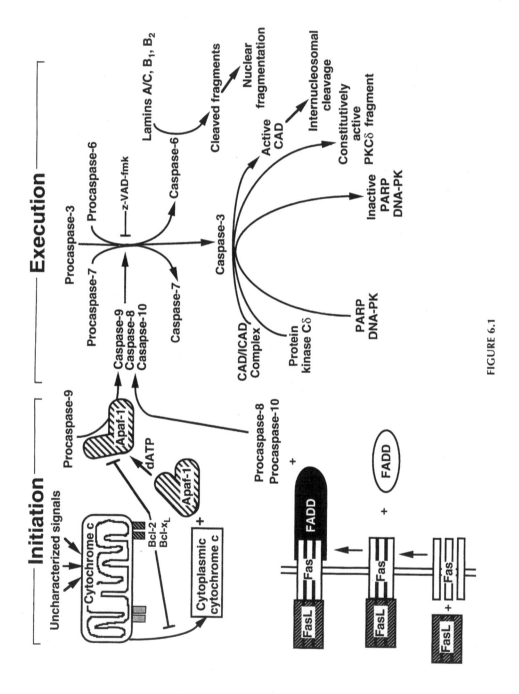

FIGURE 6.1

limited homology with the product of the *C. elegans ced-4* gene.[10] A second of these polypeptides is procaspase-9.[60] The third polypeptide required for activation of caspase-3 is cytochrome c.[29] Analysis of the interactions between these three components[10,60] has suggested that binding of cytochrome c to Apaf-1 results in a dATP-dependent change in the latter protein, thereby enabling binding of procaspase-9 and Apaf-1 through somewhat homologous amino terminal domains termed caspase recruitment domains.[60,62] This binding somehow results in conversion of procaspase-9 to active caspase-9, which then is thought to proteolytically cleave procaspase-3 to the active enzyme.[60] Earlier observations indicating that procaspase-1 is capable of catalyzing its own autoactivation, albeit at an extremely slow rate,[26,28] raise the possibility that the cytochrome c/Apaf–1 induced proteolytic activation of procaspase-9 is autocatalytic. What is unclear at this point is how binding of procaspase-9 to Apaf-1 facilitates this autoactivation.

6.2.3   Roles of Caspases in Programmed Cell Death

Once the workhorse caspases are activated, a subset of the total cellular polypeptides begin to be selectively degraded. The currently identified caspase substrates are listed in Table 11.1 of Chapter 11 by Kaufmann et al. Although it remains to be shown that any particular cleavage event is absolutely essential for the demise of the cell, the net effect of all of the cleavage events appears to be cell death.[17a,48,63]

Most cellular polypeptides remain intact during apoptosis (e.g., References 64 to 66). For those polypeptides that are cleaved, the proteolytic process can have one of three effects: (1) it can alter the assembly/disassembly properties of a structural protein; (2) it can result in inactivation of a

---

FIGURE 6.1

Two possible routes of caspase activation. In some models of PCD, currently uncharacterized signals cause release of cytochrome c from mitochondria. This cytochrome c then interacts with Apaf-1, leading to activation of caspase-9 followed by caspases -3, and -7. At present it appears that caspase-9 activates caspases-6.[170] In other models of PCD, ligation of death receptors (exemplified by Fas/CD95 in this diagram) leads to binding of procaspases-8 and -10 to membrane-associated signaling complexes. This binding results in activation of these caspases, which have in turn been shown to be capable of directly activating caspases -3, -6, and -7. Some experimental evidence suggests, that activation of caspases -8 and -10 might also result in release of cytochrome c from mitochrondria.[55] The relative contributions of cytochrome c-dependent and cytochrome c-independent pathways in activation of downstream caspases after CD95 ligation vary from cell type to cell type. Once caspases-3, -6, and -7 are activated, a variety of proteolytic events occur as illustrated on this schematic. The biochemical effects of these cleavages are summarized in the text. As also indicated in the text, antiapoptotic Bcl-2 family members have been reported to affect the caspase pathway by inhibiting cytochrome c release in some models and by competing for binding of procaspases to activation factors in other models.

particular enzyme; or (3) it can result in constitutive activation of a normally tightly regulated enzyme activity. There are multiple examples of each of these effects.

The nuclear lamins are intranuclear intermediate filament proteins that are thought to play a structural role at the interface between the peripheral chromatin and the inner nuclear membrane (reviewed in References 67,68). A number of studies have demonstrated that lamins are degraded early in the course of apoptosis.[64,65,69-74] Subsequent studies have demonstrated that caspase-6 preferentially cleaves the lamins at a site in the coiled-coil domain that is thought to be responsible for protein–protein interactions in the lamin filaments.[30,75] Interestingly, inhibition of lamin degradation prevents fragmentation of the nucleus,[73,75] suggesting that lamin degradation is required for packaging of nuclear fragments into apoptotic bodies. Additional studies have revealed that the nuclear/mitotic apparatus polypeptide NuMA, another polypeptide with extensive coiled-coil domains[76,77] that is thought to form intermediate filaments within the nucleus,[78] is also degraded during apoptosis.[74,79,80] The proteases responsible for NuMA cleavage during apoptosis have not been identified to date; and the role of NuMA cleavage in apoptotic nuclear changes remains to be determined.

Proteolytic cleavage also results in inhibition of certain cellular activities. Active caspases -3 and -7 can cleave poly(ADP-ribose) polymerase (PARP)[22,81,82] and DNA-PK,[39,83-85] two enzymes that are thought to play a role in DNA repair.[86-88] The result of this cleavage is inactivation of these enzymes[37,89,90] and, presumably, inhibition of DNA repair. Likewise, cleavage of the 70-kDa polypeptide of U1 small nuclear ribonucleoprotein particles[91] and the C1 and C2 polypeptides of heterogeneous nuclear ribonucleoprotein particles[92] is thought to inhibit RNA splicing; and cleavage of DNA replication factor C is thought to result in inhibition of DNA synthesis.[93]

Proteolytic cleavage by caspases also appears to be involved in uncontrolled activation of certain enzymes. For example, protein kinase Cδ is cleaved by caspase-3 or a closely related homolog[94] in cells treated with ionizing radiation[95] or various chemotherapeutic agents,[96] thereby generating a constitutively active 40-kDa fragment. In transient transfection assays, cDNA encoding this fragment induces apoptosis, whereas cDNA encoding full-length protein kinase Cδ does not,[94] suggesting that the 40-kDa fragment might play an active role in subsequent apoptotic events. Likewise, caspase cleavage of gelsolin generates a constitutively active fragment that cleaves actin filaments in an uncontrolled manner, possibly contributing to shape changes in apoptotic cells;[66] and cleavage of the p21-activated kinase PAK2 generates a constitutively active fragment that might be involved in subsequent fragmentation of the cell into multiple membrane-enclosed vesicles.[97] Finally, caspase activity appears to be involved in activating the endonuclease that is responsible for cleaving genomic DNA to generate the nucleosomal ladder that is characteristic of apoptotic death in many cell types.[37] Recent data indicate that a unique nuclease termed CAD (**caspase-activated deoxyribonuclease**) is synthesized as an inactive precursor that binds a second polypeptide called ICAD (inhibitor of CAD).[98] Subsequent caspase-mediated cleavage of ICAD at the sequence DEPD⇓S liberates CAD in an active form, thereby initiating internucleosomal DNA degradation.[99]

6.2.4 Regulation of Caspase Activity

Caspase precursors appear to be constitutively expressed in many — perhaps all — somatic cells (reviewed in References 100 to 105; see also Reference 106). Expression of some of these precursors (procaspases -1, -2 and -3) appears to depend on the presence of the signaling molecule Stat 1 (signal transducer and activation of transcription 1),[107] raising the possibility that levels of these caspases might be regulated, in part, at the transcriptional level. In addition, there is good evidence that the caspases are capable of autoactivation[26-28] and activating each other.[40,49-51,60,108] Given these latter observations, it is clear that there must be stringent safeguards to prevent inadvertent activation of these proteases within cells. Although the study of caspase regulation is currently in its infancy, there appear to be at least three levels of control.

First, caspase activation appears to be regulated at the level of the death receptors. As indicated above, recruitment of procaspases to these receptors results in activation of caspases -8, -10, and possibly -2. Affinity purification studies,[109] as well as cloning of cDNAs identified by their homology to caspases[110,111] or their interaction with caspases in yeast two hybrid systems,[112] have resulted in the identification of a series of mammalian polypeptides that are recruited to cell surface death receptor/adaptor molecule signaling complexes. These molecules appear to compete for procaspase binding sites or inhibit procaspase activation, thereby inhibiting subsequent apoptotic cell death.

Second, caspase activation appears to be regulated by Bcl-2 family members, a large family of polypeptides that affect the initiation of apoptosis in a variety of cell types,[7,8] including neurons.[4-6,9] Bcl-2, the founding member of this family, is localized to a variety of intracellular membranes,[113,114] including the outer mitochondrial membrane, endoplasmic reticulum, and possibly the outer nuclear membrane in the region of nuclear pores. Because Bcl-2 family members do not share significant sequence homology with other polypeptides, their functions have been somewhat enigmatic. Early studies demonstrated that Bcl-2 homologues were able to form homo- and heterodimers. Overexpression of antiapoptotic Bcl-2 family members also prevented apoptosis-associated release of calcium from the endoplasmic reticulum,[115] as well as generation of reactive oxygen species,[116] two effects that are now viewed as probably being indirect results of the inhibition of apoptosis. X-ray crystallography revealed that $Bcl-x_L$ most closely resembles the pore-forming (translocation) domain of diptheria toxin,[117] raising the possibility that Bcl-2 family members might form pores. Consistent with this possibility, several groups have demonstrated that pro- and antiapoptotic

Bcl-2 family members are capable of forming pores that can transport ions across lipid membranes (reviewed in Reference 8). Whether these ion transport capabilities contribute to the pro- and antiapoptotic effects of various Bcl-2 family members is not clear at present. Other investigations have suggested that Bcl-2 family members might affect the ability to activate caspases.[53] As discussed above, one current model suggests that translocation of cytochrome c from the intermembrane space of mitochondria to the cytosol plays a critical role in activation of some caspases (Figure 6.1).[60] Recent studies have indicated that this release of cytochrome c from mitochondria is inhibited by Bcl-2[56,118] and antiapoptotic Bcl-2 family members.[57,58] These observations, which are consistent with the reported ability of antiapoptotic Bcl-2 family members to alter transport of other polypeptides between various cellular compartments (reviewed in Reference 119), provide one potential explanation for the antiapoptotic effects of Bcl-2 and Bcl-X$_L$. Unfortunately, this explanation does not account for the observation that overexpression of Bcl-2 or Bcl-X$_L$ also inhibits the triggering of apoptosis by direct injection of cytochrome c into the cytosol of living cells.[120,121] On the other hand, it has also been demonstrated that Bcl-X$_L$ competes with certain procaspases for binding to ced-4, the *C. elegans* polypeptide with limited sequence homology to Apaf-1.[122] This latter observation raises the possibility that antiapoptotic family members might inhibit caspase activation by altering the ability of Apaf-1 to activate caspases, e.g., by altering the dATP-dependent Apaf-1 conformational change or the subsequent interaction with procaspase-9 (Figure 6.1). Further investigation of these possibilities is clearly required.

Finally, there appear to be intrinsic inhibitors of cellular caspases. The genes encoding p35 and IAP (inhibitor of **ap**optosis) were originally cloned from baculoviruses based on the ability of these gene products to inhibit virus-induced apoptosis in lepidopteran cells.[123,124] Although mammalial cells do not appear to contain any homologue of p35, a well-characterized and relatively promiscuous caspase inhibitor,[17a,125] at least four mammalian homologues of the baculovirus IAP protein, ciap-1, ciap-2, XIAP, and NAIP, have been described.[109,126-129] These polypeptides typically consist of an amino terminal domain containing three copies of a so-called BIR (**b**aculovirus **IAP r**epeat) motif[130] and a carboxyl-terminal domain containing a zinc-binding motif called a ring finger that is found in several signal transduction molecules and is thought to play a role in protein–protein interactions.[131] Interestingly, the gene encoding one of these four polypeptides, the NAIP (**n**euronal **a**poptosis **i**nhibitor **p**rotein) gene, contains mutations and deletions in patients with spinal muscular atrophy, a neurodegenerative disease characterized by excessive apoptosis in spinal cord motor neurons.[126] Although NAIP inhibits apoptosis induced by a variety of stimuli,[127] the mechanism of action of the NAIP polypeptide has not been reported to date. Recent studies on XIAP (**X** chromosome-encoded **IAP**), however, have demonstrated that this polypeptide is capable of not only inhibiting apoptosis in transfected cells, but also selectively binding to and inhibiting caspases

-3 and -7 under cell-free conditions.[132] Moreover, analysis of truncation mutants demonstrated that the BIR domain, which is conserved among the various IAP proteins, is responsible for inhibition of apoptosis.[132,133] Collectively, these observations not only indicate that mammalian cells contain polypeptides that are capable of directly inhibiting active caspases, thereby controlling their activity under physiological conditions, but also suggest that alterations in these putative caspase inhibitors can play a direct role in the pathogenesis of neurological diseases. Clearly, further studies are required to (1) examine the possibility that NAIP might be a caspase inhibitor, (2) more completely define the spectrum of proteases that are inhibited by mammalian IAP proteins, and (3) examine the regulation of these proteins — the cIAPs and XIAP as well as NAIP — in neurons under various physiological and pathological conditions.

6.3   The Potential Role of Caspases and Other Proteases in Neuronal Programmed Cell Death

In addition to the studies of NAIP described above, several other pieces of evidence suggest that caspases play an important role during neuronal PCD. Genetic analysis in the nematode *C. elegans* revealed that two genes, *ced-3* and *ced-4*, are absolutely required for all developmentally regulated cell deaths.[3] The product of the *ced-4* gene shares some similarity to Apaf-1, the docking protein that appears to be involved in caspase activation in mammalian cells.[10,60] The product of the *ced-3* gene has significant structural[11] and functional [12,134] homology to mammalian caspases. Interestingly, the supernumerary cells that survive when these genes are disrupted are typically neurons,[3] indicating that products of these genes participate in neuronal apoptosis. Conversely, the demonstration that mice lacking caspase-3 or -9 have central nervous system abnormalities as a consequence of diminished developmentally regulated neuronal apoptosis[135, 135a] suggests that these caspases play unique roles in neuronal PCD in the developing mammalian central nervous system.

Further support for the role of caspases in neuronal PCD comes from pharmacological and biochemical studies. Treatment with low molecular weight caspase inhibitors such as Z-VAD-fluoromethylketone or Boc-Asp(OMe)-fluoromethylketone not only inhibits apoptotic changes in tissue culture after application of many different stimuli,[136-139] but also diminishes central nervous system damage in animal models of stroke and epilepsy,[140] where apoptotic death is thought to play a key role. Finally, a growing number of polypeptides have been shown to be degraded by caspases in cells undergoing apoptosis (see above); and there is now evidence for cleavage of these peptides (and activation of the corresponding caspases) in several examples of neuronal PCD.

Despite the large number of studies examining caspase function during PCD, it is important to keep in mind that other proteases might also play important roles in this process. Inhibitor studies have implicated other proteases, including the proteasome,[141,142] serine proteases,[143,144] calpains,[145-147] cathepsin D,[148] and cathepsin B,[149] in apoptosis in various cell types, although the specificity of the inhibitors utilized in those studies has not been firmly established in all cases. Two observations are particularly interesting in the context of the present discussion. First, calpain I activation has been observed after administration of excitotoxic amino acids *in vivo*.[150] Second, a variety of proteasome inhibitors, including E64d, Ac-Leu-leu-norleucinal, Ac-Leu-leu-methioninal and lactacystin, delay or prevent apoptotic morphological changes in rat superior sympathetic ganglion neurons deprived of NGF.[141] These observations raise the possibility that other classes of proteases might also play a role in neuronal PCD.

6.4 Evidence for the Role of Proteases in Neuronal PCD Provoked by Various Stimuli

In the Sections that follow, we briefly review the evidence that proteases are active during neuronal PCD, summarize current controversies in this field, and identify questions that remain to be answered. Because of the overwhelming attention currently being paid to caspases, these comments focus almost exclusively on this class of proteases.

6.4.1 Trophic Factor Withdrawal-Induced Neuronal PCD

The most extensively studied type of neuronal cell death is that which occurs after withdrawal of neurotrophic factors such as NGF, NT-3, NT-4, BDNF, and CNTF (reviewed in Reference 151). In several *in vitro* models, including cultured sensory,[152] sympathetic,[153] and parasympathetic neurons,[152] motoneurons,[154,155] retinal ganglion cells,[156] and cerebellar granule neurons,[157] withdrawal of trophic support is followed by a prolonged period of time (10 to 20 h) during which the cells can be rescued by neurotrophin restoration. After this period of latency, increasing numbers of cells become committed to PCD and are no longer capable of being rescued. A variety of studies have suggested that the early stages of PCD in many such model systems require macromolecular synthesis, although the molecules essential for subsequent PCD have not been completely characterized. It is known that the synthesis of the transcription factor c-Jun and members of the Fos family are required for the occurrence of PCD in NGF-deprived sympathetic neurons at the usual time,[158,159] although it is not clear whether down-regulation of these genes with antisense oligonucleotides or dominant negative protein expression completely prevents PCD or merely delays it. Increased expression of cyclin D1 has also been reported in trophic factor-deprived sympathetic neurons,[160,161] raising the possibility that inappropriate entry into S phase might play a role in triggering neuronal PCD. Studies showing that the synthesized cyclin D1 pairs with a cyclin-dependent kinase and actually alters cell cycle distribution remain to be performed. Nonetheless, experiments demonstrating that the cyclin D1-dependent kinase inhibitor p16[INK4] and low-molecular weight inhibitors of cell cycle progression such as mimosine, olomucine, and flavopiridol can inhibit PCD in NGF-deprived sympathetic neurons or neuronally differentiated PC12 pheochromocytoma cells[162,163] provide further support for a model in which inappropriate cell cycle progression after neutrophin withdrawal contributes to PCD. Whatever the identity of the macromolecules that are synthesized during this latent period, neutrophin withdrawal ultimately results in a stereotypic series of morphological changes that include neurite withdrawal, cytoplasmic blebbing, and nuclear fragmentation in the model systems described above. Higher-order DNA fragmentation is generally demonstrable in these model systems using terminal transferase-mediated dUTP nick end-labeling (TUNEL), although cleavage of DNA into oligonucleosomal fragments is variably observed.

A number of studies have examined the potential role of caspases in neurotrophin withdrawal-induced neuronal cell death. In a seminal study, Yuan and co-workers[164] injected chicken dorsal root ganglion neurons with a control vector or a vector containing a cDNA encoding crm A, a cowpox virus polypeptide that selectively inhibits a number of different caspases.[22,165,166] Within 3 days of NGF withdrawal, ~80% of control neurons exhibited apoptotic morphology, whereas only 25% of crmA-expressing neurons were apoptotic.[164] Expression of the baculovirus protein p35, another caspase inhibitor,[125] has likewise been shown to protect rat superior cervical ganglion neurons from NGF withdrawal-induced apoptosis.[167] In subsequent studies, low-molecular weight caspase inhibitors have also been examined for their effect on neurotrophin withdrawal-induced PCD. Milligan et al. demonstrated that trophic factor withdrawal-induced death of motor neurons *in vitro* was delayed by Ac-YVAD-aldehyde, but not by leupeptin (another aldehyde protease inhibitor) or E64 (a calpain inhibitor).[168] Interestingly, administration of Ac-YVAD-aldehyde to chick embryos inhibited apoptosis occurring spontaneously in spinal cord motoneurons *in vivo*, but not apoptosis provoked by extirpation of limb buds.[168] In a similar fashion, Z-VAD-fluoromethylketone delayed the occurrence of apoptosis in PC12 cells after NGF withdrawal.[169,170] Interestingly, activation of c-Jun N-terminal kinase (JNK) occurred normally in these PC12 cells,[171] suggesting that protease activation is downstream of JNK activation. In a similar vein, studies by Deshmukh et al.[138] indicated that Boc-Aspartyl(OMe)-fluoromethylketone inhibited apoptotic morphological changes in rat sympathetic neurons deprived of NGF *in vitro*, even though this caspase inhibitor had no effect on the expression of immediate early genes.

Although the preceding studies suggest a role for caspases in neurotrophin withdrawal-induced apoptosis, identifying the caspase(s) involved has proven unexpectedly difficult. Because caspase-2 was originally identified as a mouse brain embryonic transcript down-regulated in late stages of brain development,[172] there was considerable speculation that this caspase might play a particularly important role in neuronal PCD resulting from neurotrophin deprivation. Consistent with this possibility, transfection with caspase-2 cDNA induce apoptosis in sympathetic neurons[173] and neuroblastoma cells.[174] Transfection with a variety of caspase cDNAs, however, has been observed to induce apoptosis in other cell systems, making it difficult to interpret this observation (see Chapter 11 by Kaufmann et al.). The critical issue is whether endogenously expressed caspase-2 plays a role in neuronal PCD. Two groups have reported that procaspase-2 is cleaved in sympathetic neurons undergoing NGF withdrawal-induced apoptosis.[138,175] Furthermore, Troy et al. have reported that caspase-2 antisense oligonucleotides simultaneously down-regulate caspase-2 polypeptide levels and delay PCD after NGF withdrawal in rat sympathetic neurons and PC12 cells.[176] Collectively, these observations are consistent with the view that caspase-2 plays an important role in neurotrophin withdrawal-induced apoptosis. On the other hand, caspase-2$^{-/-}$ mice do not exhibit major neurologic deformities,[176a] calling this conclusion into question.

What about other caspases? Some investigators have reported that procaspase-3 polypeptide levels are unchanged after NGF withdrawal, raising the possibility that caspase-3 fails to play a role in this process.[138,173] Other experiments, however, have suggested a different picture. In particular, Stefanis and co-workers have reported that NGF-deprived PC12 cells activate a caspase that is capable of cleaving DEVD-aminofluorocouramin,[170] a substrate that can be cleaved by caspase-3 but not caspase-2.[35] Immunoblotting and affinity labeling experiments likewise have indicated that neurotrophin withdrawal-induced apoptosis in mouse neurons and neuronal cell lines is accompanied by conversion of procaspase-3 to the active enzyme.[175,177] These results, which are difficult to reconcile with earlier reports, clearly indicate activation of caspase-3 during neutrophin withdrawal-induced apoptosis. Interestingly, the DEVD-AFC cleavage activity reportedly appears later than JNK activity, raising the possibility that caspase-3 activation follows Jun synthesis.[170]

Two groups have utilized molecular genetic approaches to examine the role of caspases in neurotrophin withdrawal-induced programmed cell death. Friedlander et al. observed diminished cell death 24 h after neutrophin withdrawal from dorsal root ganglion neurons of caspase-1$^{-/-}$ mice or mice expressing a dominant negative caspase-1 inhibitor as compared to wild-type litter mates.[178] Because a single time point was examined, it is unclear whether caspase-1 inhibition actually prevented neuronal apoptosis or merely slowed the kinetics. Consistent with the latter hypothesis, mice from these animals were viable and did not display supernumerary neurons or developmental abnormalities.[178] These results suggest that caspase-1 might play a critical role in apoptosis induced by postnatal neutrophin withdrawal but not prenatal (developmental) apoptosis. In contrast, Kuida et al.[135,135a] have observed that caspase-3$^{-/-}$ and caspase-9$^{-/-}$ mice die during development with a variety of central nervous system abnormalities that are attributable to lack of normal developmentally regulated neuronal PCD. These latter observations raise the possibility that caspases-3 and -9 might play an essential roles in neurotrophin withdrawal-induced PCD during embryonic development, but do not rule out the possibility that other caspases might also play essential roles.

Further studies are required to define the full compliment of caspases that are present in various neurons at different times during development and to identify which of these are activated during the process of neurotrophin withdrawal-induced PCD. In addition, current evidence suggests at least two possible routes of caspase activation in other cell types (Figure 6.1); and further studies are required to determine which of these contribute to caspase activation after neurotrophin withdrawal. On the one hand, the requirement for macromolecular synthesis raises the possibility that neurotrophin withdrawal is accompanied by synthesis of one of the death receptors and/or the corresponding ligand. On the other hand, the observation that reactive oxygen species are produced during neurotrophin withdrawal-induced PCD[179] raises the possibility that release of cytochrome c to the cytosol is involved in caspase activation in this setting. Although the effects of Bcl-2 on cytochrome c release and neurotrophin withdrawal-induced apoptosis (see Section 6.2) are consistent with this possibility, it remains to be determined how (or even whether) cytochrome c release is induced after neurotrophin withdrawal.

6.4.2 Excitotoxic Neuronal PCD

Studies dating to the mid 1980s indicate that excessive excitation of neurons results in cell death (e.g., Reference 180). More recently, this death has been compared and contrasted to neurotrophin withdrawal-induced PCD. Model systems that demonstrate this "excitotoxic" effect include various neuronal cells exposed to glutamate, the glutamate analog N-methyl-D-aspartate (NMDA), potassium ionophores like valinomycin, calcium channel blockers (e.g., nifedipine or verapimil),[181] low calcium medium, or 3-nitropropionic acid.[182] In addition, it has been observed that survival of cultured cerebellar granule neurons beyond the first 5 days *in vitro* is enhanced when the cells are maintained in depolarizing concentrations of potassium (20 to 25 mM).[183] Conversely, transfer of these neurons to nondepolarizing medium (5 mM potassium) results in apoptotic cell death,[157] providing one of the more widely utilized models of neuronal PCD. In contrast to neurotrophin withdrawal-induced PCD, these model systems typically cannot be rescued after the initial insult and show no requirement for macromolecular synthesis (e.g., Reference 184), the PCD observed after transfer of neurons to nondepolarizing potassium concentrations being the notable exception.[185,186] Typical morphologic features of PCD, including neurite withdrawal, chromatin condensation, plasma membrane blebbing, and nuclear fragmentation, are observed after all of these excitotoxic treatments. In addition, most excito-toxic treatments are accompanied by development of oligonucleosomal DNA fragmentation.

The role of caspases in excitotoxic PCD remains uncertain. On the one hand, proteolytic cleavage of procaspase-3 to the active protease[184,187] and appearance of corresponding DEVD-AFC cleavage activity in cytosol[187-189] has been demonstrated in several model systems. Du et al. have observed that treatment with the caspase inhibitor DEVD-aldehyde but not the inhibitor YVAD-aldehyde protects superior cervical ganglion neurons from NMDA-induced PCD.[184] Likewise, the more promiscuous caspase inhibitor Z-VAD-fluoromethylketone has been reported to inhibit valinomycin-induced PCD in cultured cortical neurons.[190] To the extent that these inhibitors are specific for caspases, the preceding observations suggest that one or more caspases might play a critical role in excitotoxic PCD. On the other hand, the same inhibitors have no effect on PCD in cerebellar granule neurons exposed to depolarizing potassium concentrations,[188,191] although other caspase inhibitors such as Z-asp-2,6-dichloroacyloxybenzylmethyl ketone do inhibit apoptosis in this model system.[189] One possibility is that the differential effect of caspase inhibitors might reflect the use of different excitotoxic stimuli. Alternatively, it has been suggested that caspase inhibitors might prevent the internucleosomal DNA degradation and morphologic manifestations of apoptosis without preventing death by a caspase-independent pathway.[188] This latter explanation is highly reminiscent of results of McCarthy et al., who demonstrated that caspase inhibitors prevent apoptotic changes without inhibiting cell death after a variety of treatments in Rat-1 fibroblasts.[192] Further studies are required to distinguish between these two possibilities. As is the case with neurotrophin withdrawal-induced apoptosis, the pathways that lead to caspase activation and the roles of individual caspases in excitotoxic PCD also remain to be clarified.

6.4.3 Neuronal PCD Subsequent to Ischemia or Hypoxia

Although the prevailing paradigm in neuropathology was that ischemia caused necrotic cell death, the demonstration of oligonucleosomal DNA fragment after transient carotid artery ligation[193,194] raised the possibility that apoptotic cell death might occur after ischemia and reperfusion. Subsequent studies in postischemic hippocampal neurons[195] and cyanide-treated PC12 cells[196] have demonstrated cell death with typical apoptotic morphology. Current models suggest that oxygen radicals formed after reperfusion set in motion the biochemical process of PCD in cells that survived the initial insult but were transiently hypoxic.

There is a paucity of information related to the role of caspases in PCD occurring under these conditions. Studies from Yuan and Moskowitz and associates indicate that treatment with the caspase inhibitor AcYVAD-chloromethylketone or Z-VAD-fluoromethylketone results in diminished infarct volume and improved neurological score after middle cerebral artery occlusion in mice.[140] Likewise, expression of a dominant negative mutant of caspase-1 reduces infarct size and/or neurological deficits after transient ischemia.[178] Subsequent studies have suggested that these effects of caspase inhibitors are indirect: It appears that caspase-1 generates interleukin-1β, which recruits inflammatory cells to the post ischemia areas and thereby worsens the cellular damage.[197] Further studies are required to define the direct role (if any) of caspases in ischemia/reperfusion-induced neuronal PCD.

6.4.4 Neurodegenerative Diseases

Premature neuronal death is observed with a number of neurodegenerative diseases, including Alzheimer's disease (AD), Parkinson's disease, Huntington's disease, and amyotrophic lateral sclerosis (ALS). Although frankly necrotic cells are observed in some instances, PCD has long been suspected to play an important role in many neurodegenerative diseases. In recent years, descriptive and experimental studies have begun to provide support for this hypothesis (reviewed in References 198-200).

AD, one of the most intensively studied forms of neurodegeneration, is associated with the intracortical formation of proteinaceous plaques containing large amounts of amyloid beta-peptide (AβP). Morphological evidence of apoptosis has been observed in AβP-treated cultured neurons[201,202] as well as the cortices of transgenic mice engineered to overexpress AβP.[203] More recently, overexpression of the familial AD gene Presenilin 2 (*PS2*) has been reported to increase apoptosis of neuronally differentiated PC12 cells after NGF withdrawal, exposure to glutamate or treatment with AβP.[204-206]

Mounting evidence suggests that excessive apoptosis plays a role in the pathogenesis of Huntington's disease, Parkinson's disease, and ALS as well. For Huntington's disease and ALS, both of which are associated with defined genetic alterations, knockout or overexpression studies indicate that the disease gene products act to regulate apoptotic cell death.[207,208] Similarly, in an *in vivo* Parkinson's disease model, apoptotic cell death has been reported.[209]

Despite accumulating evidence for the importance of apoptosis in the pathogenesis of neurodegenerative diseases, relatively little mechanistic information has been published to date. In an AD model, correlations between exposure to AβP and alterations in expression of genes that are associated with neuronal PCD in other contexts (see Section 6.4.1) have been demonstrated. Paradis et al., for example, showed down regulation of the antiapoptotic protein Bcl-2 and concomitant upregulation of the proapoptotic protein Bax in AβP-treated primary human neuron cultures.[210] In another study that examined postmortem AD tissue, DNA fragmentation (detected by TUNEL staining) was found in the same cells that displayed increased immunoreactivity of the transcription factor c-Jun, an immediate early gene product that is upregulated during trophic factor withdrawal-induced neuronal PCD.[211]

Studies exploring the role of caspases in the pathogenesis of neurodegenerative diseases have only recently begun. Although Goldberg et al. have reported that huntingtin can be cleaved by caspase-3 under cell-free conditions,[212] the cleavage of this substrate in intact cells and the contribution of this cleavage to the pathogenesis of Huntington's disease remain to be demonstrated. In contrast, the cloning of the familial AD genes *PS1* and *PS2/STM2*[213-215] resulted in analyses that suggest the direct involvement of caspase-mediated PCD pathways in the pathogenesis of AD. Vito et al., for example, reported that a cDNA homologous to *PS2/STM2* is capable of rescuing T cell hybridoma cells from Fas- and T cell receptor ligation-induced apoptosis.[205] Subsequently, Lacana et al. provided data suggesting that the antiapoptotic activity of *PS2/STM2* results from its ability to suppress ICE/CED-3-like protease activity.[216] Interestingly, caspases have also recently been implicated in the proteolytic processing of *PS1* and *PS2/STM2*, although the biological significance of such processing is currently unclear.[217] Additional studies will be required to further elucidate the importance of caspase activity in diseases like AD. Studies involving inhibitors of specific caspases or caspase knockout animals are likely to shed the most light on the involvement of caspase-mediated apoptosis in neurodegenerative diseases.

6.5   Future Directions

As indicated in the preceding sections, there is now ample evidence that caspases are activated in neurons during PCD triggered by a variety of stimuli. The identity of the caspases that are activated and the individual roles that they play in the cell death process remain to be more completely elucidated. The possibility that different caspases are involved in different types of neuronal cell death[169,176] needs to be more fully explored. There is also much to be learned about the regulation of caspase expression, caspase activation, and caspase activity in neuronal cells as well as other cell types. In addition, the pathways that lead from the proapoptotic stimuli to caspase activation are poorly understood and require further investigation.

The possible roles of other proteases in neuronal PCD also remain to be elucidated. Although calpain I has been shown to be activated in neutrophin withdrawal-induced PCD, an essential role for calpain I in this process has not been demonstrated. Likewise, experiments using proteasome inhibitors have provided indirect evidence that this protease might play a role in neuronal PCD; but more definitive molecular biologic and biochemical experiments remain to be performed.

Why study the role of proteases in neuronal PCD? Recent studies have shown that neurotrophin withdrawal, excitotoxic stimuli, hypoxia/reperfusion injury, and several degenerative processes all appear to involve apoptotic neuronal cell death. When coupled with evidence suggesting that proteases play important roles in the initiation and propagation of apoptotic cell death,[15,17,23,48,52,63,100,103,104] these observations raise the possibility that inhibition of apoptotic proteases might delay or prevent pathological neurological degeneration. Transforming this scientific hypothesis into a useful therapeutic strategy will require incisive and rigorous scientific investigation.

Acknowledgments

The authors gratefully acknowledge helpful discussions with their colleagues, including Bill Earnshaw, Greg Gores, Keith Bible, Imawati Budihardjo, and Anthony Windebank, as well as editorial assistance of Deb Strauss. We apologize to the authors of many fine articles that we were unable to cite due to space limitations.

References

1. Schwartz LM, Milligan CE: Cold thoughts of death: the role of ICE proteases in neuronal cell death. *Trends Neurosci.* 19:555, 1996.
2. Arends MJ, Wyllie AH: Apoptosis: Mechanisms and roles in pathology. *Int. Rev. Exp. Pathol.* 32:223, 1991.
3. Ellis RE, Yuan JY, Horvitz HR: Mechanisms and functions of cell death. *Annu. Rev. Cell Biol.* 7:663, 1991.
4. Garcia I, Martinou I, Tsujimoto Y, Martinou JC: Prevention of programmed cell death of sympathetic neurons by the Bcl-2 proto-oncogene. *Science* 258:302, 1992.
5. Zhong LT, Sarafian T, Kane DJ, Charles AC, Mah SP, Edwards RH, Bredesen DE: Bcl-2 Inhibits death of central neural cells induced by multiple agents. *Proc. Natl. Acad. Sci., U.S.A.* 90:4533, 1993.
6. Allsopp TE, Wyatt S, Paterson HF, Davies AM: The proto-oncogene Bcl-2 can selectively rescue neurotrophic factor-dependent neurons from apoptosis. *Cell* 73:295, 1993.
7. Yang E, Korsmeyer SJ: Molecular thanatopsis: a discourse on the Bcl2 family and cell death. *Blood* 88:386, 1996.
8. Reed JC: Double Identity for Proteins of the Bcl-2 Family. *Nature* 387:773, 1997.
9. Merry DE, Korsmeyer SJ: Bcl-2 Gene family in the nervous system. *Annu. Rev. Neurosci.* 20:245, 1997.
10. Zou H, Henzel WJ, Liu X, Lutschg A, Wang X: Apaf-1, a Human Protein Homologous to *C. elegans* CED-4, Participates in cytochrome c-dependent activation of caspase-3. *Cell* 90:405, 1997.

11. Yuan J, Shaham S, Ledoux S, Ellis HM, Horvitz HR: The *C. elegans* cell death gene and Ced-3 encodes a protein similar to mammalian interleukin-1 beta-converting enzyme. *Cell* 75:641, 1993.

12. Xue D, Shaham S, Horvitz HR: The *Caenorhabditis Elegans* cell-death protein CED-3 is a cysteine protease with substrate specificities similar to those of the human CPP32 protease. *Genes Dev.* 10:1073, 1996.

13. Alnemri ES, Livingston DJ, Nicholson DW, Salvesen G, Thornberry NA, Wong WW, Yuan J: Human ICE/CED-3 protease nomenclature. *Cell* 87:171, 1996.

14. Alnemri ES: Mammalian Cell Death Proteases: A family of highly conserved aspartate specific cysteine proteases. *J. Cell. Biochem.* 64:33, 1997.

15. Miller DK: The role of the caspase family of cysteine proteases in apoptosis. *Seminars Immunology* 9:35, 1997.

16. Nicholson DW, Thornberry NA: Caspases: killer proteases. *Trends Biochem. Sci.* 22:299, 1997.

17. Villa P, Kaufmann SH, Earnshaw WC: Caspases and caspase inhibitors. *Trends Biochem. Sci.* 22:388, 1997.

17a. Earnshaw WC, Martins LM, Kaufmann SH: Mammalian caspases: structure, activation, substrates and fuctions during apoptosis. *Ann. Rev. Biochem.* 68: in press, 1999.

18. Wilson P, Black F, Thomson A, Kim E, Griffith P, Navia A, Murcko A, Chambers P, Aldape A, Raybuck A, and Livingston J: Structure and mechanism of interleukin-1β converting enzyme. *Nature* 370:270, 1994.

19. Rotonda J, Nicholson DW, Fazil KM, Gallant M, Gareau Y, Labelle M, Peterson EP, Rasper DM, Ruel R, Vaillancourt JP, Thornberry NA, Becker JW: The three-dimensional structure of apopain/CPP32, a key mediator of apoptosis. *Nat. Struct. Biol.* 3:619, 1996.

20. Walker N, Talanian R, Brady K Dang L, Bump N, Ferenz C, Franklin S, Ghayur T, Hackett M, Hammill L, Herzog L, Hugunin M, Houy W, Mankovich J, McGuiness L, Orlewicz E, Paskind M, Pratt C, Reis P, Summani A, Terranova M, Welch J, Xiong L, Moller A, Tracey E, Kamen R and Wong W: Crystal structure of the cysteine protease interleukin-1β converting enzyme: a (p20/p10)2 homodimer. *Cell* 78:343, 1994.

21. Thornberry NA, Bull HG, Calaycay JR, Chapman KT, Howard AD, Kostura MJ, Miller DK, Molineaux SM, Weidner JR, Aunins J: A novel heterodimeric cysteine protease is required for interleukin-1 Beta processing in monocytes. *Nature* 356:768, 1992.

22. Nicholson DW, Ali A, Thornberry NA, Vaillancourt JP, Ding CK, Gallant M, Gareau Y, Griffin PR, Labelle M, Lazebnik YA: Identification and inhibition of the ICE/CED-3 protease necessary for mammalian apoptosis. *Nature* 376:37, 1995.

23. Fraser A, Evan G: A license to Kill. *Cell* 85:781, 1996.

24. Orth K, O'Rourke K, Salvesen GS, Dixit VM: Molecular ordering of apoptotic mammalian CED-3/ICE-like proteases. *J. Biol. Chem.* 271:20977, 1996.

25. Srinivasula SM, Fernandes-Alnemri T, Zangrilli J, Robertson N, Armstrong RC, Wang L, Trapani JA, Tomaselli KJ, Litwack G, Alnemri ES: The Ced-3/interleukin-1 beta converting enzyme-like homolog Mch6 and the lamin-cleaving enzyme Mch2 alpha are substrates for the apoptotic mediator CPP32. *J. Biol. Chem.* 271:27099, 1996.

26. Ramage P, Cheneval D, Chvei M, Graff P, Hemmig R, Heng R, Kocher HP, Mackenzie A, Memmert K, Revesz L, Wishart W: Expression, refolding, and autocatalytic proteolytic processing of the interleukin-1β-converting enzyme precursor. *J. Biol. Chem.* 270:9378, 1995.

27. Fernandes-Alnemri T, Litwack G, Alnemri ES: Mch2, A new member of the apoptotic Ced-3/ICE cysteine protease gene family. *Cancer Res.* 55:2737, 1995.

28. Yamin TT, Ayala JM, Miller DK: Activation of the Native 45-kDa Precursor Form of Interleukin-1-Converting Enzyme. *J. Biol. Chem.* 271:13273, 1996.

29. Liu X, Kim CN, Yang J, Jemmerson R, Wang X: Induction of apoptotic program in cell-free extracts: requirement for dATP and cytochrome C. *Cell* 86:147, 1996.

30. Takahashi A, Alnemri ES, Lazebnik YA, Fernandes-Alnemri T, Litwack G, Moir RD, Goldman RD, Poirier GG, Kaufmann SH, Earnshaw WC: Cleavage of Lamin A by Mch2α but not CPP32: multiple ICE-related proteases with distinct substrate recognition properties are active in apoptosis. *Proc. Natl. Acad. Sci., U.S.A.* 93:8395, 1996.

31. Martins LM, Kottke TJ, Mesner PW Jr., Basi GS, Sinha S, Frigon N, Jr., Tatar E, Tung JS, Bryant K, Takahashi A, Svingen PA, Madden BJ, McCormick DJ, Earnshaw WC, Kaufmann SH: Activation of Multiple Interleukin-1β Converting enzyme homologues in cytosol and nuclei of HL-60 human leukemia cell lines during etoposide-induced apoptosis. *J. Biol. Chem.* 272:7421, 1997.

32. Stennicke HR, Salvesen GS: Biochemical characteristics of caspases-3, -6, -7, and -8. *J. Biol. Chem.* 272:25719, 1997.

33. Margolin N, Raybuck SA, Wilson KP, Chen W, Fox T, Gu Y, Livingston DJ: Substrate and Inhibitor Specificity of interleukin-1β-converting enzyme and related caspases. *J. Biol. Chem.* 272:7223, 1997.

34. Thornberry NA, Rano TA, Peterson EP, Rasper DM, Timkey T, Garcia-Calvo M, Houtzager VM, Nordstrom PA, Roy S, Vaillancourt JP, Chapman KT, Nicholson DW: A combinatorial approach defines specificities of members of the caspase family and granzyme B. Functional relationships established for key mediators of apoptosis. *J. Biol. Chem.* 272:17907, 1997.

35. Talanian RV, Quinlan C, Trautz S, Hackett MC, Mankovich JA, Banach D, Ghayur T, Brady KD, Wong WW: Substrate specificities of caspase family proteases. *J. Biol. Chem.* 272:9677, 1997.

36. Sleath PR, Hendrickson RC, Kronheim SR, March CJ, Black RA: Substrate specificity of the protease that processes human interleukin-1β. *J. Biol. Chem.* 265:14526, 1990.

37. Kaufmann SH, Desnoyers S, Ottaviano Y, Davidson NE, Poirier GG: Specific proteolytic fragmentation of poly(ADP-ribose) polymerase: an early marker of chemotherapy-induced apoptosis. *Cancer Res.* 53:3976, 1993.

38. Lazebnik YA, Kaufmann SH, Desnoyers S, Poirier GG, Earnshaw WC: Cleavage of poly(ADP-ribose)polymerase by a proteinase with properties like ICE. *Nature* 371:346, 1994.

39. Song Q, Lees-Miller SP, Kumar S, Zhang Z, Chan DW, Smith GC, Jackson SP, Alnemri ES, Litwack G, Khanna KK, Lavin MF: DNA-dependent protein kinase catalytic subunit: a target for an ICE-like protease in apoptosis. *EMBO J* 15:3238, 1996.

40. Fernandes-Alnemri T, Armstrong RC, Krebs J, Srinivasula SM, Wang L, Bullrich F, Fritz LC, Trapani JA, Tomaselli KJ, Litwack G, Alnemri ES: *In Vitro* Activation of CPP32 and Mch3 by Mch4, A novel human apoptotic cysteine protease containing two FADD-like domains. *Proc. Natl. Acad. Sci., U.S.A.* 93:7464, 1996.

41. Muzio M, Chinnaiyan AM, Kischkel FC, O'Rourke K, Shevchenko A, Ni J, Scaffidi C, Bretz JD, Zhang M, Gentz R, Mann M, Krammer PH, Peter ME, Dixit VM: FLICE, a novel FADD-homologous ICE/CED-3-like protease, is recruited to the CD95 (Fas/APO-1) death-inducing signaling complex. *Cell* 85:817, 1996.

42. Boldin MP, Goncharov TM, Goltsev YV, Wallach D: Involvement of MACH, A novel MORT1/FADD-interacting protease, in Fas/APO-1 and TNF receptor-induced cell death. *Cell* 85:803, 1996.
43. Chinnaiyan AM, O'Rourke K, Tewari M, Dixit VM: FADD, A novel death domain-containing protein, interacts with the death domain of Fas and initiates apoptosis. *Cell* 81:505, 1995.
44. Kischkel FC, Hellbardt S, Behrmann I, Germer M, Pawlita M, Krammer PH, Peter ME: Cytotoxicity-dependent APO-1 (Fas/CD95)-associated proteins form a death-inducing signaling complex (DISC) with the receptor. *EMBO J* 14:5579, 1995.
45. Medema JP, Scaffidi C, Kischkel FC, Shevchenko A, Mann M, Krammer PH, Peter ME: FLICE is activated by association with the CD95 death-inducing signaling complex (DISC). *EMBO J* 16:2794, 1997.
46. Duan H, Dixit VM: RAIDD Is a new 'death' adaptor molecule. *Nature* 385:86, 1997.
47. Ahmad M, Srinivasula SM, Wang L, Talanian RV, Litwack G, Fernandes-Alnemri T, Alnemri ES: CRADD, A novel human apoptotic adaptor molecule for caspase-2, and FasL/tumor necrosis factor receptor-interacting protein RIP. *Cancer Res.* 57:615, 1997.
48. Martin SJ, Green DR: Protease activation during apoptosis: death by a thousand cuts? *Cell* 82:349, 1995.
49. Srinivasula SM, Ahmad M, Fernandes-Alnemri T, Litwack G, Alnemri ES: Molecular ordering of the Fas-apoptotic pathway: the Fas/APO-1 protease Mch5 is a CrmA-inhibitable protease that activates multiple Ced-3/ICE-like cysteine proteases. *Proc. Natl. Acad. Sci., U.S.A.* 93:14486, 1996.
50. Muzio M, Salvesen GS, Dixit VM: FLICE induced apoptosis in a cell-free system. cleavage of caspase zymogens. *J. Biol. Chem.* 272:2952, 1997.
51. Takahashi A, Hirata H, Yonehara S, Imai Y, Lee K-K, Moyer RW, Turner PC, Mesner PW, Okazaki T, Sawai H, Kishi S, Yamamoto K, Okuma M, Sasada M: Affinity labeling displays the stepwise activation of ICE-related proteases by Fas, staurosporine, and CrmA-sensitive caspase-8. *Oncogene* 14:2741, 1997.
52. Nagata S: Apoptosis by Death Factor. *Cell* 88:355, 1997.
53. Chinnaiyan AM, Orth K, O'Rourke K, Duan H, Poirier GG, Dixit VM: Molecular ordering of the cell death pathway. *J. Biol. Chem.* 271:4573, 1996.
54. Eischen CM, Kottke TJ, Martins LM, Basi GS, Tung JS, Earnshaw WC, Leibson PJ, Kaufmann SH: Comparison of apoptosis in wild-type and Fas-resistant cells: chemotherapy-induced apoptosis is not dependent on Fas/Fas ligand interactions. *Blood* 90:935, 1997.
55. Krippner A, Matsuno-Yagi A, Gottlieb RA, Babior BM: Loss of function of cytochrome C in jurkat cells undergoing Fas-mediated apoptosis. *J. Biol. Chem.* 271:21629, 1996.
56. Yang J, Liu X, Bhalla K, Kim CN, Ibrado AM, Cai J, Peng T-I, Jones DP, Wang X: Prevention of apoptosis by Bcl-2: release of cytochrome c from mitochondria blocked. *Science* 275:1129, 1997.
57. Kharbanda S, Pandey P, Schofield L, Israels S, Roncinske R, Yoshida K, Bharti A, Yuan ZM, Saxena S, Weichselbaum R, Nalin C, Kufe D: Role for Bcl-xL as an inhibitor of cytosolic cytochrome C accumulation in DNA damage-induced apoptosis. *Proc. Natl. Acad. Sci., U.S.A.* 94:6939, 1997.

58. Kim CN, Wang X, Huang Y, Ibrado AM, Liu L, Frang G, Bhalla K: Overexpression of Bcl-X(L) inhibits Ara-C-induced mitochondrial loss of cytochrome C and other perturbations that activate the molecular cascade of apoptosis. *Cancer Res.* 57:3115, 1997.

59. Marchetti P, Castedo M, Susin SA, Zamzami N, Hirsch T, Macho A, Haeffner A, Hirsch F, Geuskens M, Kroemer G: Mitochondrial Permeability transition is a central coordinating event of apoptosis. *J. Exp. Med.* 184:1155, 1996.

60. Li P, Nijhawan D, Budihardjo I, Srinivasula SM, Ahmad M, Alnemri ES, Wang X: cytochrome C and dATP-dependent formation of Apaf-1/Caspase-9 complex initiates an apoptotic protease cascade. *Cell* 91:479, 1997.

61. Wang X, Pai JT, Wiedenfeld EA, Medina JC, Slaughter CA, Goldstein JL, Brown MS: Purification of an interleukin-1 beta converting enzyme-related cysteine protease that cleaves sterol regulatory element-binding proteins between the leucine zipper and transmembrane domains. *J. Biol. Chem.* 270:18044, 1995.

62. Hofmann K, Bucher P, Tschopp J: The CARD Domain: A new apoptotic signaling motif. *Trends Biochem. Sci.* 22:155, 1997.

63. Whyte M, Evan G: Apoptosis. The last cut is the deepest. *Nature* 376:17, 1995.

64. Kaufmann SH: Induction of endonucleolytic DNA cleavage in human acute myelogenous leukemia cells by etoposide, camptothecin, and other cytotoxic anticancer drugs: a cautionary note. *Cancer Res.* 49:5870, 1989.

65. Smith GK, Duch DS, Dev IK, Kaufmann SH: Metabolic effects and kill of human T-Cell leukemia by 5-deazaacyclotetrahydrofolate, a specific inhibitor of glycineamide ribonucleotide transformylase. *Cancer Res.* 52:4895, 1992.

66. Kothakota S, Azuma T, Reinhard C, Klippel A, Tang J, Chu K, McGarry TJ, Kirschner MW, Koths K, Kwiatkowski DJ, Williams LT: Caspase-3-generated fragment of gelsolin: effector of morphological change in apoptosis. *Science* 278:294, 1997.

67. Gerace L, Burke B: Functional organization of the nuclear envelope. *Annu. Rev. Cell Biol.* 4:335, 1988.

68. Moir RD, Spann TP, Goldman RD: The dynamic properties and possible functions of nuclear lamins. *Int. Rev. Cytol.* 162B:141, 1995.

69. Ucker DS, Obermiller PS, Eckhart W, Apgar JR, Berger NA, Meyers J: Genome digestion is a dispensable consequence of physiological cell death mediated by cytotoxic T lymphocytes. *Mol. Cell. Biol.* 12:3060, 1992.

70. Lazebnik YA, Cole S, Cooke CA, Nelson WG, Earnshaw WC: Nuclear events of apoptosis *in vitro* in cell-free mitotic extracts: a model system for analysis of the active phase of apoptosis. *J. Cell Biol.* 123:7, 1993.

71. Oberhammer FA, Hochegger K, Froschl G, Tiefenbacher R, Pavelka M: Chromatin condensation during apoptosis is accompanied by degradation of lamin A+B, without enhanced activation of cdc2 kinase. *J. Cell Biol.* 126:827, 1994.

72. Neamati N, Fernandez A, Wright S, Kiefer J, McConkey DJ: Degradation of lamin B_1 precedes oligonucleosomal DNA fragmentation in apoptotic thymocytes and isolated thymocyte nuclei. *J. Immunol.* 154:3788, 1995.

73. Lazebnik YA, Takahaski A, Moir R, Goldman R, Poirier GG, Kaufmann SH, Earnshaw WC: Studies of the lamin proteinase reveal multiple parallel biochemical pathways during apoptotic execution. *Proc. Natl. Acad. Sci., U.S.A.* 92:9042, 1995.

74. Casiano CA, Martin SJ, Green DR, Tan EM: Selective cleavage of nuclear autoantigens during CD95 (Fas/APO-1)-mediated T Cell apoptosis. *J. Exp. Med.* 184:765, 1996.

75. Rao L, Perez D, White E: Lamin proteolysis facilitates nuclear events during apoptosis. *J. Cell Biol.* 135:1441, 1996.

76. Yang CH, Lambie EJ, Synder M: NuMA: An unusually long coiled-coil related protein in the mammalian nucleus. *J. Cell Biol.* 116:1303, 1992.

77. Compton DA, Cleveland DW: NuMA, A nuclear protein involved in mitosis and nuclear reformation. *Cur. Opin. Cell Biol.* 6:343, 1994.

78. Jackson DA, Cooke PR: Visualization of a filamentous nucleoskeleton with a 23 nm axial repeat. *EMBO J* 7:3667, 1988.

79. Hsu HL, Yek NH: Dynamic changes of NuMA during the cell cycle and possible appearance of a truncated form of NuMA during apoptosis. *J. Cell Sci.* 109:277, 1996.

80. Gueth-Hallonet C, Weber K, Osborn M: Cleavage of the nuclear matrix protein NuMA during apoptosis. *Exp. Cell Res.* 233:21, 1997.

81. Tewari M, Quan LT, O'Rourke K, Desnoyers S, Zeng Z, Beidler DR, Poirier GG, Salvesen GS, Dixit VM: Yama/CPP32 Beta, a mammalian homolog of CED-3, is a CrmA-inhibitable protease that cleaves the death substrate Poly(ADP-ribose) polymerase. *Cell* 81:801, 1995.

82. Fernandes-Alnemri T, Takahashi A, Armstrong R, Krebs J, Fritz L, Tomaselli KJ, Wang L, Yu Z, Croce CM, Salveson G, Earnshaw WC, Litwack G, Alnemri ES: Mch3, A novel human apoptotic cysteine protease highly related to CPP32. *Cancer Res.* 55:6045, 1995.

83. Casciola-Rosen LA, Anhalt GJ, Rosen A: DNA-dependent protein kinase is one of a subset of autoantigens specifically cleaved early during apoptosis. *J. Exp. Med.* 182:1625, 1995.

84. Han Z, Malik N, Carter T, Reeves WH, Wyche JH, Hendrickson EA: DNA-dependent protein kinase is a target for a CPP32-like apoptotic protease. *J. Biol. Chem.* 271:25035, 1996.

85. Teraoka H, Yumoto Y, Watanabe F, Tsukada K, Suwa A, Enari M, Nagata S: CPP32/Yama/apopain cleaves the catalytic component of DNA-dependent protein kinase in the holoenzyme. *FEBS Lett.* 393:1, 1996.

86. Jackson SP, Jeggo PA: DNA Double-Strand Break Repair and V(D)J Recombination: involvement of DNA-PK. *Trends Biochem. Sci.* 20:412, 1995.

87. Lindahl T, Satoh MS, Poirier GG, Klungland A: Post-translational modification of poly(ADP-ribose) polymerase induced by DNA strand breaks. *Trends Biochem. Sci.* 20:405, 1995.

88. de Murcia JM, Niedergang C, Trucco C, Ricoul M, Dutrillaux B, Mark M, Oliver FJ, Masson M, Dierich A, LeMeur M, Walztinger C, Chambon P, de Murcia G: Requirement of poly(ADP-ribose) Polymerase in recovery from DNA damage in mice and in cells. *Proc. Natl. Acad. Sci., U.S.A.* 94:7303, 1997.

89. Rosenthal DS, Ding R, Simbulan-Rosenthal CM, Vaillancourt JP, Nicholson DW, Smulson M: Intact cell evidence for the early synthesis, and subsequent late apopain-mediated suppression, of poly(ADP-Ribose) during apoptosis. *Exp. Cell Res.* 232:313, 1997.

90. Henkels KM, Turchi JJ: Induction of apoptosis in cisplatin-sensitive and -resistant human ovarian cancer cell lines. *Cancer Res.* 57:4488, 1997.

91. Casciola-Rosen LA, Miller DK, Anhalt GJ, Rosen A: Specific cleavage of the 70 kDa protein component of the U1 small nuclear ribonucleoprotein is a characteristic biochemical feature of apoptotic cell death. *J. Biol. Chem.* 269:30757, 1994.

92. Waterhouse N, Kumar S, Song Q, Strike P, Sparrow L, Dreyfuss G, Alnemri ES, Litwack G, Lavin M, Watters D: Heteronuclear Ribonucleoproteins C1 and C2, Components of the Spliceosome, Are Specific Targets of Interleukin 1β-Converting Enzyme-Like Proteases in Apoptosis. *J. Biol. Chem.* 271:29335, 1996.

93. Ubeda M, Habener JF: The Large subunit of the DNA replication complex C (DSEB/RF-C140) cleaved and inactivated by caspase-3 (CPP32/YAMA) during Fas-induced apoptosis. *J. Biol. Chem.* 272:19562, 1997.

94. Ghayur T, Hugunin M, Talanian RV, Ratnofsky S, Quinlan C, Emoto Y, Pandey P, Datta R, Huang Y, Kharbanda S, Allen H, Kamen R, Wong W, Kufe D: Proteolytic Activation of Protein Kinase C δ by an ICE/Ced-3-like protease induces characteristics of apoptosis. *J. Exp. Med.* 184:2399, 1996.

95. Emoto Y, Manome Y, Meinhardt G, Kisaki H, Kharbanda S, Robertson M, Ghayur T, Wong WW, Kamen R, Weichselbaum R, Kufe D: proteolytic activation of protein kinase C δ by an ICE-like protease in apoptotic cells. *EMBO J* 14:6148, 1995.

96. Emoto Y, Kisaki H, Manome Y, Kharbanda S, Kufe D: Activation of protein kinase Cδ in human myeloid leukemia cells treated with 1-beta-D-arabino-furanyosylcytosine. *Blood* 87:1990, 1996.

97. Rudel T, Bokoch GM: Membrane and morphological changes in apoptotic cells regulated by caspase-mediated activation of PAK2. *Science* 276:1571, 1997.

98. Enari M, Sakahira H, Yokoyama H, Okawa K, Iwamatsu A, Nagata S: A caspase-activated DNase that degrades DNA during apoptosis, and its inhibitor ICAD. *Nature* 391:43, 1998.

99. Sakahira H, Enari M, Nagata S: Cleavage of CAD Inhibitor in CAD activation and DNA degradation during apoptosis. *Nature* 391:96, 1998.

100. Nicholson DW: ICE/CED3-like proteases as therapeutic targets for the control of inappropriate apoptosis. *Nature Biotech.* 14:297, 1996.

101. Whyte M: ICE/CED-3 Proteases in apoptosis. *Trends Cell Biol.* 6:245, 1996.

102. Henkart PA: ICE family proteases: mediators of all apoptotic cell death? *Immunity* 4:195, 1996.

103. Patel T, Gores GJ, Kaufmann SH: The role of proteases during apoptosis. *FASEB J.* 10:587, 1996.

104. Takahashi A, Earnshaw WC: ICE-related proteases in apoptosis. *Curr. Opin. Genet. Dev.* 6:50, 1996.

105. Thornberry NA, Rosen A, Nicholson DW: Control of apoptosis by proteases. *Adv. Pharmacol.* 41:155, 1997.

106. Weil M, Jacobson MD, Coles HS, Davies TJ, Gardner RL, Raff KD, Raff MC: Constitutive expression of the machinery for programmed cell death. *J. Cell Biol.* 133:1053, 1996.

107. Kumar A, Commane M, Flickinger TW, Horvath CM, Stark GR: Defective TNF-α–induced apoptosis in STAT1-null cells due to low constitutive levels of caspases. *Science* 278:1630, 1997.

108. Liu X, Kim CN, Pohl J, Wang X: Purification and characterization of an interleukin-1beta-converting enzyme family protease that activates cysteine protease P32. *J. Biol. Chem.* 271:13371, 1996.

109. Rothe M, Pan MG, Henzel WJ, Ayres TM, Goeddel DV: The TNFR2-TRAF signaling complex contains two novel proteins related to baculoviral inhibitor of apoptosis proteins. *Cell* 83:1243, 1995.

110. Hu S, Vincenz C, Ni J, Gentz R, Dixit VM: I-FLICE, A novel inhibitor of tumor necrosis factor receptor-1- and CD-95-induced apoptosis. *J. Biol. Chem.* 272:17255, 1997.

111. Srinivasula SM, Ahmad M, Ottilie S, Bullrich F, Banks S, Wang Y, Fernandes-Alnemri T, Croce CM, Litwack G, Tomaselli KJ, Armstrong RC, Alnemri ES: FLAME-1, A novel FADD-like anti-apoptotic molecule that regulates Fas/TNFR1-induced apoptosis. *J. Biol. Chem.* 272:18542, 1997.

112. Goltsev YV, Kovalenko AV, Arnold E, Varfolomeev EE, Brodianskii VM, Wallach D: CASH, A novel caspase homologue with death effector domains. *J. Biol. Chem.* 272:19641, 1997.

113. Krajewski K, Tanaka S, Takayama S, Schibler MJ, Fenton W, Reed JC: Investigation of the subcellular distribution of the Bcl-2 oncoprotein: residence in the nuclear envelope, endoplasmic reticulum, and outer mitochondrial membranes. *Cancer Res.* 53:4701, 1993.

114. de Jong D, Prins FA, Mason DY, Reed JC, van Ommen GB, Kluin PM: Subcellular localization of the Bcl-2 protein in malignant and normal lymphoid cells. *Cancer Res.* 54:256, 1994.

115. Lam M, Dubyak G, Chen L, Nunez G, Miesfeld RL, Distelhorst CW: Evidence that BCL-2 represses apoptosis by regulating endoplasmic reticulum-associated Ca^{2+} Fluxes. *Proc. Natl. Acad. Sci., U.S.A.* 91:6569, 1994.

116. Hockenbery DM, Oltvai ZN, Yin X-M, Milliman CL, Korsmeyer SJ: Bcl-2 Functions in an Antioxidant pathway to prevent apoptosis. *Cell* 75:241, 1993.

117. Muchmore SW, Sattler M, Liang H, Meadows RP, Harlan JE, Yoon HS, Nettesheim D, Chang BS, Thompson CB, Wong SL, Ng SL, Fesik SW: X-ray and NMR structure of human Bcl-xL, an inhibitor of programmed cell death. *Nature* 381:335, 1996.

118. Kluck RM, Bossy-Wetzel E, Green DR, Newmeyer DD: The release of cytochrome c from mitochondria: a primary site for Bcl-2 regulation of apoptosis. *Science* 275:1132, 1997.

119. Reed JC: Bcl-2 family proteins: strategies for overcoming chemoresistance in cancer. *Adv. Pharmacol.* 41:501, 1997.

120. Li FL, Srinivasan A, Wang Y, Armstrong RC, Tomaselli KJ, Fritz LC: Cell-specific induction of apoptosis by microinjection of cytochrome C. *J. Biol. Chem.* 272:30299, 1997.

121. Zhivotovsky B, Orrenius S, Brustugun OT, Doskeland SO: Injected cytochrome C induces apoptosis. *Nature* 391:449, 1998.

122. Chinnaiyan AM, O'Rourke K, Lane BR, Dixit VM: Interaction of CED-4 with CED-3 and CED-9: a molecular framework for cell death. *Science* 275:1122, 1997.

123. Clem RJ, Fechheimer M, Miller LK: Prevention of apoptosis by a baculovirus gene during infection of insect cells. *Science* 254:1388, 1991.

124. Crook NE, Clem RJ, Miller LK: An apoptosis-inhibiting baculovirus gene with a zinc finger-like motif. *J. Virol.* 67:2168, 1993.

125. Bump NJ, Hackett M, Hugunin M, Seshagiri S, Brady K, Chen P, Ferenz C, Franklin S, Ghayur T, Li P: Inhibition of ICE family proteases by baculovirus antiapoptotic protein p35. *Science* 269:1885, 1995.

126. Roy N, Mahadevan MS, McLean M, Shutler G, Yaraghi Z, Farahani R, Baird S, Besner-Johnston A, Lefebvre C, Kang X, Salih M, Augry H, Tamai K, Guan X, Ioannou P, Crawford TO, de Jong PJ, Surh L, Ikeda J-E, Korneluk RG, MacKenzie A: The gene for neuronal apoptosis inhibitory protein is partially deleted in individuals with spinal muscular atrophy. *Cell* 80:167, 1995.

127. Liston P, Roy N, Tamai K, Lefebvre C, Baird S, Cherton-Horvat G, Farahani R, McLean M, Ikeda JE, MacKenzie A, Korneluk RG: Suppression of apoptosis in mammalian cells by NAIP and a related family of IAP genes. *Nature* 379:349, 1996.

128. Uren AG, Pakusch M, Hawkins CJ, Puls KL, Vaux DL: Cloning and expression of apoptosis inhibitory protein homologs that function to inhibit apoptosis and/or bind tumor necrosis factor receptor-associated factors. *Proc. Natl. Acad. Sci., U.S.A.* 93:4974, 1996.

129. Duckett CS, Nava VE, Gedrich RW, Clem RJ, Van Dongen JL, Gilfillan MC, Shiels H, Hardwick JM, Thompson CB: A conserved family of cellular genes related to the baculovirus IAP Gene and encoding apoptosis inhibitors. *EMBO J* 15:2685, 1996.

130. Birnbaum MJ, Clem RJ, Miller LK: An apoptosis-inhibiting gene from a nuclear polyhedrosis virus encoding a polypeptide with Cys/His sequence motifs. *J. Virol.* 68:2521, 1994.

131. Lovering R, Hanson IM, Borden KL, Martin S, O'Reilly NJ, Evan GI, Rahman D, Pappin DJ, Trowsdale J, Freemont PS: Identification and preliminary characterization of a protein motif related to the zinc finger. *Proc. Natl. Acad. Sci., U.S.A.* 90:2112, 1993.

132. Deveraux QL, Takahashi R, Salvesen GS, Reed JC: X-linked iap is a direct inhibitor of cell-death proteases. *Nature* 388:300, 1997.

133. Hay BA, Wassarman DA, Rubin GM: Drosophila homologs of baculovirus inhibitor of apoptosis proteins function to block cell death. *Cell* 83:1253, 1995.

134. Xue D, Horvitz HR: Inhibition of the *Caenorhabditis Elegans* Cell-death protease CED-3 by a CED-3 cleavage site in baculovirus P53 protein. *Nature* 377:248, 1995.

135. Kuida K, Zheng TS, Na S, Kuan C, Yang D, Karasuyama H, Rakic P, Flavell RA: decreased apoptosis in the brain and premature lethality in CPP32-deficient mice. *Nature* 384:368, 1996.

135a. Kuida K, Hayder TF, Kuan CY, Gu Y, Taya C, Karasuyzma H, Su MS, Rakie P, Florell RA: Reduced apoptosis and cytochrane c- mediated caspase activation in mice lacking caspase 9, Cell 94: 325, 1998.

136. Zhu H, Fearnhead HO, Cohen GM: An ICE-like protease is a common mediator of apoptosis induced by diverse stimuli in human monocytic THP.1 cells. *FEBS Lett.* 374:303, 1995.

137. Mashima T, Naito M, Kataoka S, Kawai H, Tsuruo T: Aspartate-based inhibitor of interleukin-1- beta-converting enzyme prevents antitumor agent-induced apoptosis in human myeloid leukemia U937 cells. *Biochem. Biophys. Res. Commun.* 209:907, 1995.

138. Deshmukh M, Vasilakos J, Deckwerth TL, Lampe PA, Shivers BD, Johnson EM: genetic and metabolic status of NGF-deprived sympathetic neurons saved by an inhibitor of ICE family proteases. *J. Cell Biol.* 135:1341, 1996.

139. Sarin A, Williams MS, Alexander-Miller MA, Berzofsky JA, Zacharchuk CM, Henkart PA: Target cell lysis by CTL granule exocytosis is independent of ICE/Ced-3 family proteases. *Immunity* 6:209, 1997.

140. Hara H, Friedlander RM, Gagliardini V, Ayata C, Fink K, Huang Z, Shimizu-Sasamata M, Yuan J, Moskowitz MA: Inhibition of interleukin-1 Beta converting enzyme family proteases reduces ischemic and excitotoxic neuronal damage. *Proc. Natl. Acad. Sci., U.S.A.* 94:2007, 1997.

141. Sadoul R, Fernandez PA, Quiquerez AL, Martinou I, Maki M, Schroter M, Becherer JD, Irmler M, Tschopp J, Martinou JC: Involvement of the proteasome in the programmed cell death of NGF-Deprived Sympathetic Neurons. *EMBO J* 15:3845, 1996.
142. Grimm LM, Goldberg AL, Poirier GG, Schwartz LM, Osborne BA: Proteasomes play an essential role in thymocyte apoptosis. *EMBO J* 15:3835, 1996.
143. McConkey DJ: Calcium-Dependent, Interleukin 1-converting enzyme inhibitor-insensitive degradation of lamin B1 and DNA fragmentation in isolated thymocyte nuclei. *J. Biol. Chem.* 271:22398, 1996.
144. Stefanis L, Troy CM, Qi H, Greene LA: Inhibitors of trypsin-like serine proteases inhibit processing of the caspase Nedd-2 and Protect PC12 cells and sympathetic neurons from death evoked by withdrawal of trophic support. *J. Neurochem.* 69:1425, 1997.
145. Sarin A, Adams DH, Henkart PA: Protease inhibitors selectively block T cell receptor-triggered programmed cell death in a murine T cell hybridoma and activated peripheral T cells. *J. Exp. Med.* 178:1693, 1993.
146. Sarin A, Clerici M, Blatt SP, Hendrix CW, Shearer GM, Henkart PA: Inhibition of activation-induced programmed cell death and restoration of defective immune response of HIV+ donors by cysteine protease inhibitors. *J. Immunol.* 153:862, 1994.
147. Squier MK, Miller AC, Malkinson AM, Cohen JJ: Calpain activation in apoptosis. *J. Cell. Physiol.* 159:229, 1994.
148. Deiss LP, Galinka H, Berissi H, Cohen O, Kimchi A: Cathepsin D protease mediates programmed cell death induced by interferon-gamma, Fas/APO-1 and TNF-alpha. *EMBO J* 15:3861, 1996.
149. Lotem J, Sachs L: Differential Suppression by protease inhibitors and cytokines of apoptosis induced by wild-type p53 and cytotoxic agents. *Proc. Natl. Acad. Sci., U.S.A.* 93:12507, 1996.
150. Siman R, Noszek JC: Excitatory amino acids activate calpain I and induce structural protein breakdown *in Vivo. Neuron* 1:279, 1988.
151. Deshmukh M, Johnson EM, Jr.: Programmed cell death in neurons: focus on the pathway of nerve growth factor deprivation-induced death of sympathetic neurons. *Mol. Pharmacol.* 51:897, 1997.
152. Scott SA, Davies AM: Inhibition of protein synthesis prevents cell death in sensory and parasympathetic neurons deprived of neurotrophic factor *in vitro. J. Neurobiol.* 21:630, 1990.
153. Martin DP, Schmidt RE, DiStefano PS, Lowry OH, Carter JG, Johnson EM: Inhibitors of protein synthesis and RNA synthesis prevent neuronal death caused by nerve growth factor deprivation. *J. Cell Biol.* 106:829, 1988.
154. Henderson CE, Camu W, Mettling C, Gouin A, Poulsen K, Karihaloo M, Rullamas J, Evans T, McMahon SB, Armanini MP, Berkemeier L, Phillips HS, Rosenthal A: Neurotrophins Promote motor neuron survival and are present in embryonic limb bud. *Nature* 363:266, 1993.
155. Hughes RA, Sendtner M, Thoenen H: Members of several gene families influence survival of rat motoneurons *in vitro* and *in vivo. J. Neurosci. Res.* 36:663, 1993.
156. Meyer-Franke A, Kaplan MR, Pfrieger FW, Barres BA: Characterization of the signaling interactions that promote the survival and growth of developing retinal ganglion cells in culture. *Neuron* 15.805, 1995.

157. D'Mello SR, Galli C, Ciotti T, Calissano P: Induction of apoptosis in cerebellar granule neurons by low potassium: inhibition of death by insulin-like growth factor I and cAMP. *Proc. Natl. Acad. Sci., U.S.A.* 90:10989, 1993.

158. Estus S, Zaks WJ, Freeman RS, Gruda M, Bravo R, Johnson EM, Jr.: Altered gene expression in neurons during programmed cell death: identification of c-*jun* as necessary for neuronal apoptosis. *J. Cell Biol.* 127:1717, 1994.

159. Ham J, Babij C, Whitfield J, Pfarr CM, Lallemand D, Yaniv M, Rubin LL: A c-*Jun* dominant negative mutant protects sympathetic neurons against programmed cell death. *Neuron* 14:3125, 1995.

160. Freeman RF, Estus S, Johnson EM, Jr.: Analysis of cell-related gene expression in postmitotic neurons: selective induction of cyclin d1 during programmed cell death. *Neuron* 12:343, 1994.

161. Kranenburg O, van der Eb AJ, Zantema A: Cyclin D1 is an essential mediator of apoptotic neuronal cell death. *EMBO J* 15:46, 1996.

162. Park DS, Farinelli SE, Greene LA: Inhibitors of cyclin-dependent kinases promote survival of post-mitotic neuronally differentiated PC12 cells and sympathetic neurons. *J. Biol. Chem.* 271:8161, 1996.

163. Park DS, Levine B, Ferrari G, Greene LA: Cyclin dependent kinase inhibitors and dominant negative cyclin dependent kinase 4 and 6 promote survival of NGF-deprived sympathetic neurons. *J. Neurosci.* 17:8975, 1997.

164. Gagliardini V, Fernandez PA, Lee RK, Drexler HC, Rotello RJ, Fishman MC, Yuan J: Prevention of vertebrate neuronal death by the crmA gene. *Science* 263:826, 1994.

165. Ray CA, Black RA, Kronheim SR, Greenstreet TA, Sleath PR, Salvesen GS, Pickup DJ: Viral Inhibition of inflammation: cowpox virus encodes an inhibitor of the interleukin-1 beta converting enzyme. *Cell* 69:597, 1992.

166. Zhou Q, Snipas S, Orth K, Muzio M, Dixit VM, Salvesen GS: Target protease specificity of the viral serpin CrmA. analysis of five caspases. *J. Biol. Chem.* 272:7797, 1997.

167. Martinou I, Fernandez P-A, Missotten M, White E, Allet B, Sadoul R, Mertinou J-C: Viral proteins E1B19K and p35 protect sympathetic neurons from cell death induced by NGF deprivation. *J. Cell Biol.* 128:201, 1995.

168. Milligan CE, Prevette D, Yaginuma H, Homma S, Cardwell C, Fritz LC, Tomaselli KJ, Oppenheim RW, Schwartz LM: Peptide inhibitors of the ICE protease family arrest programmed cell death of motoneurons *in vivo* and *in vitro*. *Neuron* 15:385, 1995.

169. Troy CM, Stefanis L, Prochiantz A, Greene LA, Shelanski M: The contrasting roles of ICE family proteases and interleukin-1β in apoptosis induced by trophic factor withdrawal and by copper/zinc superoxide dismutase down-regulation. *Proc. Natl. Acad. Sci., U.S.A.* 93:5635, 1996.

170. Stefanis L, Park DS, Yan CYI, Farinelli SE, Troy CM, Shelanski ML, Greene LA: Induction of CPP32-like activity in PC12 cells by withdrawal of trophic support. *J. Biol. Chem.* 271:30663, 1996.

171. Park DS, Stefanis L, Yan CYI, Farinelli SE, Greene LA: Ordering of cell death pathway. *J. Biol. Chem.* 271:21898, 1996.

172. Kumar S, Tomooka Y, Noda M: Identification of a set of genes with developmentally down-regulated expression in the mouse brain. *Biochem. Biophys. Res. Commun.* 185:1155, 1992.

173. Allet B, Hochmann A, Martinou I, Berger A, Missotten M, Antonsson B, Sadoul R, Martinou J-C, Bernasconi L: Dissecting processing and apoptotic activity of a cysteine protease by mutant analysis. *J. Cell Biol.* 135:479, 1996.
174. Kumar S, Kinoshita M, Noda M, Copeland NG, Jenkins NA: Induction of apoptosis by the mouse Nedd2 Gene, which encodes a protein similar to the product of the *Caenorhabditis elegans* cell death gene Ced-3 and the mammalian IL-1 beta-converting enyzme. *Genes Dev.* 8:1613, 1994.
175. Srinivasan A, Foster LM, Testa M-P, Ord T, Keane RW, Bredesen DE, Kayalar C: Bcl-2 expression in neural cells blocks activation of ICE/CED-3 family proteases during apoptosis. *J. Neurosci.* 16:5654, 1996.
176. Troy CM, Stefanis L, Greene LA, Shelanski ML: Nedd2 is required for apoptosis after trophic factor withdrawal, but not superoxide dismutase (SOD1) Downregulation, In sympathetic neurons and PC12 cells. *J. Neurosci.* 17:1911, 1997.
176a. Bergeron L, Perez GI, Macdonald G, Shi L, Sun L, Sun Y, Jurisicova A, Varuza S, Latham KE, Flaws JA, Salter JC, Hara H, Moskowitz MA, Li E, Greenberg A, Tilly JL, Yuan J: Defects in Regulation of apoptosis in caspase-2deficient mice. Genes Dev. 12:1304, 1998.
177. Keane RW, Srinivasan A, Foster LM, Testa M-P, Örd T, Nonner D, Wang H-G, Reed JC, Bredesen DE, Kayalar C: Activation of CPP32 during apoptosis of neurons and astrocytes. *J. Neurosci. Res.* 48:168, 1997.
178. Friedlander RM, Gagliardini V, Hara H, Fink KB, Li W, MacDonald G, Fishman MC, Greenberg AH, Moskowitz MA, Yuan J: Expression of a dominant negative mutant of interleukin-1β converting enzyme in transgenic mice prevents neuronal cell death induced by trophic factor withdrawal and ischemic brain injury. *J. Exp. Med.* 185:933, 1997.
179. Bredesen DE: Neuronal apoptosis: genetic and biochemical modulation, in Tomei LC, Cope FO (Eds): *Apoptosis II: The Molecular Basis of Apoptosis in Disease.* Cold Spring Harbor, NY: Cold Spring Harbor Laboratory, 1994, p. 397.
180. Olney JW: Inciting excitotoxic cytocide amoung central neurons. *Adv. Exper. Med. Biol.* 203:631, 1986.
181. Koh JY, Cotman CW: Programmed cell death: its possible contribution to neurotoxicity mediated by calcium channel antagonists. *Brain Res.* 587:233, 1992.
182. Behrens MI, Koh J, Canzoniero LM, Sensi SL, Csernansky CA, Choi DW: 3-Nitropropionic acid induces apoptosis in cultured straital and cortical neurons. *NeuroReport* 6:545, 1995.
183. Gallo V, Kingsbury A, Balazs R, Jorgensen OS: The role of depolarization in the survival and differentiation of cerebellar granule cells in culture. *J. Neurosci.* 7:2203, 1987.
184. Du Y, Bales KR, Dodel RC, Hamilton-Byrd E, Horn JW, Czilli DL, Simmons LK, Ni B, Paul SM: Activation of a caspase 3-related cysteine protease is required for glutamate-mediated apoptosis of cultured cerebellar granule neurons. *Proc. Natl. Acad. Sci., U.S.A.* 94:11657, 1997.
185. Schulz JB, Weller M, Klockgether T: Potassium deprivation-induced apoptosis of cerebellar granule neurons: a sequential requirement for new mRNA and protein synthesis, ICE-like protease activity, and reactive oxygen species. *J. Neurosci.* 16:4696, 1996.
186. Ni B, Wu X, Du Y, Su Y, Hamilton-Byrd E, Rockey PK, Rosteck P, Poirier GG, Paul SM: Cloning and expression of a rat brain interleukin-1β-converting enzyme (ICE)-related protease (IRP) and its possible role in apoptosis of cultured cerebellar granle neurons. *J. Neurosci.* 17:1561, 1997.

187. Armstrong RC, Aja TJ, Hoang KD, Gaur S, Bai X, Alnemri ES, Litwack G, Karanewsky DS, Fritz LC, Tomaselli KJ: Activation of the CED3/ICE-Related protease CPP32 in Cerebellar granule neurons undergoing apoptosis but not necrosis. *J. Neurosci.* 17:553, 1997.
188. Miller TM, Moulder KL, Knudson CM, Creedon DJ, Deshmukh M, Korsmeyer SJ, Johnson EM: Bax deletion further orders the cell death pathway in cerebellar granule cells and suggests a caspase-independent pathway to cell death. *J. Cell Biol.* 139:205, 1997.
189. Nath R, Raser KJ, McGinnis K, Nadimpalli R, Stafford D, Wang KKW: Effects of ICE-Like Protease and calpain inhibitors on neuronal apoptosis. *NeuroReport* 8:249, 1996.
190. Yu SP, Yeh C-H, Sensi SL, Gwag BJ, Canzoniero LMT, Farhangrazi ZS, Ying HS, Tian M, Dugan LL, Choi DW: Mediation of neuronal apoptosis by enhancement of outward potassium current. *Science* 278:114, 1997.
191. Taylor J, Gatchalian CL, Keen G, Rubin LL: Apoptosis in cerebellar granule neurones: involvement of interleukin-1β converting enzyme-like proteases. *J. Neurochem.* 68:1598, 1997.
192. McCarthy NJ, Whyte MK, Gilbert CS, Evan GI: Inhibition of Ced-3/ICE-related proteases does not prevent cell death induced by oncogenes, DNA Damage, or the Bcl-2 Homologue Bak. *J. Cell Biol.* 136:215, 1997.
193. Okamoto M, Matsumoto M, Ohtsuki T, Taguchi A, Mikoshiba K, Yanagihara T, Kamada T: Internucleosomal DNA Cleavage involved in ischemia-induced neuronal death. *Biochem. Biophys. Res. Commun.* 196:1356, 1993.
194. MacManus JP, Buchan AM, Hill IE, Rasquinha I, Preston E: Global Ischemia can cause DNA fragmentation indicative of apoptosis in rat brain. *Neurosci. Lett.* 164:89, 1993.
195. Nitatori T, Sato N, Waguri S, Karasawa Y, Araki H, Shibanai K, Kominami E, Uchiyama Y: Delayed neuronal death in the CA1 pyramidal cell layer of the gerbil hippocampus following transient ischemia is apoptosis. *J. Neurosci.* 15:1001, 1995.
196. Mills EM, Gunasekar PG, Pavlakovic G, Isom GE: Cyanide-Induced apoptosis and oxidative stress in differentiated PC12 Cells. *J. Neurochem.* 67:1039, 1996.
197. Friedlander RM, Brown RH, Gagliardini V, Wang J, Yuan J: Inhibition of ICE Slows ALS in Mice. *Nature* 388:31, 1997.
198. Nishimoto I, Okamoto T, Giambarella U, Iwatsubo T: Apoptosis in neurodegenerative diseases. *Adv. Pharmacol.* 41:337, 1997.
199. Ellerby HM, Martin SJ, Ellerby LM, Naiem SS, Rabizadeh S, Salvesen GS, Casiano CA, Cashman NR, Green DR, Bredesen DE: Establishment of a cell-free system of neuronal apoptosis: comparison of premitochondrial, mitochondrial, and postmitochondrial phases. *J. Neurosci.* 17:6165, 1997.
200. Stefanis L, Burke RE, Greene LA: Apoptosis in neurodegenerative disorders. *Curr. Opin. Neurol.* 10:299, 1997.
201. Loo DT, Copani A, Pike CJ, Whittemore ER, Walencewicz AJ, Cotman CW: Apoptosis is induced by beta-amyloid in cultured central nervous system neurons. *Proc. Natl. Acad. Sci., U.S.A.* 90:7951, 1993.
202. Forloni G, Chiesa R, Smiroldo S, Verga L, Salmona M, Tagliavini F, Angeretti N: Apoptosis mediated neurotoxicity induced by chronic application of beta amyloid fragment 25-35. *NeuroReport* 4:523, 1993.
203. LaFerla FM, Tinkle BT, Bieberich CJ, Haudenschild CC, Jay G: The Alzheimer's A beta peptide induces neurodegeneration and apoptotic cell death in transgenic Mice. *Nat. Genet.* 9:21, 1995.

204. Wolozin B, Iwasaki K, Vito P, Ganjei JK, Lacana E, Sunderland T, Zhao B, Kusiak JW, Wasco W, D'Adamio L: Participation of presenilin 2 in apoptosis: enhanced basal activity conferred by an alzheimer mutation. *Science* 274:1710, 1996.
205. Vito P, Wolozin B, Ganjei JK, Iwasaki K, Lacana E, D'Adamio L: Requirement of the familial alzheimer's disease gene PS2 for apoptosis. opposing effect of ALG-3. *J. Biol. Chem.* 271:31025, 1996.
206. Guo Q, Sopher BL, Furukawa K, Pham DG, Robinson N, Martin GM, Mattson MP: Alzheimer's presenilin mutation sensitizes neural cells to apoptosis induced by trophic factor withdrawal and amyloid beta-peptide: involvement of calcium and oxyradicals. *J. Neurosci.* 17:4212, 1997.
207. Zeitlin S, Liu JP, Chapman DL, Papaioannou VE, Efstratiadis A: Increased apoptosis and early embryonic lethality in mice nullizygous for the huntington's disease gene homologue. *Nat. Genet.* 11:155, 1995.
208. Rabizadeh S, Gralla EB, Borchelt DR, Gwinn R, Valentine JS, Sisodia S, Wong P, Lee M, Hahn H, Bredesen DE: Mutations associated with amyotrophic lateral sclerosis convert superoxide dismutase from an antiapoptotic gene to a proapoptotic gene: studies in yeast and neural cells. *Proc. Natl. Acad. Sci., U.S.A.* 92:3024, 1995.
209. Ziv I, Melamed E, Nardi N, Luria D, Achiron A, Offen D, Barzilai A: Dopamine induces apoptosis-like cell death in cultured chick sympathetic neurons — a possible novel pathogenetic mechanism in Parkinson's disease. *Neurosci. Lett.* 170:136, 1994.
210. Paradis E, Douillard H, Koutroumanis M, Goodyer C, LeBlanc A: Amyloid beta peptide of alzheimer's disease downregulates Bcl-2 and upregulates Bax expression in human neurons. *J. Neurosci.* 16:7533, 1996.
211. Anderson AJ, Su JH, Cotman CW: DNA Damage and apoptosis in Alzheimer's disease: colocalization with c-Jun immunoreactivity, relationship to brain area, and effect of postmortem delay. *J. Neurosci.* 16:1710, 1996.
212. Goldberg YP, Nicholson DW, Rasper DM, Kalchman MA, Koide HB, Graham RK, Bromm M, Kazemi-Esfarjani P, Thornberry NA, Vaillancourt JP, Hayden MR: Cleavage of Huntingtin by apopain, a proapoptotic cysteine protease, is modulated by the polyglutamine tract. *Nat. Genet.* 13:442, 1996.
213. Sherrington R, Rogaev EI, Liang Y, Rogaeva EA, Levesque G, Ikeda M, Chi H, Lin C, Li G, Holman K: Cloning of a gene bearing missense mutations in early-onset familial alzheimer's disease. *Nature* 375:754, 1995.
214. Rogaev EI, Sherrington R, Rogaeva EA, Levesque G, Ikeda M, Liang Y, Chi H, Lin C, Holman K, Tsuda T: Familial Alzheimer's disease in kindreds with missense mutations in a gene on chromosome 1 Related to the Alzheimer's disease Type 3 gene. *Nature* 376:775, 1995.
215. Levy-Lahad E, Wasco W, Poorkaj P, Romano DM, Oshima J, Pettingell WH, Yu C, Jondro PD, Schimdt SD, Wang K, Crowley AC, Fu Y-H, Guenette SY, Galas D, Nemens E, Wijsman EM, Bird TD, Schellenberg GD, Tanzi RE: Candidate gene for the chromosome 1 familial Alzheimer's disease locus. *Science* 269:973, 1995.
216. Lacana E, Ganjei JK, Vito P, D'Adamio L: Dissociation of apoptosis and activation of IL-1beta-converting enzyme/Ced-3 proteases by ALG-2 and the truncated Alzheimer's gene ALG-3. *J. Immunol.* 158:5129, 1997.
217. Loetscher H, Deuschle U, Brockhaus M, Reinhardt D, Nelboeck P, Mous J, Grunberg J, Haass C, Jacobsen H: Presenilins are processed by caspase-type proteases. *J. Biol. Chem.* 272:20655, 1997.

7

Molecular Mechanisms in the Activation of Apoptotic and Antiapoptotic Pathways by Ceramide

Rick T. Dobrowsky

CONTENTS

7.1 Introduction

Ceramide is the common moiety forming the hydrophobic backbone of glycosphingolipids and sphingomyelin (SM). Over the last 8 years, numerous reports have shown that cytokines,[1-9] hormones,[10-12] neurotrophins,[13,14] the T cell costimulatory molecule CD 28,[15] chemotherapeutic agents,[16-20] and environmental stresses[21,22] increase cellular ceramide levels primarily via the hydrolysis of SM (Figure 7.1). Collectively, these studies support a role for ceramide as a bioactive sphingolipid metabolite involved in the regulation of numerous cellular responses. Foremost among these responses is the ability of ceramide to affect growth suppression pathways leading to cell growth arrest, cell differentiation, and programmed cell death (apoptosis).

Several lines of evidence support a role for ceramide as an important mediator in apoptosis. First, ceramide production occurs in response to agents which induce apoptosis, i.e., tumor necrosis factor-α (TNF),[23-26] Reaper,[27] nerve growth factor,[28] daunorubicin,[16,17] vincristine,[19] arabinofuranosylcytosine,[18] FAS

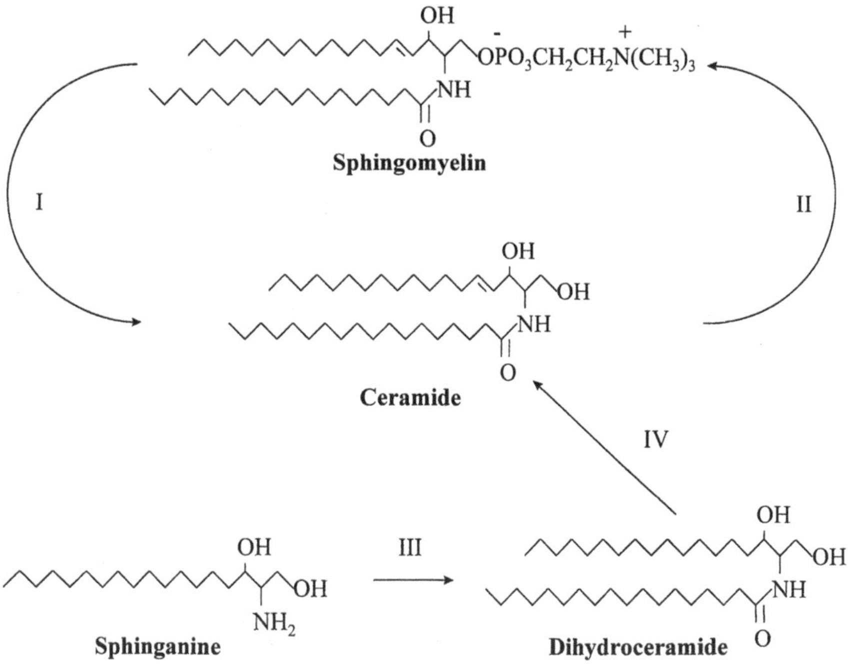

FIGURE 7.1
Schematic of ceramide production via sphingomyelin breakdown or *de novo* synthesis. The relevant enzymes are (I) acid or neutral sphingomyelinase, (II) phosphatidylcholine:ceramide cholinephosphotransferase, (III) sphinganine N-acyltransferase (dihydroceramide synthase), and (IV) dihydroceramide desaturase.

antigen,[29-33] antibody crosslinking,[34,35] and ionizing radiation.[22,26,39] Second, the kinetics of ceramide accumulation correlates closely with the kinetics of apoptosis induced by these agents.[36] Finally, exogenous ceramide mimics the ability of the aforementioned agents to induce apoptosis in hematopoietic cells,[23,37] fibroblasts,[25,38] lung endothelium,[39] pheochromocytoma cells,[40] and oligodendroglial cells.[28,41]

In contrast to glial cells, primary cultures of cortical neurons are relatively insensitive to ceramide-induced apoptosis,[41] except at very high concentrations of ceramide.[42] Indeed, ceramide prevents or at least delays apoptosis of sympathetic neurons induced by growth factor withdrawal.[43] Ceramide also protects hippocampal[44] and cortical neurons (RTD, unpublished data) from apoptosis induced by amyloid-β peptides. These results indicate that in postmitotic neurons ceramide may activate antiapoptotic pathways and suggest that the ceramide signal is interpreted in a very distinct manner in mature neurons vs. glial cells.

The intent of this review is to describe the molecular mechanisms whereby ceramide may mediate apoptotic or antiapoptotic responses. It is hoped that these discussions will serve as overview for scientists new to the field and provide a framework for the design of future investigations into how ceramide may activate similar molecular targets but promote very dichotomous biologies in different cell types.

7.2 Ceramide as a Mediator of Apoptosis

7.2.1 Ceramide Generation

7.2.1.1 Sphingomyelinase Activation

Numerous cell culture studies have identified that sphingomyelinase (SMase) activity is increased following stimulation with TNF,[1,4] FAS antigen,[29,33,45] ionizing radiation,[39] and chemotherapeutics.[17,18] SMase is a type of phospholipase C which cleaves phosphocholine from SM generating ceramide (Figure 7.1). Further, at least two types of SMases have been implicated in generating a bioactive pool of ceramide. These enzymes can be readily distinguished, based upon their neutral or acidic pH optima for activity. Of these two enzymes, the gene encoding for acid SMase has been cloned and knockout animals generated.

Compelling evidence for a critical role of acidic SMase and ceramide in radiation-induced apoptosis comes from studies using acid SMase-deficient knockout mice.[39] Following irradiation of wild-type animals, ceramide levels in lung tissue increased 2-fold over control levels within 10 min and remained elevated for at least 1 hour. Ten hours following irradiation, histologic analysis of tissues from the wild-type animals indicated extensive apoptosis in lung and heart endothelium as well as mesothelium of the pleura and pericardium. In contrast, the lack of acid SMase in the knockout animals completely abolished ceramide production and the induction of an apoptotic phenotype in lung tissue. The conclusion from these studies is that the activation of acid SMase and the production of ceramide may be critical to the induction of an apoptotic phenotype in lung tissue in response to ionizing radiation.

Interestingly, cells from the thymic cortex of acid SMase deficient mice were not as resistant to radiation-induced apoptosis. Although p53 is a critical component in the apoptosis of thymic cells, lung tissue from p53-deficient mice showed a similar extent of radiation-induced apoptosis to that in wild-type mice. Together, these data suggest that p53 is not involved in mediating apoptosis within lung endothelium and that the role of acid SMase in apoptosis may be very cell specific.[39] This cell specificity may underlie the reported differences in the activation of either acid[33] or neutral[29] SMases in various cell lines in response to FAS antigen. Unfortunately, the mechanisms regulating which SMase may become activated in response to a given stimulus remain poorly defined.

In contrast to the acidic SMase, establishing a definitive role for the Mg^{2+}-dependent neutral SMase in apoptosis has been difficult, since this enzyme has not been purified or cloned, nor have any specific inhibitors of its activity been identified. Despite these shortcomings, exciting progress has been evident on the potential role of this enzyme in apoptosis. A most striking example is the correlation between increased cell death, the induction of neutral SMase activity, and a 15-fold increase in cellular ceramide levels in Molt-4 cells in response to serum deprivation.[46]

Previous studies have indicated that the neutral SMase is preferentially activated by FAS[29] and is very susceptible to proteolytic degradation.[47] In this regard, the site-specific inhibitor of certain ICE-like proteases (YVAD-CMK), significantly attenuated FAS-dependent ceramide production and decreased apoptosis by 60% in Jurkat cells.[31] Although an assessment of the specific SMase which was being inhibited by YVAD-CMK was not performed, these results suggest that an ICE-like protease may regulate neutral SMase activation. Indeed, a dominant negative mutant of the FAS-binding protein, FADD, abolished FAS-induced ceramide generation and apoptosis.[48] Since an ICE-like protease lies downstream of FADD,[49,50] these results also place this or a closely related protease upstream of ceramide production. Similarly, inhibition of ICE proteases blocked ceramide production and apoptosis induced by ectopic expression of Reaper (a cell death protein in *Drosophila*) in mammalian cells.[27]

The cytokine response modifier A (CrmA) is a viral protein which also inhibits members of the ICE family of cysteine proteases.[51] Similar to YVAD.CMK, transfection of CrmA into MCF-7 cells blocked TNF-induced ceramide generation and apoptosis.[24] However, exogenous ceramide circumvented the inhibition of apoptosis by CrmA, indicating that the cell death machinery was intact and that ceramide-induced apoptosis occurred downstream of the CrmA-sensitive protease.[24] Collectively, the above reports strongly suggest that an ICE-like protease activity is upstream of FAS- and TNF-induced SMase activation and ceramide generation. Although not definitely established, this SMase activity may be the Mg^{2+}-dependent neutral pH enzyme. More important, however, in conjunction with the genetic studies from the acid SMase knockout mice, these data strongly support evidence that ceramide is a critical mediator in the induction phase of apoptosis by these varied stimuli.

7.2.1.2 De Novo Synthesis

Daunorubicin is an anthracycline antibiotic that induces apoptosis and is widely used in cancer chemotherapy. Not surprisingly, ceramide mimicked the effect of daunorubicin in inducing apoptosis in U937 human monoblastic leukemia cells and HL-60 cells.[16,17,20] However, despite the fact that daunorubicin increased intracellular ceramide levels, a concomitant decrease in SM levels was not observed. These results suggested that ceramide may be a mediator in the apoptotic response to this chemotherapeutic agent, but was not generated through SM catabolism.[16,20] Although most studies have focused on the activation of SMases as the enzymes responsible for ceramide generation, ceramide may also be formed *de novo* from the acylation of sphinganine by sphinganine N-acyltransferase.[52] Dihydroceramide subsequently serves as substrate for the enzyme dihydroceramide desaturase which introduces the 4,5-*trans* double bond present in ceramide[53,54,54a] (Figure 7.1). Indeed, 10 μM daunorubicin induced a 70% increase in the V_{max} of sphinganine N-acyltransferase after 4 to 6 h of drug treatment. Similar results have also been reported for the effect of daunorubicin on dihydroceramide synthase activity in HL60 cells.[20] Additionally, increased *de novo* ceramide synthesis precedes apoptosis induced by palmitate after inhibition of the mitochondrial enzyme, carnitine palmitoyltransferase.[57] Although it is unknown whether dihydroceramide desaturase activity is also subject to regulation, this enzyme is critical for the synthesis of an apoptogenic pool of lipid, since dihydroceramide is ineffective at inducing apoptosis.[23,25]

An important tool in defining the role of this sphingolipid biosynthetic pathway in apoptosis was the discovery that the acylation of sphinganine to dihydroceramide is inhibited specifically by fumonisin B_1.[55] Fumonisin B_1 is a member of a family of fungal metabolites which can induce a wide range of toxicities in both animals and humans that are due to alterations in sphingolipid metabolism (for an excellent review, see Reference 56). Significantly, pretreatment of U937 leukemia cells with 25 μM fumonisin B_1 dramatically attenuated increases in ceramide levels and apoptosis in response to 10 μM daunorubicin[16]; supporting the conclusion that *de novo* synthesis of ceramide may be involved in the cellular actions of daunorubicin.

A caveat to these studies, however, concerns the physiologic relevance of *de novo* ceramide synthesis in the clinical efficacy of daunorubicin. At usual pharmacologic doses of daunorubicin (0.1 to 1 μM), ceramide levels also increased, but through activation of neutral SMase.[17] Indeed, fumonisin B_1

was without effect on apoptosis or ceramide generation induced by 0.1 to 1 μM daunorubicin; suggesting that stimulation of *de novo* synthesis may be inconsequential at therapeutic levels of the drug. Further, other chemotherapeutic agents such as vincristine, methotrexate, etoposide, and *cis*-platinum did not increase dihydroceramide synthase activity. Unfortunately, other anthracyclines such as doxorubicin, idarubucin, epirubicin, and mitoxantrone were not tested.[16] Regardless of the mechanism of ceramide production, however, these studies agree that ceramide is a likely mediator in the apoptotic action of this important chemotherapeutic agent.

7.2.1.3 Localization of Ceramide Production and Relationship to Apoptosis

The bulk of cellular SM localizes to the external leaflet of the plasma membrane.[58] This orientation has raised the question of where the signal-sensitive pool of SM used in the generation of ceramide is located. Linardic and Hannun[59] first addressed this question by treating HL-60 cells with bacterial SMase, thereby depleting 48% of cellular SM. Since the cells are impermeable to the bacterial enzyme, it can only hydrolyze externally oriented SM and is inaccessible to internal pools of SM (Figure 7.2A). If a ligand such as vitamin D_3 was hydrolyzing only an internal pool of SM, then SM hydrolysis induced by co-treatment of the cells with bacterial SMase and the ligand should be additive. Accordingly, treatment of HL-60 cells with vitamin D_3 alone induced a 15.4% decrease in SM, whereas treatment with Vitamin D_3 plus bacterial SMase resulted in a loss of 65.3% (48% + 15%) of total SM. These results strongly suggested that vitamin D_3 was hydrolyzing an internal pool of SM. Similar results have also been obtained in cells treated with TNF or nerve growth factor.[14,60] Subsequent cell fractionation experiments suggest that the internal signal-sensitive pool of SM localizes to the plasma membrane,[59] possibly within specialized microdomains called caveolae.[61,62]

The importance of this internal pool of SM in producing an apoptogenic pool of ceramide has been elegantly demonstrated by Zhang et al.[63] These investigators cloned the neutral SMase from *Bacillus cereus* and placed it in an inducible mammalian expression vector under the control of the *lac* operator. Upon induction of the enzyme with isopropylthiogalactoside (IPTG) in stably transfected Molt-4 cells, cellular ceramide levels increased greater than 2-fold and the extent of cell death increased 6-fold.[63] Significantly, treatment of the cells with exogenous bacterial SMase failed to induce death despite increasing cellular ceramide levels (Figure 7.2B). Thus, hydrolysis of the outer leaflet SM, while increasing cellular ceramide levels, does not induce an apoptogenic pool of ceramide. These results provide strong biologic evidence suggesting that the reservoir of SM for production of the bioactive pool of ceramide is distinct from that present on the outer leaflet pool of the plasma membrane. However, it is unclear why treatment with bacterial SMase has reportedly induced apoptosis in some cell lines[25,38,42] but not others.[40,63]

7.2.2 The Effect of Ceramide on Mitochondrial Function and Its Potential Role in Apoptosis

7.2.2.1 *Inhibition of Mitochondrial Respiration*

Several recent reports have provided convincing evidence that ceramide can inhibit complex III of the mitochondrial respiratory chain and increase the generation of H_2O_2 both *in vivo*[50,64] and in isolated mitochondria (Figure 7.3).[64,65] Indeed, rotenone (complex I inhibitor) and thenoyltrifluoroacetone (complex II inhibitor) blocked electron transfer prior to complex III and inhibited ceramide-induced increases in H_2O_2. However, Antimycin A (complex III inhibitor at the level of electron transfer to cytochrome c) increased ceramide-induced H_2O_2 production, suggesting that ceramide was acting at or near the same site of action as Antimycin A (Figure 7.3).[50,64,65] Interestingly, dihydroceramide was ineffective at increasing H_2O_2 in whole cells,[50] but had a similar potency to ceramide in inducing H_2O_2 production in isolated mitochondria.[65] However, other structurally related lipids were ineffective at inducing H_2O_2 generation. These results suggest that ceramide may affect cellular processes not only through direct interaction with protein targets[67-69] but also indirectly through the generation of reactive oxygen species.

7.2.2.2 *Ceramide-Induced Production of H_2O_2 and the Mitochondrial Permeability Transition*

A recent development in elucidating the sequence of events mediating the apoptotic cascade was the realization that changes in mitochondrial permeability and disruption of the mitochondrial transmembrane potential marks a critical point in the execution of apoptosis from which recovery is irreversible.[66] The mitochondrial transmembrane potential is a product of the respiratory activity of mitochondria. During oxidative phosphorylation, electrons are passed from one carrier to another along the four complexes of the respiratory chain. (Figure 7.3). However, elements such as coenzyme Q within the electron transport chain may carry both protons and electrons. In passing electrons from coenzyme Q to a cytochrome, for instance, the protons must be ejected. These protons collect in the intermembrane space of the mitochondria and form a pH and electrical gradient across the inner mitochondrial membrane ($\Delta\Psi$, transmembrane potential). The mitochondrial permeability transition results from a rapid increase in the permeability of the inner mitochondrial membrane which occurs through the opening of a transmembrane pore.[70] This process leads to a collapse of the ion gradient, mitochondrial depolarization (loss of $\Delta\Psi$), generation of reactive oxygen species, and loss of oxidative phosphorylation.[66,71] The nature of the mitochondrial permeability transition pore is unknown, but it may involve a mitochondrial cyclophilin which can be inhibited from opening by cyclosporine and sphingosine.[70]

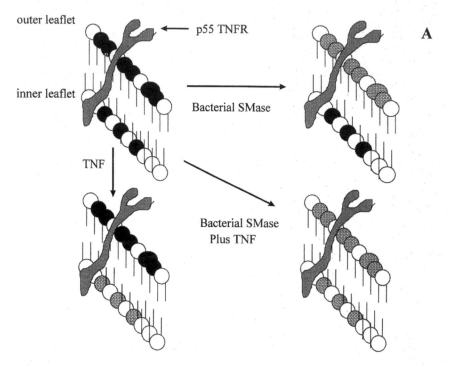

FIGURE 7.2

The signal-sensitive pool of sphingomyelin is internally oriented and involved in apoptosis. (A) Treatment of cells with exogenous bacterial SMase hydrolyzes SM ● oriented on the outer leaflet of the plasma membrane to ceramide ● . However, the enzyme is inaccessible to an internal pool of SM which is hydrolyzed by TNF (see text for more details). (B) Treatment of cells with exogenous bacterial SMase hydrolyzes SM ● oriented on the outer leaflet of the plasma membrane to ceramide ● . However, the resulting increase in cellular ceramide does not create an apoptogenic pool of lipid. In contrast, after transfection of bacterial SMase into cells and induction with IPTG, the enzyme has access to internally oriented of SM which upon hydrolysis, creates an apoptogenic pool of ceramide.

Ceramide did not induce a change in the mitochondrial $\Delta\Psi$ nor inhibit oxidative phosphorylation. These results indicate that neither ceramide nor ceramide-induced H_2O_2 directly affect the permeability transition. Moreover, ceramide-induced increases in H_2O_2 preceded the mitochondrial permeability transition, suggesting that the generation of H_2O_2 by ceramide is not a consequence of increased permeability of the inner mitochondrial membrane.[50,64,65] Since the permeability transition increases the production of reactive oxygen species,[72] the accumulation of ceramide within mitochondria and its inhibition of electron transport may have a role in activating the pore and/or amplifying the production of reactive oxygen species.[65] Interestingly, TNF treatment of cultured hepatocytes led to a 2- to 3-fold increase in ceramide levels in mitochondria isolated from these cells.[65] If this result is reproducible, it will be important to determine how ceramide may accumulate in mitochondria and the necessity of this accretion to ceramide-mediated apoptosis.

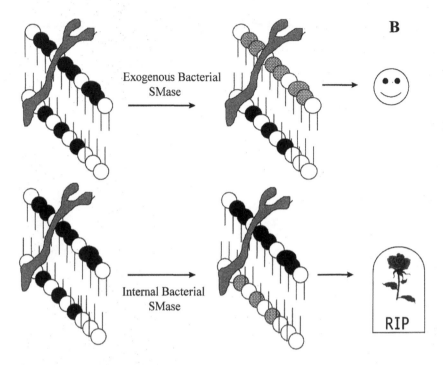

Exogenous Bacterial
SMase

Internal Bacterial
SMase

B

RIP

FIGURE 7.2 (continued)

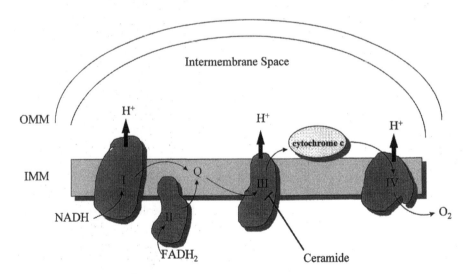

FIGURE 7.3
Ceramide inhibits mitochondrial respiration by blocking electron transport. The four complexes (I-IV) of the respiratory chain are located on the inner mitochondrial membrane (IMM). As electrons are transferred along the complexes (indicated by the arrows), protons are ejected into the intermembrane space, creating an electrochemical gradient ($\Delta\Psi$) across the IMM. Ceramide inhibits electron transport within complex III of the respiratory chain. Q represents the ubiquinone complex and OMM is the outer mitochondrial membrane.

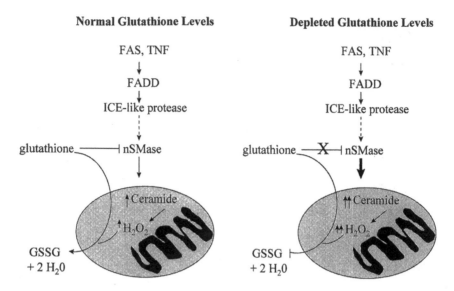

FIGURE 7.4

Glutathione depletion may amplify ceramide-induced H_2O_2 production. In response to stimuli such as FAS and TNF, mitochondrial levels of ceramide may increase, stimulating H_2O_2 production. The oxidation of the reduced form of glutathione to GSSG with the concomitant reduction of H_2O_2 to water is the primary defense of mitochondria for peroxide removal. In cells depleted of glutathione, peroxide metabolism is severely compromised in the mitochondria leading to increased H_2O_2 levels. Additionally, depletion of cellular glutathione may relieve the allosteric inhibition of neutral SMase activity by glutathione thus elevating ceramide levels and further amplifying H_2O_2 production.

   Not surprisingly, antioxidants inhibited ceramide-induced H_2O_2 production and subsequent apoptosis.[50] The level of mitochondrial peroxides is primarily regulated via the oxidation of glutathione (γ-glutamycysteinylglycine) by glutathione peroxidase. Accordingly, glutathione-depleted mitochondria isolated from animals treated with buthionine-L-sulfoximine (BSO, a specific inhibitor of γ-glutamylcysteine synthase), were more sensitive to ceramide-induced H_2O_2 generation than mitochondria whose glutathione levels had been repleted.[65] Moreover, Liu and Hannun[73] demonstrated that physiologic concentrations of glutathione may serve as a negative allosteric regulator of neutral SMase *in vitro*. Indeed, treatment of Molt-4 cells with BSO for 48 h reduced cellular glutathione to 9% of control levels and induced a 2.5-fold increase in ceramide. Thus, depletion of cellular glutathione may create a double insult to the cell by elevating (mitochondrial?) ceramide levels and amplifying the level of H_2O_2 through increased production and decreased degradation[73] (Figure 7.4).

7.2.2.3   Ceramide Activation of Caspases and Inhibition by Bcl-2

Ceramide activates caspase-3 and induces proteolysis of poly(ADP)ribose polymerase.[74] Although the mechanism of ceramide activation of caspase-3 is unknown, it may relate to the release of cytochrome c. Cytochrome c is located in the intermembrane space of the mitochondria peripherally bound to the inner mitochondrial membrane (Figure 7.3). Recent data have identified the release of cytochrome c from mitochondria as a critical component of apoptosis and in the activation of caspase-3,[75] at least in some cell systems.[76] Although discrepancies exist regarding whether cytochrome c is released before[77,78] or after the permeability transition,[71,79] these results raise the possibility that ceramide may affect the release of cytochrome c and related factors or be a component with apoptotic protease-activating factor 1[75] and/or apoptosis-inducing factor[79] in activating caspases.

Bcl-2 is well recognized for its ability to inhibit apoptosis induced by TNF or ceramide.[80] Importantly, Bcl-2 does not inhibit ceramide production, indicating that it acts downstream of ceramide,[19,24] most likely at the level of the mitochondria by protecting against unidentified ceramide-elicited inducers of the permeability transition.[71] Although the mechanism whereby Bcl-2 inhibits apoptosis is unknown, it may relate to its ability to inhibit caspase activation,[74,81] to inhibit the release of cytochrome c and the apoptosis-inducing factor,[77-79] to function as an antioxidant protein,[82] to effect Ca^{2+} fluxes across mitochondrial membranes, to bind Raf-1,[77,80,83] and to antagonize the activation of stress-activated protein kinases.[84]

Although overexpression of Bcl-2 inhibits ceramide-induced apoptosis, it is tempting to speculate that ceramide may affect Bcl-2 in the induction of apoptosis. Recent data have shown that Bcl-2 binds Raf-1 and that this interaction facilitates resistance to apoptosis,[83] potentially through the phosphorylation and inactivation of the proapoptotic protein BAD by Raf-1.[85] Ceramide activates a 97-kDa protein kinase[68] recently identified as the kinase suppressor of ras (KSR).[86] KSR is a membrane-associated Raf-1 kinase which phosphorylates Raf-1 on a unique motif in the N-terminal domain[86,87] which is not involved in BAD phosphorylation nor the interaction with Bcl-2.[83,85] Similar to Bcl-2, KSR lacks a signal sequence and may bind to membranes through interaction with its hydrophobic carboxyl terminal domain.[80,86] Although it is unknown whether KSR may localize to mitochondrial membranes, increases in mitochondrial pools of ceramide[65] may activate KSR and affect the antiapoptotic activity of the Bcl-2-Raf–1 complex. On the other hand, one study has suggested that ceramide-activated protein kinase (KSR) may not be involved in apoptosis.[88]

7.2.3 Effect of Ceramide on Stress-Activated Protein Kinases

Stress-activated protein kinases (SAPKs or c-Jun N-terminal kinases, JNK) and p38 kinase are members of the MAP kinase family, which are activated by a phosphorylation cascade involving upstream kinase–kinase activities (reviewed in Reference 89). In many cells, one biologic consequence of SAPK and p38 kinase activation is apoptosis. There is significant overlap between the stimuli known to activate SAPK and induce ceramide production,[90] suggesting that SAPK activation may be a downstream effector of ceramide signaling. Indeed, addition of C_2-ceramide to HL-60 cells activated SAPKs and increased *c-jun* mRNA levels.[91] These results are consistent with the reports that the activation of *c-jun*/AP-1 is necessary for ceramide-induced apoptosis in HL-60 cells[37] and that inhibition of SAPK by dominant-negative SEK1 (the immediate upstream kinase activator of SAPKs) or TAM-67 (a dominant-negative c-Jun mutant lacking the N-terminal transactivation domain) blocked apoptosis induced by growth factor withdrawal[92] or ceramide.[26] However, a separate study found that Tam-67 alone failed to protect cells from ceramide-induced apoptosis, although the combination of Tam-67 and an inhibitor of p38 kinase was effective.[93] These results suggest that cross talk between SAPKs and p38 kinase may be important in regulating the response to ceramide, and that the role of these MAP kinase family members in ceramide-induced apoptosis may vary depending upon the cell type. Although the mechanism whereby ceramide leads to SAPK activation is unknown, it may involve the generation of reactive oxygen species.[94] Alternatively, the activation of GTPases such as Rac1 and Ras[30,93] may regulate an upstream MAP kinase kinase kinase such as TAK1, which may be necessary for ceramide activation of SAPKs.[94] Indeed, dominant negative mutants of Rac1 and ras inhibited ceramide-induced activation of SAPKs and p38 kinase and blocked apoptosis.[30,93]

7.3 Ceramide, a Potential Mediator of Antiapoptotic Signals in Post-Mitotic Neurons

Primary cultures of sympathetic neurons are dependent upon nerve growth factor (NGF) for survival.[43] However, the addition of the cell-permeable ceramide analogue, C_2-ceramide, concomitant with NGF withdrawal increased neuronal survival by up to 84% after 48 h. Moreover, the time necessary to kill 50% of the neurons increased from 27 to 81 h. Similarly, ceramide protected hippocampal neurons against cell death induced by amyloid-beta peptides, glutamate, and $FeSO_4$ toxicity.[44] The protective effects of ceramide in the latter study were only seen upon preincubation of the cells with ceramide and were blocked by inhibitors of protein and RNA synthesis. Significantly, in both studies, ceramide concentrations exceeding 25 μM induced cell death. These results are similar to studies showing that neurons are much less sensitive than glial cells to ceramide-induced death.[28,41] For example, chick cortical neurons required 50 to 100 μM ceramide in serum free medium to achieve significant levels of cell death[42]; these concentrations are at least 10 to 20 fold greater than those required for death in glial cells.[28,41] These data suggest that ceramide may temporarily activate antiapoptotic pathways in neurons and that a fundamental difference exists between how the ceramide signal is interpreted by postmitotic neurons vs. glial cells.

7.3.1 Potential Molecular Mechanisms of the Protective Effects of Ceramide

7.3.1.1 *Inhibition of Cell Cycle Reentry by Postmitotic Neurons*

Although the discussion to this point has focused on ceramide as a mediator of apoptosis, ceramide also induces cell growth arrest at the G_0/G_1 phase of the cell cycle.[46] Importantly, ceramide can induce a G_0/G_1 cell cycle arrest without inducing apoptosis.[46,95] In many cells, apoptosis occurs from late G_1 to early S phase, while cell cycle arrest in G_0 or early G_1 can suppress apoptosis.[96,97] The cell cycle/apoptosis hypothesis predicts that neuronal apoptosis may be initiated by an abortive attempt of postmitotic neurons to reenter the cell cycle.[98] Several studies support the view that once neurons attempt to reenter the cell cycle at the G_1/S checkpoint, they are committed to die.[99,100] Indeed, inhibitors which block progression through G_2/M do not protect postmitotic sympathetic neurons from undergoing apoptosis in response to NGF withdrawal, indicating that neurodegenerative pathways are already irreversibly activated.[98,100]

Several studies have shown that cyclin D1 expression and cyclin-dependent kinase 4 (cdk4) activity are necessary for neuronal apoptosis. Since the cyclin D–cdk4 complex is the first cdk known to be activated and is necessary for the progression of cells from G_1,[101] activation of this complex in postmitotic neurons may be an early signal for apoptosis.[102] Additionally, hyperphosphorylation of retinoblastoma (Rb) protein by cyclin D1–cdk4 and the neuron specific cdk5 may accelerate apoptosis.[103,104] These results raise the possibility that the antiapoptotic effect of ceramide may relate to the ability of ceramide to induce Rb dephosphorylation,[105] inhibit cdk activity,[95] and arrest cells at the G_0/G_1 checkpoint,[46] thereby blocking an attempt at cell cycle reentry. Prevailing evidence suggests that ceramide-induced growth arrest is regulated at least in part by changes in the phosphorylation state of Rb[19,105] and the inhibition of cyclin-dependent kinases.[95] The importance of Rb in effecting ceramide-induced cell cycle arrest is underscored by the attenuated ability of ceramide to inhibit the proliferation of cells either lacking Rb or in which Rb is sequestered by binding to the adenoviral E1A antigen.[105] Further, ceramide dramatically attenuates the activation of cdk4 induced by treatment of primary cultures of cortical neurons with amyloid-beta peptides (RTD, unpublished data). However, neurons may also undergo apoptosis through pathways independent of attempts at cell cycle reentry, i.e., peroxynitrite-induced cell death. It will be interesting to determine if ceramide may also be protective against these inducers of apoptosis.

7.3.1.2 *Ceramide Activation of NF-κB and Protection against Apoptosis*

NF-κB is an inducible transcription factor classically implicated in the regulation of inflammatory response genes.[106] NF-κB is activated by numerous stimuli including amyloid-beta peptides,[107] reactive oxygen species,[106,107] TNF,[108] chemotherapeutic agents,[20] NGF[109] and ceramide.[2,47,109] Intriguingly, the activation of NF-κB may elicit both apoptotic and antiapoptotic responses in neurons vs. glial cells.[107,110-112] For example, activation of NF-κB in glial cells may be involved in apoptosis induced by amyloid-beta peptides, potentially through the activation of proinflammatory or cytotoxic genes such as nitric oxide synthase and interleukin-6.[106,107] In contrast, activation of NF-κB in neurons may increase the expression of antioxidant genes and decrease neuronal apoptosis induced by amyloid-beta peptides[44] or growth factor withdrawal.[114] Moreover, previous studies had suggested that TNF could protect neurons from amyloid-beta-induced apoptosis by decreasing H_2O_2 production and activating NF-κB.[115] Further, ceramide pretreatment mimicked the effect of TNF in protecting neurons from cell death induced by amyloid-beta peptides or glutamate.[44] However, as previously discussed, ceramide can increase H_2O_2 production by directly inhibiting mitochondrial electron flow.[65] These results raise the possibility that fundamental differences may exist in the effects of ceramide on mitochondria in different cell types. Alternatively, low levels of ceramide may induce sufficient generation of H_2O_2 to activate NF-κB and induce the expression of antioxidant genes but not be apoptotic. Indeed, recent data suggest that the production of H_2O_2 by low concentrations of Aβ (0.1 μM) may induce a protective response in neurons through a transient activation of NF-κB.[107] In contrast, doses of amyloid-beta peptides typically used to induce apoptosis (10 μM) generate significantly more H_2O_2 which inhibits NF-κB activation.[107] Thus, depending upon the cell type and its genetic program, the nature of the specific activator, the magnitude and duration of the activation signal, as well an interaction with other signaling pathways, similar effector proteins may be involved in eliciting neurodegenerative or neuroprotective responses.

In summary, the dichotomous responses of neurons to ceramide are consistent with the hypothesis first put forth by Hannun and co-workers that ceramide may serve as a sensor of cellular stress involved in gauging the extent of cellular injury and initiating a response.[19,36] If the stress is short term, transient ceramide production may promote an antiapoptotic response by affecting cell cycle progression and afford the cell an opportunity to recover from the stress. In contrast, if the stress endures, prolonged ceramide production in combination with other signals may initiate entry into apoptotic pathways. Understanding the circumstances and mechanisms whereby ceramide may activate antiapoptotic signals will undoubtedly provide basic insight into the relationships between lipid signaling pathways, cell cycle progression, and apoptosis. These efforts may provide novel approaches to regulating or slowing neurodegenerative processes.

Acknowledgments

This work was supported by grant MCB 9513596 from the National Science Foundation, a Career Development Award from the Juvenile Diabetes Foundation International, and by funds from the Higuchi Biosciences Center at the University of Kansas.

References

1. Kim, M., Linardic, C., Obeid, L., Hannun, Y. Identification of sphingomyelin turnover as an effector mechanism for the action of tumor necrosis factor-α and γ-interferon, *J Biol Chem*, 266, 484, 1991.
2. Schutze, S., Potthoff, K., Machleidt, T., Berkovic, D., Wegmann, K., Kronke, M. TNF activates NF-κB by phosphatidylcholine-specific phospholipase C-induced "acidic" sphingomyelin breakdown, *Cell*, 71, 765, 1992.
3. Mathias, S., Younes, A., Kan, C., Orlow, I., Joseph, C., Kolesnick, R.N. Activation of the sphingomyelin signaling pathway in intact EL4 cells and in a cell-free system by IL-1β, *Science*, 259, 519, 1993.
4. Dressler, K.A., Mathias, S., Kolesnick, R.N. Tumor necrosis factor-α activates the sphingomyelin signal transduction pathway in a cell-free system, *Science*, 255, 1715, 1992.
5. Masamune, A., Igarashi, Y., Hakomori, S. Regulatory role of ceramide in interleukin (IL)-1β-induced E-selectin expression in human umbilical vein endothelial cells, *J Biol Chem*, 271, 9368, 1996.
6. Belka, C., Weigmann, K., Dieter, A., Holland, R., Neuloh, M., Herrmann, F., Kronke, M., Brach, M.A. Tumor necrosis factor (TNF)-α activates c-*raf*-1 kinase via the p55 TNF receptor engaging neutral sphingomyelinase, *EMBO J*, 14, 1156, 1995.
7. Ballou, L.R., Chao, C.P., Holness, M.A., Barker, S.C., Raghow, R. Interleukin-1-mediated PGE$_2$ production and sphingomyelin metabolism, *J Biol Chem*, 267, 20044, 1992.
8. Chen, J., Nikolova-Karakashian, M., Merrill, A.H., Morgan, E.T. Regulation of cytochrome P450 2C11 (CYP2C11) gene expression by interleukin-1, sphingomyelin hydrolysis, and ceramides in rat hepatocytes, *J Biol Chem*, 270, 25233, 1995.
9. Yanaga, F., Watson, S.P. Tumor necrosis factor a stimulates sphingomyelinase through the 55 kDa receptor in HL-60 cells, *FEBS Lett*, 314, 297, 1992.
10. Okazaki, T.O., Bell, R.M., Hannun, Y.A. Sphingomyelin turnover induced by vitamin D$_3$ in HL-60 cells, *J Biol Chem*, 264, 19076, 1989.
11. Geilen, C.C., Bektas, M., Weider, T., Kodelja, V., Goerdt, S., Orfanos, C.E. 1α,25-Dihydroxyvitamin D$_3$ induces sphingomyelin hydrolysis in HaCaT cells via tumor necrosis factor α, *J Biol Chem*, 272, 8997, 1997.

12. Okazaki, T.O., Bielawska, A., Bell, R.M., and Hannun, Y.A. Role of ceramide as a lipid mediator of $1\alpha,25$-dihydroxyvitamin D_3-induced HL-60 cell differentiation, J *Biol Chem*, 265, 15823, 1985.

13. Dobrowsky, R.T., Werner, M.H., Castellino, A.M., Chao, M.V., Hannun, Y.A. Activation of the sphingomyelin cycle through the low-affinity neurotrophin receptor, *Science*, 265, 1596, 1994.

14. Dobrowsky, R.T., Jenkins, G.M., Hannun, Y.A. Neurotrophins induce sphingomyelin hydrolysis: modulation by co-expression with trk receptors, *J Biol Chem*, 270, 22135, 1995.

15. Boucher, L., Wiegmann, K., Futterer, A., Pfeffer, K., Mahleidt, T., Schutze, S., Mak, T.W., Kronke, M. CD28 signals through acidic sphingomyelinase, *J Exp Med*, 181, 2059, 1995.

16. Bose, R., Verheji, M., Haimovitz-Friedman, A., Scotto, K., Fuks, Z., Kolesnick, R.N. Ceramide synthase mediates danorubicin-induced apoptosis: an alternative mechanism for generating death signals, *Cell*, 82, 405, 1995.

17. Jaffrezou, J., Levade, T., Bettaieb, A., Andrieu, N., Bezombes, C., Maestre, N., Vermeersch, S., Rousse, A., Laurent, G. Daunorubicin-induced apoptosis: triggering of ceramide generation through sphingomyelin hydrolysis, *EMBO J*, 15, 2417, 1996.

18. Strum, J.C., Small, G.W., Pauig, S.B., Daniel, L.W. 1-β-D-Arabinofuranosylcytosine stimulates ceramide and diglyceride formation in HL-60 cells, *J Biol Chem*, 269, 15493, 1994.

19. Zhang, J., Alter, N., Reed, J.C., Borner, C., Obeid, L.M., Bcl-2 interrupts the ceramide-mediated pathway of cell death, *Proc Natl Acad Sci U.S.A.*, 93, 5325, 1996.

20. Boland, M.P., Foster, S.J., O'Neill, L.A.J. Daunorubicin activates NF-κB and induces κB-dependent gene expression in HL-60 promyelocytic and Jurkat T lymphoma cells, *J Biol Chem*, 272, 12952, 1997.

21. Chang, Y., Abe, A., Shayman, J.A. Ceramide formation during heat shock: a potential mediator of alpha B-crystallin transcription, *Proc Natl Acad Sci U.S.A.*, 92, 12275, 1995.

22. Haimovitz-Friedman, A., Kan, C.C., Ehleiter, D., Persaud, R.S., McLoughlin, M., Fuks, Z., Kolesnick, R.N. Ionizing radiation acts on cellular membranes to generate ceramide and initiate apoptosis, *J Exp Med*, 180, 525, 1994.

23. Obeid, L.M., Linardic, C.M., Karolak, L.A., Hannun, Y.A. Programmed cell death induced by ceramide, *Science*, 259, 1769, 1993.

24. Dbaibo, G.S., Perry, D.K., Gamard, C.J., Platt, R., Poirier, G.G., Obeid, L.M., Hannun, Y.A. Cytokine response modifier A (CrmA) inhibits ceramide formation in response to tumor necrosis factor (TNF)-α: CrmA and Bcl-2 target distinct components in the apoptotic pathway, *J Exp Med*, 185, 481, 1997.

25. Jarvis, W.D., Kolesnick, R.N., Fornari, F.A., Traylor, R.S., Gewirtz, D.A., Grant, S. Induction of apoptotic DNA damage and cell death by activation of the sphingomyelin pathway, *Proc Natl Acad Sci U.S.A.*, 91, 73, 1994.

26. Verheij, M., Bose, R., Lin, X.H., Bei, Y., Jarvis, W.D., Grant, S., Birrer, M.J., Szabo, E., Zon, L.I., Kyriakis, J.M., Haimovitz-Friedman, A., Fuks, Z., Kolesnick, R.N. Requirement for ceramide-initiated SAPK/JNK signaling in stress-induced apoptosis, *Nature*, 380, 75, 1996.

27. Pronk, G.J., Ramer, K., Amiri, P., Williams, L.T. Requirement of an ICE-like protease for induction of apoptosis and ceramide generation by REAPER, *Science*, 271, 808, 1996.

28. Cassacia-Bonnefil, P., Carter, B.D., Dobrowsky, R.T., Chao, M.V. Nerve growth factor-mediated death of oligodendrocytes by the p75 neurotrophin receptor, *Nature*, 383, 716, 1996.

29. Tepper, C.G., Jayadev, S., Liu, B., Bielawska, A., Wolff, R., Yonehara, S., Hannun, Y.A., Seldin, M.F. Role for ceramide as an endogenous mediator of fas-induced cytotoxicity, *Proc Natl Acad Sci U.S.A.*, 92, 8443, 1995.

30. Gulbins, E., Bissonnette, R., Mahboubi, A., Martin, S., Nishioka, W., Brunner, T., Baier, G., Baier-Bitterlich, G., Byrd, C., Lang, F., Kolesnick, R., Altman, A., Green, D. FAS-induced apoptosis is mediated via a ceramide-initiated RAS signaling pathway, *Immunity*, 2, 341, 1995.

31. Gamard, C., Dbaibo, G.S., Obeid, L.M., Hannun, Y.A. Selective involvement of ceramide in cytokine-induced apoptosis, *J Biol Chem*, 272, 16474, 1997.

32. Sillence, D.J., Allan, D. Evidence against an early signaling role for ceramide in FAS-mediated apoptosis, *Biochem J*, 324, 29, 1997.

33. Cifone, M.G., De Maria, R., Roncaioli, P., Rippo, M.R., Azuma, M., Lanier, L.L., Santoni, A., Testi, R. Apoptotic signaling through CD95 (FAS/Apo-1) activates an acidic sphingomyelinase, *J Exp Med*, 177, 1547, 1993.

34. Quintans, J., Kilkus, C.L., McShan, A.R., Dawson, G. Ceramide mediates the apoptotic response of WEHI 231 cells to anti-immunoglobulin, corticosteroids and irradiation, *Biochem Biophys Res Comm*, 202, 710, 1994.

35. Weisner, D.A., Kilkus, J.P., Gottschalk, A.R., Quintans, J., Dawson, G. Anti-immunoglobulin-induced apoptosis in WEHI 231 Cells involves the slow formation of ceramide from sphingomyelin and is blocked by Bcl-x_L, *J Biol Chem*, 272, 9868, 1997.

36. Hannun, Y.A. Functions of ceramide in coordinating cellular response to stress, *Science*, 274, 1855, 1996.

37. Sawai, H., Okazaki, T., Yamamoto, H., Okano, H., Takeda, Y., Tashima, M., Sawada, H., Okuma, M., Ishikura, H., Umehara, H., Domae, N. Requirement of AP-1 for ceramide-induced apoptosis in human leukemia HL-60 cells, *J Biol Chem*, 270, 27326, 1995.

38. Jarvis, W.D., Fornari, F.A.J., Browning, J.L., Gewirtz, D.A. Attenuation of ceramide-induced apoptosis by diglyceride in human myeloid leukemia cells, *J Biol Chem*, 269, 31685, 1994.

39. Santana, P., Pena, L.A., Haimovitz-Friedman, A., Martin, S., Green, D., McLoughlin, E.H., Cordon-Cardo, C., Schuchman, E.H., Fuks, Z., Kolesnick, R. Acid sphingomyelinase-deficient human lymphoblasts and mice are defective in radiation-induced apoptosis, *Cell*, 86, 189, 1996.

40. Hartfield, P.J., Mayne, G.C., Murray, A.W. Ceramide induces apoptosis in PC12 cells, *FEBS Lett*, 401, 148, 1997.

41. Casaccia-Bonnefil, P., Aibel, L., Chao, M.V. Central glial and neuronal populations display differential sensitivity to ceramide-dependent cell death, *J Neurosci Res*, 43, 382, 1996.

42. Wiesner, D.A., Dawson, G. Staurosporine induces programmed cell death in embryonic neurons and activation of the ceramide pathway, *J Neurochem*, 66, 1418, 1996.

43. Ito, A., Horigome, K. Ceramide prevents neuronal programmed cell death by NGF deprivation, *J Neurochem*, 65, 463, 1995.

44. Goodman, Y., Mattson, M.P. Ceramide protects hippocampal neurons against excitotoxic and oxidative insults, and amyloid β-peptide toxicity, *J Neurochem*, 66, 869, 1996.

45. Grazia Cifone, M., Roncaioli, P., De Maria, R., Camarda, G., Santoni, A., Ruberti, G., Testi, R. Multiple pathways originate at the FAS/APO-1 (CD95) receptor: sequential involvement of phosphatidylcholine-specific phospholipase C and acidic sphingomyelinase in the propagation of the apoptotic signal. *EMBO J*, 14, 5859, 1995.

46. Jayadev, S., Liu, B., Bielawska, A.E., Lee, J.Y., Nazaire, F., Pushkareva, M.Y., Obeid, L.M., Hannun, Y.A. Role for ceramide in cell cycle arrest, *J Biol Chem*, 270, 2047, 1995.

47. Weigmann, K., Schutze, S., Machleidt, T., Witte, D., Kronke, M. Functional dichotomy of neutral and acidic sphingomyelinases in tumor necrosis factor signaling, *Cell*, 78, 1005, 1994.

48. Chinnaiyan, A.R., Tepper, C.G., Seldin, M.F., O'Rourke, K., Kischkel, F.C., Hellbardt, S., Krammer, P.H., Peter, M.E., Dixit, V.M. FADD/MORT1 is a common mediator of CD95 (FAS/APO-1) and tumor necrosis factor receptor-induced apoptosis, *J Biol Chem*, 271, 4961, 1996.

49. Chinnaiyan, A.M., Orth, K., O'Rourke, K., Duan, H., Poirer, G.G., Dixit, V.M. Molecular ordering of the cell death pathway, *J Biol Chem*, 271, 4573, 1996.

50. Quillet-Mary, A., Jaffrezou, J., Mansat, V., Bordier, C., Naval, J., Laurent, G. Implication of mitochondrial hydrogen peroxide generation in ceramide-induced apoptosis, *J Biol Chem*, 272, 21388, 1997.

51. Tewari, M., Telford, W.G., Miller, R.A., Dixit, V.M. CrmA, a poxvirus-encoded serpin, inhibits cytotoxic T-lymphocyte-mediated apoptosis, *J Biol Chem*, 270, 22705, 1995.

52. Merrill, A.H.J., Jones, D.D. An update of the enzymology and regulation of sphingomyelin metabolism, *Biochim Biophys Acta*, 1044, 1, 1990.

53. Michel, C., van Echten-Deckert, G., Rother, J., Sandhoff, K., Wang, E., Merrill, A.H.J. Characterization of ceramide synthesis, *J Biol Chem*, 272, 22432, 1997.

54. Kok, J.W., Nikolova-Karakashian, M., Klappe, K., Alexander, C., Merrill, A.H.J. Dihydroceramide biology, *J Biol Chem*, 272, 21128, 1997.

55. Merrill, A.H.J., van Echten, G., Wang, E., Sandhoff, K. Fumonisin B_1 inhibits sphingosine (sphinganine) N-acyltransferase and *de novo* sphingolipid biosynthesis in cultured neurons *in situ*, *J Biol Chem*, 268, 27299, 1993.

56. Merrill, A.H.J., Schmelz, E., Dillehay, D.L., Spiegel, S., Shayman, J.A., Schroeder, J.J., Riley, R.T., Voss, K.A., Wang, E. Sphingolipids — the enigmatic lipid class: biochemistry, physiology and pathophysiology, *Toxicol Appl Pharmacol*, 142, 208, 1996.

57. Paumen, M.B., Ishida, Y., Muramatsu, M., Yamamoto, M., Honjo, T. Inhibition of carnitine palmitoyltransferase I augments sphingolipid synthesis and palmitate-induced apoptosis, *J Biol Chem*, 272, 3324, 1997.

58. Barenholz, Y., Thompson, T.E. Sphingomyelins in bilayers and biological membranes, *Biochim Biophys Acta*, 604, 129, 1980.

59. Linardic, C.M., Hannun, Y.A. Identification of a distinct pool of sphingomyelin involved in the sphingomyelin cycle, *J Biol Chem*, 269, 23530, 1994.

60. Andrieu, N., Salvayre, R., Levade, T. Comparative study of the metabolic pools of sphingomyelin and phosphatidylcholine sensitive to tumor necrosis factor, *Eur J Biochem*, 236, 738, 1996.

61. Liu, P., Ying, Y., Ko, Y., Anderson, R.G.W. Localization of platelet-derived growth factor-stimulated phosphorylation to caveolae, *J Biol Chem*, 271, 10299, 1996.

62. Bilderback, T.R., Grigsby, R.J., and Dobrowsky, R.T. Association of p75^{NTR} with caveolin and localization of neurotrophin-induced sphingomyelin hydrolysis to caveolae. *J Biol Chem*, 272, 10922-7, 1997.
63. Zhang, P., Liu, B., Jenkins, G.M., Hannun, Y.A., Obeid, L.M. Expression of neutral sphingomyelinase identifies a distinct pool of sphingomyelin involved in apoptosis, *J Biol Chem*, 272, 9609, 1997.
64. Gudz, T.I., Tserng, K., Hoppel, C.L. Direct inhibition of mitochondrial respiratory chain complex III by cell-permeable ceramide, *J Biol Chem*, 272, 24154, 1997.
65. Garcia-Ruiz, C., Colell, A., Maris, M., Morales, A., Fernandez-Checa, J.C. Direct effects of ceramide on the mitochondrial electron transport chain leads to generation of reactive oxygen species, *J Biol Chem*, 272, 11369, 1997.
66. Zamzami, N., Marchetti, P., Castedo, M., Zanin, C., Vayssiere, J., Petit, P.X., Kroemer, G. Reduction in mitochondrial potential constitutes an early irreversible step of programmed lymphocyte death *in vivo*, *J Exp Med*, 181, 1661, 1995.
67. Dobrowsky, R.T., Kamibayashi, C., Mumby, M.C., Hannun, Y.A. Ceramide activates heterotrimeric protein phosphatase 2A, *J Biol Chem*, 268, 15523, 1993.
68. Mathias, S., Dressler, K.A., Kolesnick, R.N. Characterization of a ceramide-activated protein kinase: stimulation by tumor necrosis factor α, *Proc Natl Acad Sci U.S.A.*, 88, 10009, 1991.
69. Lozano, J., Berra, E., Muncio, M.M., Diaz-Meco, M.T., Dominguez, I., Sanz, L., Moscat, J. Protein kinase C ζ isoform is critical for κB-dependent promoter activation by sphingomyelinase, *J Biol Chem*, 269, 19200, 1994.
70. Bernardi, P., Broekemeier, K.M., Pfeiffer, D.R. Recent progress on regulation of the mitochondrial permeability transition pore; A cyclosporin-sensitive pore in the inner mitochondrial membrane, *J Bioenerget Biomembr*, 26, 509, 1994.
71. Susin, S.A., Zamzami, N., Castedo, M., Daugas, E., Wang, H., Geley, S., Fassy, F., Reed, J.C., Kroemer, G. The central executioner of apoptosis: multiple connections between protease activation and mitochondria in FAS/APO-1/CD95- and ceramide-induced apoptosis, *J Exp Med*, 186, 25, 1997.
72. Zamzami, N., Marchetti, P., Castedo, M., Decaudin, D., Macho, A., Hirsch, T., Susin, S.A., Petit, P.X., Mignotte, B., Kroemer, G. Sequential reduction of mitochondrial membrane transmembrane potential and generation of reactive oxygen species in early programmed cell death, *J Exp Med*, 182, 367, 1995.
73. Liu, B., Hannun, Y.A. Inhibition of the neutral magnesium-dependent sphingomyelinase by glutathione, *J Biol Chem*, 272, 16281, 1997.
74. Smyth, M.J., Perry, D.K., Zhang, J., Poirier, G.G., Hannun, Y.A., Obeid, L.M. prICE:a downstream target for ceramide-induced apoptosis and for the inhibitory action of Bcl-2, *Biochem J*, 316, 25, 1996.
75. Liu, X., Kim, C.N., Yang, J., Jemmerson, R., Wang, X. Induction of apoptotic program in cell free extracts: requirement for dATP and cytochrome c, *Cell*, 86, 147, 1996.
76. Adachi, S., Cross, A.R., Babior, B.M., Gottlieb, R.A. Bcl-2 and the outer mitochondrial membrane in the inactivation of cytochrome c during FAS-mediated apoptosis, *J Biol Chem*, 272, 21878, 1997.
77. Kluck, R.M., Bossy-Wetzel, E., Green, D.R., Newmeyer, D.D. The release of cytochrome c from mitochondria: a primary site for Bcl-2 regulation of apoptosis, *Science*, 275, 1132, 1997.
78. Yang, J., Liu, X., Bhalla, K., Kim, C.N., Ibrado, A.M., Cai, J., Peng, T., Jones, D.P., Wang, X. Prevention of apoptosis by Bcl-2: release of cytochrome c from mitochondria blocked, *Science*, 275, 1129, 1997.

79. Susin, S.A., Zamzami, N., Castedo, M., Hirsch, T., Marchetti, P., Macho, A., Daugas, E., Geuskens, M., Kroemer, G. Bcl-2 inhibits the mitochondrial release of an apoptogenic protease, *J Exp Med*, 184, 1331, 1996.

80. Reed, J.C. Double identity for proteins of the Bcl-2 family, *Nature*, 387, 773, 1997.

81. Perry, D.K., Smyth, M.J., Wang, H., Reed, J.C., Duriez, P., Poirier, G.G., Obeid, L.M., Hannun, Y.A. Bcl-2 acts upstream of the PARP protease and prevents its activation, *Cell Death Differ*, 4, 29, 1997.

82. Hockenberry, D.M., Oltvai, Z.N., Yin, X., Milliman, C.L., Korsmeyer, S.J. Bcl-2 functions in an antioxidant pathway to prevent apoptosis, *Cell*, 75, 241, 1993.

83. Wang, H., Rapp, U.R., Reed, J.C. Bcl-2 targets the protein kinase Raf-1 to mitochondria, *Cell*, 87, 629, 1996.

84. Park, J., Kim, I., Jun Oh, Y., Lee, K., Han, P., Choi, E. Activation of c-Jun N-terminal kinase antagonizes an antiapoptotic action of Bcl-2, *J Biol Chem*, 272, 16725, 1997.

85. Zha, J., Harada, H., Yang, E., Jockel, J., Korsmeyer, S. Serine phosphorylation of death agonist BAD in response to survival factor results in binding to 14-3-3 not Bcl-X_L, *Cell*, 87, 619, 1996.

86. Zhang, Y., Yao, B., Delikat, S., Bayoumy, S., Lin, X., Basu, S., McGinley, M., Cahn-hui, P., Lichenstein, H., Kolesnick, R. Kinase suppressor of Ras is ceramide-activated protein kinase, *Cell*, 89, 63, 1997.

87. Yao, B., Zhang, Y., Delikat, S., Mathias, S., Basu, S., Kolesnick, R. Phosphorylation of Raf by ceramide-activated protein kinase, *Nature*, 378, 307, 1995.

88. Karasavvas, N., Erukulla, R.K., Bittman, R., Lockshin, R., Zakeri, Z. Stereospecific induction of apoptosis in U937 cells by N-octanoyl-sphingosine stereoisomers and N-octyl-sphingosine, *Eur J Biochem*, 236, 729, 1996.

89. Robinson, M.J., Cobb, M.H. Mitogen-activated protein kinase pathways, *Curr Opin Cell Biol*, 9, 181, 1997.

90. Kyriakis, J.M., Banerjee, P., Nikolakaki, E., Dal, T., Rubie, E., Ahmad, M.F., Avruch, J., Woodgett, J.R. The stress-activated protein kinase subfamily of c-Jun kinases, *Nature*, 369, 156, 1994.

91. Westwick, J.K., Bielawska, A., Dbaibo, G., Hannun, Y.A., Brenner, D.A. Ceramide activates the stress-activated protein kinase, *J Biol Chem*, 270, 22689, 1995.

92. Xia, Z., Dickens, M., Raingeaud, J., Davis, R.J., Greenberg, M.E. Opposing effects of ERK and JNK-p38 MAP kinases on apoptosis, *Science*, 270, 1326, 1996.

93. Brenner, B., Koppenhoefer, U., Weinstock, C., Linderkamp, O., Lang, F., Gulbins, E. Fas- or ceramide-induced apoptosis is mediated by a Rac1-regulated activation of Jun N-terminal kinase/p38 kinases and GADD 153, *J Biol Chem*, 272, 22173, 1997.

94. Shirakabe, K., Yamaguchi, K., Shibuya, H., Irie, K., Matsuda, S., Moriguchi, T., Gotoh, Y., Matsumoto, K., Nishida, E. TAK1 mediates the ceramide signaling to stress-activated protein kinase/c-Jun N-terminal kinase, *J Biol Chem*, 272, 8141, 1997.

95. Rani, C.S.S., Abe, A., Chang, Y., Rosenzweig, N., Saltiel, A.R., Radin, N.S., Shayman, J.A. Cell cycle arrest induced by an inhibitor of glucosylceramide synthase, *J Biol Chem*, 270, 2859, 1995.

96. Evan, G.I., Brown, L., Whyte, M., Harrington, E. Apoptosis and the cell cycle, *Curr Opin Cell Biol*, 7, 825, 1995.

97. Deshaies, R.J. The self-destructive personality of a cell cycle in transition, *Curr Opin Cell Biol*, 7, 781, 1995.

98. Park, D.S., Farinelli, S.E., Greene, L.A. Inhibitors of cyclin-dependent kinases promote survival of postmitotic neuronally differentiated PC12 cells and sympathetic neurons, *J Biol Chem*, 271, 8161, 1996.

99. Rukenstein, A., Rydel, R.E., Greene, L.A. Multiple agents rescue PC12 cells from serum-free cell death by translation- and transcription-independent mechanisms, *J Neurosci*, 11, 2552, 1991.

100. Farinelli, S.E., Greene, L.A. Cell cycle blockers mimosine, ciclopirox, and deferoxamine prevent the death of PC12 cells and postmitotic sympathetic neurons after removal of trophic support, *J Neurosci*, 16, 1150, 1996.

101. Sherr, C.J. G1 phase progression:cycling on cue, *Cell*, 79, 551, 1994.

102. Freeman, R.S., Estus, S., Johnson, E.M.J. Analysis of cell cycle-related gene expression in postmitotic neurons: selective induction of *cyclin D1* during programmed cell death, *Neuron*, 12, 343, 1994.

103. Kranenburg, O., van der Eb, A.J., Zantema, A. Cyclin D1 is an essential mediator of apoptotic neuronal cell death, *EMBO J*, 15, 46, 1996.

104. Lee, K., Helbing, C.C., Choi, K., Johnston, R.N., Wang, J.H. Neuronal Cdc-2like kinase (Nclk) binds and phosphorylates the retinoblastoma protein, *J Biol Chem*, 272, 5622, 1997.

105. Dbaibo, G., Pushkareva, M.Y., Jayadev, S., Schwarz, J.K., Horowitz, J.M., Obeid, L.M., Hannun, Y.A. Retinoblastoma gene product as a downstream target for a ceramide-dependent pathway of growth arrest, *Proc Natl Acad Sci U.S.A.*, 92, 1347, 1995.

106. Baeuerle, P., Henkel, T. Function and activation of NF-κB in the immune system, *Annu Rev Immunol*, 12, 141, 1994.

107. Kaltschmidt, B., Uherek, M., Volk, B., Baeuerle, P., Kalschmidt, C. Transcription factor NF-κB is activated in primary neurons by amyloid β peptides and in neurons surrounding early plaques from patients with Alzheimer disease, *Proc Natl Acad Sci U.S.A.*, 94, 2642, 1997.

108. Hsu, H., Shu, H., Pan, M., Goeddel, D.V. TRADD-TRAF2 and TRADD-FADD interactions define two distinct TNF receptor 1 signal transduction pathways, Cell, 84, 299, 1996.

109. Carter, B.D., Kaltschmidt, C., Kaltschmidt, B., Offenhauser, N., Bohm-Matthaei, R., Baeuerle, P.A., Barde, Y. Selective activation of NF-κB by nerve growth factor through the neurotrophin receptor p75, *Science*, 272, 542, 1996.

110. Beg, A.A., Baltimore, D. An essential role for NF-κb in preventing TNF-α-induced cell death, *Science*, 274, 782, 1996.

111. Wang, C., Mayo, M.W., Baldwin, A.S.J. TNF and cancer therapy-induced apoptosis: potentiation by inhibition of NF-κB, *Science*, 274, 784, 1996.

112. Van Antwerp, D.J., Martin, S.J., Kafri, T., Green, D.R., Verma, I.M. Suppression of TNF-α-induced apoptosis by NF-κB, *Science*, 274, 787, 1996.

113. Bayaert, R., Fiers, W. Molecular mechanism of tumor necrosis factor-induced cytotoxicity, *FEBS Letts*, 340, 9, 1994.

114. Greenlund, L.L.S., Deckwerth, T.L., Johnson, E.M.J. Superoxide dismutase delays neuronal apoptosis: a role for reactive oxygen species in programmed neuronal death, *Neuron*, 14, 303, 1995.

115. Barger, S.W., Horster, D., Furukawa, K., Goodman, Y., Krieglstein, J., Mattson, M.P. Tumor necrosis factor α and β protect neurons against amyloid β-peptide toxicity: Evidence for involvement of a κB-binding factor and attenuation of peroxide and Ca^{2+} accumulation, *Proc Natl Acad Sci U.S.A.*, 92, 9328, 1995.

Section B

Methods

8

Assessment of Cell Viability and Histochemical Methods in Apoptosis

Kasturi L. Puranam and Rose-Mary Boustany

CONTENTS

8.1   Introduction

Historically, the distinction between various types of cell death has been based on the morphology of the cells as they undergo the process. The classification of cell death as apoptosis or necrosis now depends on morphological and biochemical criteria, in addition to the circumstances that lead to these processes.[1-3] The necrosis phenomenon involves swelling of the plasma membrane and other organelles, eventually leading to their rupture. Cells undergoing the apoptotic process shrink and increase in density while maintaining organelle integrity until relatively late in the course of events. The plasma membrane of apoptotic cells forms bud-like protrusions that eventually separate as membrane-bound vesicles containing remnants of the cell: these are referred to as apoptotic bodies. The chromatin in the nucleus of these cells goes through remarkable changes leading to condensation and finally, to a breakdown of the nucleus. The nuclear fragments are emitted by the cell as apoptotic bodies. Biochemically, the nuclear DNA in apoptotic cells gets cleaved by activated endonucleases into 200 to 300 kb and 30 to 50 kb fragments, which may get further cleaved into internucleosomal, 180 to 200-bp fragments.[4,5] The apoptosis detection methods discussed in this chapter are based mainly on the profound changes in cell membrane and nuclear morphology that occur during the course of apoptosis. The methods will first be discussed briefly, followed by short protocols for each of the methods.

Detection methods that assess the viability of cells undergoing apoptosis are based on (a) the intactness of the plasma membrane; (b) the morphological state of the nucleus; or, (c) certain cellular enzymes and/or their biochemical function. Membrane impermeant dyes such as trypan blue, ethidium bromide, or propidium iodide are used in assays that are based on the principle of dye exclusion.[6-8] Cells with damaged plasma membrane allow entry of these dyes, whereas viable cells with intact plasma membrane exclude them. Membrane-permeable nuclear dyes such as acridine orange and the Hoechst 33258 dye are used for detecting apoptotic cells, based only on their ability to stain the nucleus by chelating DNA and thus differentiating a normal nucleus from one that has condensed due to apoptosis.[9,10]

Viability assays based on the quantitation of a biochemical function that are to be discussed in this chapter include the MTT (3-(4,5-dimethythiazol-2-yl)-2,5-diphenyl tetrazolium bromide) assay and the LDH (lactate dehydrogenase) release assay. The MTT assay is based on the evaluation of cell viability by quantitating the metabolic product, formazan, of the tetrazolium salt, MTT.[11] The MTT formazan has a dark blue color and can be easily quantitated using a spectrophotometer. The LDH assay is functionally based on the observation that injured or nonintact cells tend to leak the cytosolic enzyme LDH into the surrounding media.[12] The LDH enzyme converts pyruvate to lactate in the presence of NADH (nicotinamide adenine dinucleotide). LDH release is then measured by spectrophotometrically monitoring the decrease in the absorption of NADH.

Histochemical methods used to detect apoptosis are based on *in situ* labeling of cleaved nuclear DNA.[13] The TUNEL (TdT-mediated dUTP nick end labeling) method utilizes the ability of the TdT (terminal deoxynucleotidyl transferase) enzyme to incorporate deoxyuridine (dUTP) at the 3'-OH sites of DNA breaks.[14] The dUTP is conveniently labeled with a tag which is then detected, depending on the chemical nature of the tag. The final detection can be using a color development reagent such as DAB (3, 3'-diaminobenzidine) or by fluorescence. Double-labeling techniques incorporating the TUNEL method have also proved helpful in associating apoptosis with the expression of various proteins in the cell.[15] Finally, the old standby method of detection of fragmented DNA by gel electrophoresis or DNA laddering is discussed.[16]

8.2   Methods to Assay Cell Viability

8.2.1   Assays Based Only on Dye Exclusion Principle

Assessment of cell viability using high molecular weight dyes such as trypan blue has become a widely used method to differentiate viable from damaged cells.[8,17] Normal cells or those that are in the early phase of apoptosis, have intact plasma membrane, and thus exclude trypan blue; whereas, dead cells or those that are in late apoptotic stages with lysed membranes are permeable to the dye. The trypan blue assay for cell viability is inexpensive and reliable and is thus one of the most commonly used viability assays. Both adherent as well as suspension cells can be easily stained with trypan blue and counted using a hemocytometer under a light microscope. Viable cells exclude the dye and appear unstained or white, whereas dead cells take up the dye and are stained blue.

The usefulness of viability assays using trypan blue, particularly for some cell types, is somewhat limited due to the fact that once cells are exposed to the dye, the number of dead cells tends to increase with incubation time.[18,19] As time goes on, the percentage of dead cells artificially increases during this assay. One can overcome this problem by incubating cell samples for different experimental time points with trypan blue immediately prior to counting it, thus ensuring reliable results.

8.2.2   Assays Based on Dye Exclusion and Nuclear Staining

Viability assay by staining for apoptotic cells with the fluorescent dye ethidium bromide is also based on the principle of dye exclusion. The ethidium bromide homodimer can permeate dead cells and fluorescently stain the nuclei as it chelates DNA; intact cells remain unstained.[7,20] When viewed under a fluorescence microscope, the nuclei of dead cells have a bright red fluorescence.[21] The other fluorescent dye frequently used in similar experiments is propidium iodide (PI). PI is also a membrane impermeable nuclear dye, under a fluorescence microscope the stained dead cells emit red fluorescence upon green excitation.[6,22,23] The advantage of using both these dyes for viability assay is that in addition to distinguishing between live and dead cells based on membrane integrity, they also stain the nucleus, and thus differentiate a normal nucleus from a condensed, apoptotic one.

One advantage of using fluorescent dyes for cell viability assay over trypan blue is that the count of stained cells remains constant with time.[24] However, while counting several hundred cells under a fluorescence microscope, the fluorescent dye tends to bleach. This problem is overcome by photographing various fields and counting stained cells from the photographs.

8.2.3 Assays Based on Nuclear Morphology

Apoptotic nuclei have characteristic condensed chromatin forming crescent shaped clumps at the periphery of the nuclear membrane.[1] These distinctive morphological changes in the nucleus of the apoptotic cell can be visualized at a very gross level by staining with membrane-permeable (or inclusion) dyes such as acridine orange (AO) and Hoechst 33258.[9,10,25] These dyes bind to nuclear DNA and emit fluorescence when excited by ultraviolet or visible light. When bound to DNA, AO emits a yellow-green fluorescence at concentrations lower than 20 μM, whereas Hoechst 33258 dye emits blue fluorescence.[10,24] Condensed nuclei in apoptotic cells are clearly distinguishable with these staining methods as they are highly fluorescent, visibly shrunk, and homogenous bodies when compared to normal nuclei.

Viability assays based on nuclear staining can also be carried out with membrane-impermeant dyes such as propidium iodide, provided the entire population of cells being analyzed is made permeant prior to incubation with the dye. This is commonly achieved either by fixation of cells followed by staining, or by staining with PI in the presence of a detergent such as Triton X-100.[17] Furthermore, as an extension of the use of these fluorescent dyes for viability assay, one can use a combination of exclusion and inclusion dyes such as ethidium bromide or PI (exclusion) and AO (inclusion). This permits a more detailed analysis of viable, early apoptotic, and late apoptotic/necrotic cells in the same population.[17,24,26] In such experiments, viable cells are stained with only the membrane permable dye (AO), which emits a green fluorescence. At the same excitation wavelength, late apoptotic or dead cells will emit a bright red fluorescence; even though both AO and PI enter such cells, the emission of PI is stronger than AO. Cells that are in the early stages of apoptosis will have mostly green fluorescence (AO) and a very low level of red fluorescence.

8.2.4 Assays Based on Biochemical Function

8.2.4.1 MTT Assay

Measurement of cell viability by the MTT method is commonly used to assay cell survival during apoptosis. This method was originally described to measure cell proliferation in cytotoxic assays.[11] It has been very widely used in drug screening and drug sensitivity testing for various tumor cell lines.[27] The main advantage of this assay is that it is a rapid and quantitative (colorimetric) method to measure the number of surviving cells. This assay is based on the ability of live cells to reduce a light yellow compound, MTT, into a dark blue product, formazan MTT, that is quantitated spectrophotometrically by absorbance measurement at 570 nm.[11] The cellular reduction of MTT is now believed to occur both in the mitochondria by the enzyme succinate dehydrogenase and also extra-mitochondrially involving the pyridine nucleotide cofactors, NADH and NADPH.[28] The MTT assay is easily performed with cells grown in microtiter assay plates. The MTT formazan crystals generated from the reduction reaction are best dissolved in an organic solvent such as DMSO and then quantitated by absorbance measurement at 570 nm.[29] Only live cells in a given cell population reduce MTT; hence, only viable cells contribute to the absorbance measurement. Therefore, an increase or decrease in the number of live cells will show up as a corresponding increase or decrease in the absorbance value at 570 nm.

The reduction of MTT to the formazan product is influenced by many parameters, only some of which can be controlled by changing the cellular growth and assay parameters.[29,30] The assay cannot be applied to all cell lines, since some do not metabolize MTT (for example, breast cancer cell lines VACC 732 and VACC 812) and some lose the ability to do so at high density (for example, colon carcinoma cell line HCT-8).[31,32] The absorbance at 570 nm which is directly related to the level of the product, MTT formazan, varies significantly among different cell lines; therefore, assay conditions for each cell line should be initially established before a detailed use of this method.

8.2.4.2 LDH Assay

The LDH efflux assay originally developed for measuring cytotoxicity is also used to quantitate damaged cells.[33] The LDH assay is a sensitive enzyme release assay that is based on the fact that cytosolic LDH leaks out of cells with damaged membranes. The released LDH can be rapidly and spectrophotometrically monitored by one of the following two ways: the LDH enzyme reduces pyruvate to lactate in the presence of NADH which is oxidized to NAD^+. The level of LDH is most commonly measured by monitoring the reduction of absorbance of NADH at 340 nm.[12,34] Instead of measuring the absorbance of NADH, one can also measure its fluorescence at 460 nm with an excitation at 360 nm.[35] The increased sensitivity of fluorescence measurement over that of absorbance, has been cited as the major advantage of this detection method. The second method of detecting LDH is based on the ability of LDH to catalyze the reverse reaction, i.e., oxidation of lactate to pyruvate using NAD^+, which in turn is reduced to NADH. This reaction is coupled to another enzymatic reaction wherein NADH reduces INT ((2-(p-iodophenyl)-3-(p-nitrophenyl)-5-phenyltetrazolium chloride)) to a red formazan product that can be detected and spectrometrically quantitated between 450 to 490 nm.[34]

Both the MTT and LDH assays for measurement of cell viability are rapid and reliable methods. In testing for drug sensitivity, a good correlation has been found between these two methods.[27] The MTT method cannot be used for all cell types as described earlier. A major advantage of the LDH assay is that it is very sensitive. Neurons that have a relatively high level of the LDH enzyme are commonly assayed for viability by this method.[12]

8.3 Histochemical Methods to Detect Apoptosis

8.3.1 TUNEL Method

Profound changes involving the digestion of nuclear DNA in apoptosis were demonstrated in early experiments done on thymocytes.[16] It is now recognized that the digestion of DNA occurs in two phases, one that cleaves the genomic DNA into large 200- to 300-kb and 30- to 50-kb fragments, and the second that further cleaves the DNA to produce fragments 180 to 200 bp in size.[36,37] The final DNA fragments produced as a result of the cleavage, most often, are the nucleosomal units that, by gel electrophoresis, show up as a DNA ladder, referred to as an oligonucleosomal ladder. The identity of the nucleases involved in DNA fragmentation during apoptosis is still very controversial and is the topic of much discussion.[38,39] Nucleases involved in the generation of the internucleosomal, 200-bp fragments have been variously identified as DNaseI, DNaseII, and several other proteins.[40-42]

However, an observation that is not controversial is the generation of 3′-OH groups on the fragmented DNA.[43] These new 3′-OH groups can be identified by *in situ* labeling and forms the basis of the histochemical ISEL (in situ end labeling) method to detect apoptotic cells. The ISEL technique relies on the labeling of DNA by terminal deoxynucleotidyl transferase (TdT) or DNA polymerase. Based on the enzyme that is used to label the DNA, the method is either called TUNEL (TdT mediated dUTP nick end labeling) or ISNT (in situ nick translation).[14,44,45] Significant labeling differences have been observed upon comparison of the two ISEL techniques. The TUNEL method has been found to be superior to the ISNT method, as it has a higher sensitivity and also labels a larger number of cells. This was found to be the case in tissues of the nervous system, tissues from biopsies of breast cancer, head and neck squamous cell carcinoma, and non-Hodgkin's lymphoma, and also cells such as HL60.[13,46] The difference in labeling is based on the activity of the two enzymes involved. The DNA polymerase enzyme only recognizes 3'-OH recessed ends, whereas the TdT enzyme recognizes the 3'-OH at 3' recessed, 5' recessed as well as blunt ends.[46,47] Therefore, TdT labeling tends to be stronger than that with DNA polymerase. In addition, some tissue samples (breast cancer) that are TUNEL-positive do not even label with the ISNT method leading to the hypothesis that DNA breaks in this case of apoptosis probably do not produce 3' recessed OH groups.[46] The *in situ* labeling method described in this chapter will be restricted to the TUNEL method.

The labeling technique in the TUNEL method is functionally based on the activity of the TdT enzyme that transfers dUTP to the 3'-OH groups of the fragmented DNA.[14] The dUTP moiety is chemically tagged, thus allowing the visualization of cells that are apoptotic. Normal nuclei have very low levels of 3'-OH breaks and do not produce any appreciable signal as opposed to apoptotic nuclei. In the original method, dUTP was labeled with biotin, and detection was achieved with avidin-peroxidase followed by color development.[14] Subsequently, several modifications of the original method have been made. One such change involves the chemical modification of dUTP to digoxigenin-dUTP, which is then detected by using an antibody that recognizes the digoxigenin group.[48] This antibody can itself be fluorescently labeled or can be detected by the addition of a horseradish peroxidase (HRP) conjugated secondary antibody followed by DAB color development that results in brown apoptotic nuclei. A more recent and simple technique involves the direct tagging of dUTP with the fluorescent molecule, fluorescein.[49] This modification drastically reduces the time required to label apoptotic cells. In addition, the direct fluorescent TUNEL method produces significantly lower levels of background fluorescence when compared to the method that employs fluorescently labeled antidigoxigenin antibody. In experiments where the detection is based on a color reaction, the cells are counterstained with a staining reagent such as methylene blue.

The results of a TUNEL staining experiment with digoxigenin-dUTP, followed by detection using HRP-conjugated antidigoxigenin antibody and DAB are shown in Figure 8.1. Brain tissue sections from a patient with juvenile Batten disease (JNCL) were subjected to TUNEL staining using the ApopTag kit from Oncor (Gaithersburg, MD).[50,51] Apoptotic cells are identifiable as those containing brown nuclei; normal nuclei stain blue. Figure 8.2 shows the results from a TUNEL staining experiment with indirect fluorescent detection. A human brain tissue section (treated with DNaseI to artificially induce DNA fragmentation) was labeled with digoxigenin-dUTP and then visualized by addition of a rhodamine-conjugated antidigoxigenin antibody that fluoresces red. The small, dense nuclei (white arrows) can be easily recognized as apoptotic nuclei. The results from a direct fluorescent TUNEL staining experiment performed on cells that were made apoptotic by withdrawal of growth factors is shown in Figure 8.3. Neuronal precursor cells, NT2, were grown under conditions of serum withdrawal for 20 h and fixed

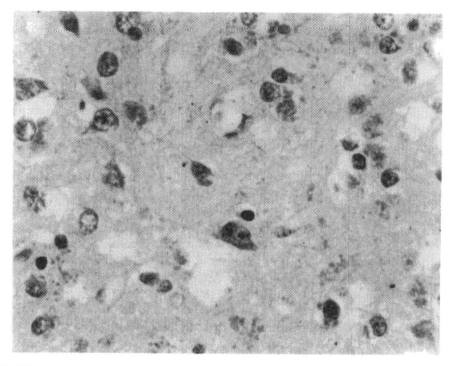

FIGURE 8.1

TUNEL staining of brain tissue section using digoxigenin-dUTP and DAB color development. Paraffin-embedded brain section from a Batten patient (JNCL) was deparaffinized in xylene, rehydrated, and stained for apoptosis by the TUNEL method. Digoxigenin-dUTP was detected with HRP-conjugated antidigoxigenin antibody and DAB. When viewed under a light microscope, apoptotic cells are identified as those with brown nuclei, whereas normal cells have blue nuclei. See color plates following page 148.

prior to staining with ApopTag Direct (Fluorescein) kit from Oncor (Gaithersburg, MD).[52] In this case the detection method was direct, since fluorescein-labeled dUTP was used in the TdT labeling reaction.[49] The cells were counterstained with propidium iodide, which is taken up by every cell since all cells were made membrane permeable during fixation. Cells with apoptotic nuclei are those that stain yellow-green; normal nuclei counterstain red.

8.3.2 Double Labeling for TUNEL and Other Proteins

A variation of the TUNEL technique involves the double labeling for apoptosis and other proteins under investigation.[15,53] This method can be used to elegantly demonstrate whether or not apoptosis and the protein of interest are colocalized to the same cell. This double labeling procedure can be performed using either DAB or the fluorescent method of detection for both apoptosis and the protein being studied. This method is now commonly used in many different systems. For example, TUNEL has been combined

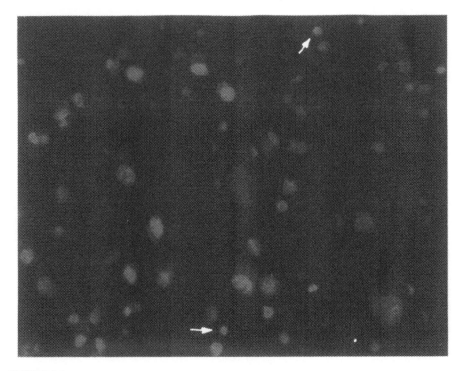

FIGURE 8.2
Fluorescent TUNEL (indirect) staining of brain tissue section. Paraffin-embedded section of human brain was deparaffinized, rehydrated, and treated with DNase prior to TUNEL staining using digoxigenin-dUTP. The dUTP-labeled fragmented DNA was then visualized by using a rhodamine labeled antidigoxigenin antibody that has red fluorescence. Apoptotic cells with the condensed, round, red nuclei (white arrows) are easily distinguishable from normal nuclei. See color plates following page 148.

with immunocytochemistry to stain for GFAP (glial fibrillary acidic protein) in astrocytes and transferrin in oligodendrocytes following peripheral nerve injury in the CNS.[54]

The result from a double-labeling experiment as described above is shown in Figure 8.4. Paraffin-embedded tissue sections of a diseased juvenile Batten brain were subjected to the double-labeling method in order to determine if apoptotic neurons overexpress the Bcl-2 protein.[55] The tissue sections were labeled for apoptotic nuclei using the ApopTag kit from Oncor. Immunolabeling for Bcl-2 was carried out using a polyclonal anti-Bcl-2 antibody followed by biotin conjugated secondary antibody and finally, HRP-streptavidine. DAB was used for color development for both TUNEL and immunolabeling. A brown nucleus indicates an apoptotic neuron. Brown cytoplasm identifies neurons overexpressing Bcl-2. Labeling of affected (Batten) tissue showed three types of cells: (1) those that were neither apoptotic (blue nuclei) nor Bcl-2 overexpressors (blue cytoplasm); (2) those that were either only apoptotic (brown nucleus, thin white arrow) or only Bcl-2 overexpressors (brown cytoplasm, broad white arrow); and, finally, (3) those

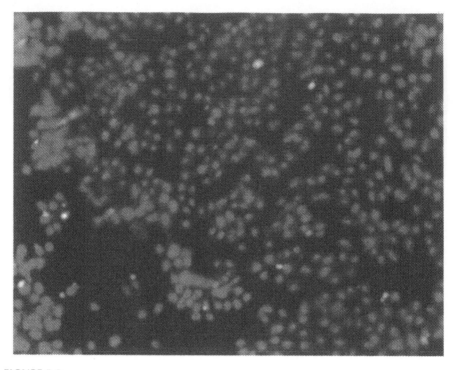

FIGURE 8.3

Direct fluorescent TUNEL staining of cells. Neuronal NT2, precursor cells, were grown on glass coverslips and cultured for 20 h in serum-free medium. Cells were then fixed in 2% formaldehyde, 0.2% glutaraldehyde, and stained by TUNEL method. Fluorescein-labeled dUTP was used for direct visualization of fragmented nuclei. Propidium iodide was used to counterstain the cells. Apoptotic nuclei are those with green or yellow-green fluorescence. See color plates following page 148.

neurons that were both apoptotic and Bcl-2 overexpressors (brown nucleus and cytoplasm, white arrowhead).

Double labeling results with fluorescently labeled antibodies for TUNEL and the Bcl-2 protein are shown in Figure 8.5. Paraffin-embedded tissue section from a diseased (Batten) brain was first labeled for apoptosis with TdT and digoxigenin-dUTP and detected by using fluorescein-labeled anti-digoxigenin antibody. The protein of interest, Bcl-2, was detected by immunolabeling with a polyclonal anti-Bcl-2 antibody followed by rhodamine-labeled secondary antibody. The green fluorescein labeling (A) identifies apoptotic nuclei only, whereas the red rhodamine labeling (B) indicates the expression of Bcl-2 protein in the cytoplasm. The apoptotic cells can be clearly identified as those with the small homogenous, dense, green rounded nuclei (Figure 8.5A) some of which also express the Bcl-2 protein (white arrows in Figure 8.5B). In addition, there are cells that are not apoptotic but which express Bcl-2 as indicated by a broad white arrow in Figure 8.5B. It is important to note that the intrinsic green autofluorescence of some tissues, such

8.5.1.2   Ethidium Bromide or Propidium Iodide Staining

Materials and Equipment

- Ethidium bromide (20 µg/ml)
- Propidium iodide (20 µg/ml)
- Phosphate buffered saline (PBS)
- Fluorescence microscope

Procedure

1. Wash and resuspend cells in PBS at an approximate density of 0.5 to 1×10^6/ml and then add ethidium bromide to a final concentration of 2 µg/ml. Incubate cells in dye for 30 to 60 min at room temperature. For propidium iodide staining, add the dye to a final concentration of 3 to 5 µg/ml and incubate the cells at 4°C for 5 min. Wash cells twice in PBS and resuspend in PBS for microscopic analysis.

2. Load 10 µl of the cell suspension in the counting chamber of a hemocytometer and view under a fluorescence microscope. For ethidium bromide-stained cells, use excitation wavelength of 360 nm and emission filter for 600 nm (red fluorescence). In order to count propidium iodide stained cells, use excitation wavelength of 525 nm with an emission filter for 600 nm.

3. Count dead cells as those that fluoresce red and the total cells in any chamber under phase contrast conditions. Calculate cell density and viability as described above.

8.5.1.3   Nuclear Staining with Acridine Orange or Hoechst 33258

Materials and Equipment

- Acridine orange (100 µg/ml)
- Hoechst 33258 (2.5 µg/ml)
- Phosphate buffered saline (PBS)
- Fluorescence microscope

Procedure

1. Wash cells and resuspend in PBS at a density of 0.5 to 1×10^6/ml. Add acridine orange to achieve a final concentration of 4 to

5 µg/ml. Incubate cells at room temperature for 10 to 30 min. For Hoechst staining, add dye to a final concentration of 0.05 µg/ml and incubate cells for 30 min at room temperature.

2. Wash cells three times with PBS and proceed to microscopic analysis as described above. View acridine orange-stained cells at 515 to 535 nm following excitation at 488 nm. For Hoechst-stained cells, use excitation wavelength of 365 nm and emission wavelength of 460 nm.

3. Apoptotic cells are distinguished from normal cells based on nuclear staining. Normal cells will display nucleus with homogenous fluorescent chromatin, whereas apoptotic cells are those with condensed chromatin and therefore, stongly fluorescent nuclei. Viability is calculated from the number of dead and live cells as described above.

8.5.1.4   MTT Assay

Materials and Equipment

- MTT dissolved in PBS at 5 mg/ml, filter sterilized
- Lysis buffer (DMSO or 0.04 N HCl in isopropanol or 10% SDS in 0.01 N HCl)
- Spectrometer with a microtiter (ELISA) plate reader

Procedure

1. Incubate 1 to 2 ×10^4 cells in 0.1 ml culture medium in a 96-well microtiter plate with 10 µl of the stock MTT solution at 37°C for 2 to 4 h.

2. Add 100 µl of lysis buffer to each well of the plate and mix thoroughly to dissolve the dark blue crystals. Check for complete solubilization before measuring the absorbance of the formazan product.

3. Measure absorbance at 570 nm, using the plate reader on the spectrometer. The absorbance correlates directly to the number of viable cells in the sample.

8.5.1.5   LDH Assay

Materials and Equipment

- 0.1 M potassium phosphate buffer (pH 7.2 at 25°C)
- 7.5 mM sodium pyruvate (prepared fresh in phosphate buffer)

- 1 mM NADH (freshly made in phosphate buffer)
- Spectrophotometer with automatic recording capability

Procedure

1. Remove media (0.1 to 0.5 ml) from cells being assayed. Add 1/10 volume of the sodium pyruvate solution (final concentration of 0.75 mM). Add phosphate buffer to get a final volume of 3 ml. For background value, use plain culture medium for assay.

2. Start the enzymatic reaction by addition of 1/10 volume of NADH (0.1 mM final concentration) to the sample.

3. Immediately start recording the absorbance of the sample at 340 nm at 2 s intervals. The concentration of LDH, automatically calculated from the slope of the absorbance curve, is expressed as U/ml (units per ml). One unit is defined as the amount of LDH required to cause a decrease of 0.001/min in the absorbance of a 3-ml reaction.

4. Measure total LDH from the cells being assayed by rupturing cells at –70°C (freeze-thawing). Express degree of apoptosis as a ratio of LDH from the sample (minus background) over the total LDH in the cells being assayed.

8.5.2 Histochemical Assay

8.5.2.1 TUNEL Staining

Materials and Equipment

- Xylene
- Ethyl alcohol (absolute, 95%, 70% diluted in distilled water)
- Phosphate buffered saline (PBS)
- Hydrogen peroxide (30% solution)
- Proteinase K (20 mg/ml)
- Equilibration buffer (100 mM sodium cacodylate, 0.1 M dithiothreitol, 5 mM cobalt chloride, 30 mM Trizma base, pH 7.2)
- TdT enzyme (1U/μl)
- Reaction buffer (0.001 mM digoxigenin-dUTP in equilibration buffer)
- Stop solution (300 mM sodium chloride, 30 mM sodium citrate)
- Peroxidase conjugated antidigoxigenin antibody (for indirect TUNEL staining)
- Counterstain (1% w/v methyl green in 0.1 M sodium acetate)
- DAB (0.025% in PBS, 0.1% hydrogen peroxide freshly added)

- Coplin jars
- Light microscope

Procedure

1. Deparaffinize and rehydrate tissue sections by washing specimen in two changes of xylene for 5 min each, two changes of absolute ethanol for 5 min each, one wash in 95% and 70% ethanol for 3 min each and finally, one wash in PBS for 5 min.

2. Apply proteinase K (100 μl of 2.5 to 5 μg/ml) directly to the tissue sample, incubate for 15 min at room temperature, wash four times in distilled water for 2 min each.

3. Quench endogenous peroxide by incubation in 3% hydrogen peroxide in PBS for 5 min at room temperature. Wash twice in PBS for 5 min each.

4. Remove all excess liquid from slide, apply 50 μl of equilibration buffer to each sample, cover with a plastic coverslip and incubate at room temperature for up to 30 min.

5. Remove equilibration buffer by tapping off liquid, apply 25 μl of TdT solution (20 U/100 μl) made up in reaction buffer. Cover with plastic coverslip and incubate in a humid chamber at 37°C for 1 h.

6. Terminate labeling reaction by incubating the section in stop buffer for 15 min at room temperature. Wash thrice in PBS for 5 min each wash.

7. For indirect TUNEL staining, apply peroxidase conjugated anti-digoxigenin antibody and incubate sections for 30 min at room temperature. Wash samples thrice with PBS for 5 min each.

8. Add 150 μl of DAB solution to sections and let the color development proceed for 5 min at room temperature. Wash samples four times with distilled water for 5 min each.

9. Counterstain sections for 3 min in methyl green solution. Wash sections with three changes of water followed by three changes of 100% butanol.

10. Wash in three changes of xylene and mount sections under glass coverslips, using an appropriate mounting medium. View samples under light microscope.

Notes:

1. While staining fixed cells, omit steps 1 and 2; start with step 3.
2. For a negative control, omit the TdT enzyme from the reaction mix in step 5.

3. For a positive control, treat a tissue section with DNase I (10 to 20 U/μl) for 10 min at room temperature, followed by four washes in distilled water, 5 min each. This treatment is carried out between steps 3 and 4.

8.5.2.2 TUNEL Staining and Immunolabeling for a Protein

Materials and Equipment

- All the materials for TUNEL staining as described above
- Bovine serum albumin (BSA), IgG free (5% w/v in PBS)
- Antibody for the protein to be detected
- Biotin-labeled secondary antibody
- HRP-streptavidine
- Hematoxylin stain

Procedure

1. Proceed with the steps for TUNEL staining until step 6 (termination of TdT reaction).
2. Incubate sections with BSA (to block nonspecific binding) for 1 h at room temperature.
3. Remove the BSA solution and apply primary antibody (appropriately diluted in BSA solution) to sections, incubate overnight (18 to 20 h) at room temperature. Wash thrice with PBS, 5 min each.
4. Apply biotin-labeled secondary antibody (diluted 1:300 with PBS) to sections, incubate at room temperature for 1 h, and wash thrice with PBS, 5 min each.
5. Incubate sections with HRP-streptavidine (diluted 1:300 with PBS) for 15 min at room temperature. Wash with PBS once for 5 min.
6. Continue with step 7 from the TUNEL procedure described above. Include hematoxylin blue in the counterstaining step.

8.5.3 DNA Laddering

Materials and Equipment

- Phosphate buffered saline (PBS)
- Lysis buffer (5 mM Tris-HCl, pH 7.4, 20 mM EDTA, 0.5% Triton X-100)

- Buffer-saturated phenol, pH 8.0
- Chloroform-isoamylalcohol (24:1, v/v)
- Sodium acetate (3 M, pH 5.2)
- Ethanol, absolute
- TE buffer (10 mM Tris-HCl, pH 8.0, 1 mM EDTA)
- RNase (1 mg/ml)
- TBE buffer (89 mM Tris-OH, 89 mM boric acid, 2 mM EDTA)

Procedure

1. Wash cells twice with ice-cold PBS and resuspend in lysis buffer (0.5 ml for 1×10^6 cells). For adherent cells, collect by scraping in lysis buffer. Incubate for 20 min at 4°C with gentle agitation.
2. Remove cellular debris by centrifugation for 2 min at full speed in a microfuge.
3. Extract supernatant twice with buffered phenol and once with chloroform-isoamylalcohol.
4. Precipitate DNA by addition of 0.1 volume of sodium acetate and 2 volumes of ethanol. Incubate for 30 min at –20°C, spin DNA at 4°C for 20 min. Wash once with 70% ethanol, air dry DNA pellet.
5. Resuspend DNA in TE buffer and digest with RNase (50 µg/ml) at 37°C for 30 min.
6. Analyze DNA on a 1.8% agarose gel in TBE buffer.

Note: This method is only for the extraction and detection of low molecular weight DNA. Genomic DNA can be extracted by following procedure described elsewhere.[58]

Acknowledgments

We would like to thank Yusuf Hannun and Ram Puranam for critical reading of the manuscript. We also appreciate the technical assistance of Wei-Hua Qian and Jerome Fennel. This work was supported by grant R01 NS 30170 from NINDS.

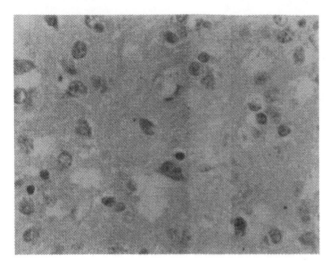

FIGURE 8.1
TUNEL staining of brain tissue section using digoxigenin-dUTP and DAB color development. Paraffin-embedded brain section from a Batten patient (JNCL) was deparaffinized in xylene, rehydrated, and stained for apoptosis by the TUNEL method. Digoxigenin-dUTP was detected with HRP-conjugated antidigoxigenin antibody and DAB. When viewed under a light microscope, apoptotic cells are identified as those with brown nuclei, whereas normal cells have blue nuclei.

FIGURE 8.2
Fluorescent TUNEL (indirect) staining of brain tissue section. Paraffin-embedded section of human brain was deparaffinized, rehydrated, and treated with DNase prior to TUNEL staining using digoxigenin-dUTP. The dUTP-labeled fragmented DNA was then visualized by using a rhodamine labeled antidigoxigenin antibody that has red fluorescence. Apoptotic cells with the condensed, round, red nuclei (white arrow) are easily distinguishable from normal nuclei.

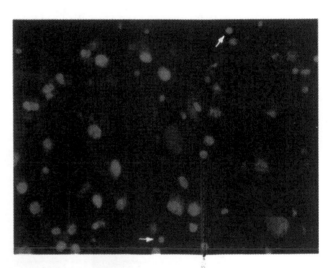

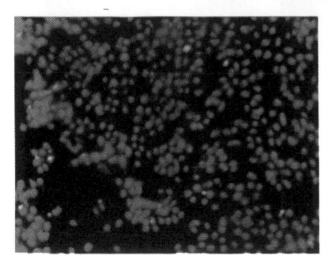

FIGURE 8.3
Direct fluorescent TUNEL staining of cells. Neuronal precursor cells, NT2, were grown on glass coverslips and cultured for 20 h in serum-free medium. Cells were then fixed in 2% formaldehyde, 0.2% glutaraldehyde, and stained by TUNEL method. Fluorescein labeled dUTP was used for direct visualization of fragmented nuclei. Propidium iodide was used to counterstain the cells. Apoptotic cells are those with green or yellow-green fluorescence.

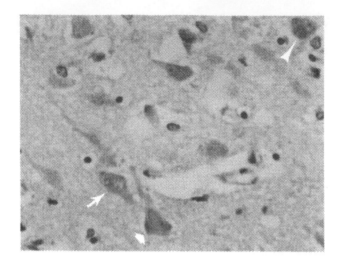

FIGURE 8.4
Double labeling for apoptosis by TUNEL method and Bcl-2 protein by immunocytochemical method. Paraffin-embedded brain section from Batten patient (JNCL) was labeled with digoxigenin-dUTP followed by HRP-conjugated antidigoxigenin antibody. Bcl-2 detection was carried out by immunolabeling with a polyclonal anti-Bcl-2 antibody followed by biotin labeled secondary antibody and, finally, HRP-streptavidin. Color development was done using DAB. A brown nucleus identifies an apoptotic cell (thin white arrow), while a brown cytoplasm indicates Bcl-2 overexpression (broad white arrow).

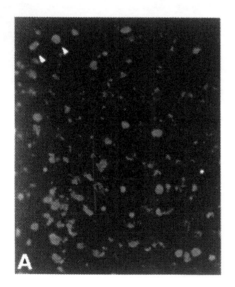
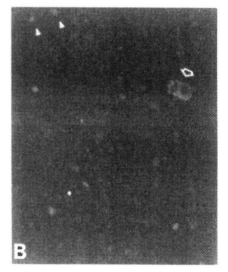

FIGURE 8.5
Fluorescent double labeling for apoptosis and Bcl-2. Paraffin-embedded brain tissue section was stained by the TUNEL method with digoxigenin-dUTP followed by fluorescein labeled antidigoxigenin antibody. Bcl-2 detection was done using a polyclonal anti-Bcl-2 antibody and rhodamine labeled secondary antibody. Green fluorescence (A) shows TUNEL-positive nuclei as condensed, rounded bodies (arrow heads). Red fluorescence (B) indicates Bcl-2 overexpression (arrow and arrow heads).

References

1. Wyllie, A. H., Kerr, J. F. R., and Currie, A. R., Cell death: the significance of apoptosis, *Int. Rev. Cytol.*, 68, 251, 1980.
2. Earnshaw, W. C., Nuclear changes in apoptosis, *Curr. Op. Cell Biol.*, 7, 337, 1995.
3. Majno, G. and Joris, I., Apoptosis, oncosis and necrosis. An overview of cell death, *Am. J. Pathol.*, 146, 3, 1995.
4. Arends, M. J., Morris, R. G., and Wyllie, A. H., Apoptosis, the role of the endonuclease, *Am. J. Pathol.* 136, 593, 1990.
5. Bortner, C. D., Oldenburg, N. B. E., and Cidlowski, J. A., The role of DNA fragmentation in apoptosis, *Trends Cell Biol.*, 5, 21, 1995.
6. Yeh, C.-J. G., Hsi, B.-L., and Faulk, W. P., Propidium iodide as a nuclear marker in immunofluorescence. II. Use with cellular identification and viability studies, *J. Immunol. Methods*, 43, 269, 1981.
7. Tymianski, M., Charlton, M. P., Carlen, P. L., and Tator, C. H., Secondary Ca^{2+} overload indicates early neuronal injury which precedes staining with viability indicators, *Brain Res.*, 607, 319, 1993.
8. Perry, S. W., Epstein, L. G., and Gelbard, H. A., *In situ* trypan blue staining of monolayer cell cultures for permanent fixation and mounting, *BioTechniques*, 22, 1020, 1997.
9. Oberhammer, F., Wilson, J. W., and Dive, C., Apoptotic death in epithelial cells: cleavage of DNA to 300 and/or 50 kb fragments prior to or in the absence of internucleosomal fragmentation, *EMBO J.*, 12, 3679, 1993.
10. Figiel, I. and Kaczmarek, L., Cellular and molecular correlates of glutamamte-evoked neuronal programmed cell death in the *in vitro* cultures of rat hippocampal dentate gyrus, *Neurochem. Int.*, 31, 229, 1997.
11. Mossmann, T., Rapid colorimetric assay for cellular growth and survival: application to proliferation and cytotoxicity assays, *J. Immunol. Methods*, 65, 55, 1983.
12. Koh, J. Y. and Choi, D. W., Quantitative determination of glutamate mediated cortical neuronal injury in cell culture by lactate dehydrogenase efflux assay, *J. Neurosci. Methods*, 20, 83, 1987.
13. Migheli, A., Cavalla, P., Marino, S., and Schiffer, D., A study of apoptosis in normal and pathological nervous tissue after *in situ* end-labeling of DNA strand breaks, *J. Neuropath. Exp. Neurol.*, 53, 606, 1994.
14. Gavrieli, Y., Sherman, Y., and Ben-Sasson, S. A., Identification of programmed cell death *in situ* via specific labeling of nuclear DNA fragmentation, *J. Cell Biol.*, 119, 493, 1992.
15. Sgonc, R., Boeck, G., Dietrich, H., Gruber, J., Recheis, H., and Wick, G., Simultaneous determination of cell surface antigens and apoptosis, *Trends Genet.*, 10; 41, 1994.
16. Wyllie, A. H., Glucocorticoid-induced thymocyte apoptosis is associated with endogenous endonuclease activation, *Nature*, 284, 555, 1980.
17. Petit, P. X., Lecoeur, H., Zorn, E., Dauguet, C., Mignotte, B. and Gougeon, M. L., Alterations in mitochondrial structure and function are early events of dexamethasone-induced thymocyte apoptosis, *J. Cell Biol.*, 130, 157, 1995.
18. Black, L. and Berenbaum, M. C., Factors affecting the dye exclusion test for cell viability, *Exp. Cell Res.*, 35, 9, 1964.

19. Tennant, J. R., Evaluation of the trypan blue technique determination of cell viability, *Transplantation*, 2, 685, 1964.
20. Edidin, M., A rapid, quantitative fluorescence assay for cell damage by cytotoxic antibodies, *J. Immunol.*, 104, 1303, 1970.
21. Beletsky, I. P. and Umansky, S. R., A new assay for cell death, *J. Immunol. Methods*, 134, 201, 1990.
22. Tanke, H. J., Van Der Linden, P. W. G., and Langerak, J., Alternative fluorochromes to ethidium bromide for automated read out of cytotoxicity tests, *J. Immunol. Methods*, 52, 91, 1982.
23. Manev, H., Kharlamov, E., Uz, T., Mason, R. P., and Cagnoli, C. M., Characterization of zinc-induced neuronal death in primary cultures of rat cerebellar granule cells, *Exp. Neurol.*, 146, 171, 1997.
24. Bank, H. L., Assessment of islet cell viability using fluorescent dyes, *Diabetologia*, 30, 812, 1987.
25. Boix, J., Llecha, N., Yuste, V.-J., and Comella, J. X., Characterization of the cell death process induced by staurosporine in human neuroblastoma cell lines, *Neuropharmacology*, 36, 811, 1997.
26. Vasconcelos, A. C. and Lam, K. M., Apoptosis induced by infectious bursal disease virus, *J. Gen. Virol.*, 75, 1803, 1994.
27. Coley, H. M., Lewandowicz, G., Sargent, J. M., and Verrill, M. W., Chemosensitivity testing of fresh and continuous tumor cell cultures using lactate dehydrogenase, *Anticancer Res.*, 17, 231, 1997.
28. Berridge, M. V. and Tan, A. S., Characterization of the cellular reduction of 3-(4,5-dimethythiazol-2-yl)-2,5-diphenyltetrazolium bromide (MTT): subcellular localization, substrate dependence, and involvement of mitochondrial electron transport in MTT reduction, *Arch. Biochem. Biophys.*, 303, 474, 1993.
29. Twentyman, P. R. and Luscombe, M., A study of some variables in a tetrazolium dye (MTT) based assay for cell growth and chemosensitivity, *Br. J. Cancer*, 56, 279, 1987.
30. Vistica, D. T., Skehan, P., Scudiero, D., Monks, A., Pittmen, A., and Boyd, M. R., Tetrazolium-based assays for cellular viability: a critical examination of selected parameters affecting formazan production, *Cancer Res.*, 51, 2515, 1991.
31. Einsphar, J., Alberts, D. S., Gleason, M., Dalton, W. S., and Leibovitz, A., Pharmacological pitfalls in the use of the MTT (vs. human tumour clonogenic-HCTA) assay to quantitate chemosensitivity of human tumour cell lines, *Proc. Am. Assoc. Cancer Res.*, 29, 492, 1988.
32. Finlay, G. J., Wilson, W. R., and Baguley, B. C., Comparison of *in vitro* activity of cytotoxic drugs towards human carcinoma and leukaemia cell lines, *Eur. J. Cancer Clin. Oncol.*, 22, 655, 1986.
33. Korzeniewski, C. and Callewaert, D. M., An enzyme-release assay for natural cytotoxicity, *J. Immunol. Methods*, 64, 313, 1983.
34. Decker, T. and Lohmann-Matthes, M.-L., A quick and simple method for the quantitation of lactate dehydrogenase release in measurements of cellular cytotoxicity and tumor necrosis factor (TNF) activity, *J. Immunol. Methods*, 15, 61, 1988.
35. Moran, J. H. and Schnelmann, R. G., A rapid β-NADH-linked fluorescence assay for lactate dehydrogenase in cellular death, *J. Pharm. Toxicol. Methods*, 36, 41, 1996.

36. Dusenbury, C. E., Davis, M. A., Lawrence, T. S., and Maybaum, T. S., Induction of megabase DNA fragments by 5-fluorodeoxyuridine in human colorectal tumor (HT29) cells, *Mol. Pharmacol.*, 39, 285, 1991.

37. Walker, P. R., Smith, C., Youdale, T., Leblanc, J., Whitfield, J. F. and Sikorska, M., Topoisomerase II-reactive chemotherapeutic drugs induce apoptosis in thymocytes, *Cancer Res.*, 51, 1078, 1991.

38. Walker, P. R., Weaver, V. M., Lach, B., LeBlanc, J., and Sikorska, M., Endonuclease activities associated with high molecular weight and internucleosomal DNA fragmentation in apoptosis, *Exp. Cell Res.*, 213, 100, 1994.

39. Cohen, G. M., Sun, X.-M., Fearnhead, H., MacFarlane, M., Brown, D. G., Snowden, R. T., and Dinsdale, D., Formation of large molecular weight fragments of DNA is a key commited step of apoptosis, *J. Immunol.*, 153, 507, 1994.

40. Barry, M. A. and Eastman, A., Identification of deoxyribonuclease II as an endonuclease involved in apoptosis, *Arch. Biochem. Biophys.*, 300, 440, 1993.

41. Peitsch, M. C., Polzar, B., Stephan, H., Crompton, T., MacDonald, H. R., Mannherz, H. G., and Tschopp, J., Characterization of the endogenous deoxyribonuclease involved in nuclear DNA degradation during apoptosis (programmed cell death), *EMBO J.*, 12, 371, 1993.

42. Montague, J. W., Gaido, M. L., Frye, C., and Cidlowski, J. A., A calcium-dependent nuclease from apopotic rat thymocytes is homologous with cyclophilin, *J. Biol. Chem.*, 269, 18877, 1994.

43. Wyllie, A. H., Morris, R. G., Smith, A. L., and Dunlop, D., Chromatin cleavage in apoptosis: association with condensed chromatin morphology and dependence on macromolecular synthesis, *J. Pathol.*, 142, 67, 1984.

44. Gold, R., Schmied, M., Rothe, G., Zischler, H., Breitschopf, H., Wekerle, H., and Lassmann, H., Detection of DNA fragmentation in apoptosis: application of *in situ* nick translation to cell culture systems and tissue sections, *J. Histochem. Cytochem.*, 41, 1023, 1993.

45. Wijsman, J. H., Jonker, R. R., Keijzer, R., Van De Velde, C. J. H., Cornelisse, C. J., and Ven Dierendonck, J. H., A new method to detect apoptosis in paraffin sections: In situ end-labeling of fragmented DNA, *J. Histochem. Cytochem.*, 41, 7, 1993.

46. Mundle, S. D., Gao, X. Z., Khan, S., Gregory, S. A., Preisler, H. D., and Raza, A., Two *in situ* labeling techniques reveal different patterns of DNA fragmentation during spontaneous apoptosis *in vivo* and induced apoptosis *in vitro*, *Anticancer Res.*, 15, 1895, 1995.

47. Roychoudhary, R., Jay, E., and Wu, R., Terminal labeling and addition of homopolymer tracts to duplex DNA fragments by terminal deoxunucleotidyl transferase, *Nucleic Acid Res.*, 3, 863, 1976.

48. Schmitz, G. G., Walter, T., Seibl, R., and Kessler, C., Nonradioactive labeling of oligonucleotides *in vitro* with the hapten digoxigenin by tailing with terminal transferase, *Anal. Biochem.*, 192, 222, 1991.

49. Sgonc, R. and Wick, G., Methods for the detection of apoptosis, *Int. Arch. Allergy Immunol.*, 105, 327, 1994.

50. Boustany, R.-M. and Kolodny, E. H., Neurology of the neuronal ceroid-lipofuscinoses: late infantile and juvenile types, *Am. J. Med. Genet.*, 42, 533, 1992.

51. Lane, S. C., Jolly, R. D., Schmechel, D. E., Alroy, J., and Boustany, R.-M., Apoptosis is the mechanism of neurodegeneration in Batten disease, *J. Neurochem.*, 67, 677, 1996.

52. Pleasure, S. J. and Lee, V. M.-Y., NTera 2 cells: a human cell line which displays characteristics expected of a human committed neuronal progenitor cell, *J. Neurosci. Res.*, 35, 585, 1993.

53. Tornusciolo, D. R. Z., Schmidt, R. E., and Roth, K. A., Simultaneous detection of TdT-mediated dUTP-biotin nick end-labeling (TUNEL)-positive cells and multiple immunohistochemical markers in single tissue sections, *BioTechniques*, 19, 800, 1995.

54. Gehrmann, J. and Banati, R. B., Microglial turnover in the injured CNS: activated microglia undergo delayed DNA fragmentation following peripheral nerve injury, *J. Neuropathol. Exp. Neurol.*, 54, 680, 1995.

55. Puranam, K., Qian, W.-H., Nikbakht, K., Venable, M., Obeid, L., Hannun, Y. A. and Boustany, R.-M., Upregulation of Bcl-2 and elevation of ceramide in Batten disease, *Neuropediatrics*, 28, 37, 1997.

56. Obeid, L. M., Linardic, C. M., Karolak, L. A. and Hannun, Y. A., Programmed cell death induced by ceramide, *Science*, 259, 1769, 1993.

57. Rosenbaum, D. M., Michaelson, M., Batter, D. K., Doshi, P., and Kessler, J. A., Evidence for hypoxia-induced, programmed cell death of cultured neurons, *Ann. Neurol.*, 36, 864, 1994.

58. Strauss, W. M., Preparation of genomic DNA from mammalian tissue, in *Current Protocols in Molecular Biology*, Ausubel, F. M., Brent, R., Kingston, R. E., Moore, D. D., Seidman, J. G., Smith, J. A. and Struhl, K., Eds., Wiley, New York, 1992, Chap. 2, Section 2.

9

Assessment of Ultrastructural Changes Associated with Apoptosis

Donald E. Schmechel

CONTENTS

9.1 Introduction

Practical assessment of morphological changes associated with programmed cell death or apoptosis relies in practice on light microscopy. Examination of tissue sections should include looking for pycnotic cells or apoptotic bodies, looking for histological evidence of nuclear fragmentation, using dyes and internucleosomal DNA cleavage, using TUNEL staining, and other screening methods for cell death and tissue reaction. These approaches are quite reliable in most defined experimental contexts for detecting and quantitating cell death. For routine screening and assessment, light microscopy is to be preferred for the rapid ability to survey large areas. However, in new experimental paradigms or complex tissues with multiple cell classes, adequate documentation of the cellular changes of apoptosis must also be based on ultrastructural demonstration of characteristic features of apoptotic cell death. Likewise, there must be careful description of all forms of cell death observed in the tissue and specific inventory of features that suggest necrosis or forms of cell death other than apoptosis. In this chapter, we will focus on the ultrastructural delineation of apoptosis using transmission elecron microscopy in nervous tissue.

Apoptosis or programmed cell death was first defined by observations on cell death in developing tissues in young organisms.[1-4] Even in these studies, the delineation of apoptosis or programmed cell death, as contrasted to necrosis or accidental cell death, was not always easily made.[2-4] The most compelling evidence for apoptosis in complex tissues would be the complete observation of an individual cell's lifespan from birth to programmed death. This has been achieved in the nematode *Caenorhabditis elegans*, where 131 of 1090 somatic cells undergo programmed cell death.[5,6] In *C. elegans*, programmed cell death can be precisely contrasted with degenerative cell death.[6] In programmed cell death, there may be early recognition of the doomed cell by neighboring cell processes, followed by condensation of the cytoplasm, dilatation of the nuclear membrane, chromatin aggregation, and alteration of nuclear and nucleolar morphology. Phagocytosis splits the cell into membrane-bound fragments, the nucleus breaks up into fragments, and the cell disappears within as short a time as 1 hour. Programmed cell death during development is contrasted with degenerative cell death of normal cells induced by mutations in *C. elegans*. Necrotic-like degenerative cell death is manifested by swelling of the cytoplasm and nucleus, marked vacuolated appearance, and eventual lysis and disappearance. This process is longer than apoptosis and may take up to 8 h or more. Of note is the fact that effective disposal of cells dying of apoptotic or degenerative cell death can also be influenced or even arrested by specific genetic mutations.[6]

In studies of apoptosis in the vertebrate nervous system, observations are made more complex by the abundance of cell classes and the complexity of nervous tissue.[7-9] Major advances have been achieved by studying cell death during development or in genetically defined degenerations where the degree of programmed cell death is extensive and/or the tissue cell classes affected are well defined such as the examples of development of motor neurons, retinal cells, or in inherited retinal degeneration.[7-10] These studies suggest that the degree of apoptosis during development is commonly underappreciated, perhaps because of the rapidity of cell death.

When evidence for apoptotic cell death is sought in neurodegenerative disease in adult animals or man or in the context of injury to the mature nervous system, there needs to be a careful review of the experimental methods applied to demonstration of apoptosis. Many studies now use a combination of histological analysis, TUNEL staining for demonstration of DNA cleavage, electrophoretic analysis of DNA for ladder pattern support- ive of internucleosomal DNA breaks, and finally, ultrastructural analysis using electron microscopy.[11-13] Ultrastructural examination of tissue is an important component of such a multidimensional approach and a sensitive probe for the early morphological equivalents to the various steps of pro- grammed cell death and of degenerative cell death or cell necrosis.

The dialectic between the extreme of early nuclear changes with cell con- densation observed in programmed cell death and the extreme of early energetic/membrane integrity failure with cell swelling in necrotic or degen- erative cell death has existed since the earliest studies on cell death.[1-3] Early studies of cell death during development proposed three types of cell death:

Type 1 — apoptosis (other synonyms — shrinkage necrosis, preco- cious pycnosis, nuclear type of cell death) with pronounced nuclear condensation, blebbing of cytoplasmic membrane, condensation of cytoplasm to electron-dense state with loss of ribosomes from RER, and prominent heterophagic elimination;

Type 2 — autophagic cell death with some nuclear changes; cell mem- brane changes; abundant autophagocytotic vacuoles with dilata- tion of ER, mitochondria, and Golgi; and inconsistent and late heterophagic disposal;

Type 3A — nonlysosomal disintegration with late nuclear changes; disrupted cell membrane; general disintegration of the cytoplasm with swelling, abnormal organelles, and "empty space," and no particular phagocytosis; and

Type 3B — cytoplasmic type of cell death with late nuclear change; rounding up and swelling of cell; dilatation of ER, nuclear enve- lope, mitochondria, and Golgi; and heterophagocytosis by neigh- boring cells.[2-4]

The relative proportion of these forms of cell death has been related to age of the organism, developmental stage of the tissue, and actual amount or load of dying cells contrasted to ability of remaining cells to participate in phagocytosis.[4] This level of complexity contrasts with the simple extremes of apoptosis or necrosis and is similar to the more complex descriptions of early workers.

Recent reports emphasize the ultrastructural delineation of multiple features of nuclear and cytoplasmic morphology to allow creating multiple possible categories of cell death.[4] The main distinction is between (1) *necrotic cells* showing swelling of cytoplasm and nucleus with retention of the shape of the nucleus; and vacuolation and loss of ultrastructural integrity of cell membrane, organelles, and cytoplasmic membranes, and nucleus; and (2) *apoptotic cells* showing condensation of nucleus and cytoplasm with early nuclear changes in chromatin and convoluted nuclear membrane, and relative preservation of cell membrane, of organelles such as mitochondria and of cytoplasmic structures (apoptosis). A third type of degeneration was noted with shared features between apoptosis and necrosis, namely, early nuclear changes and vacuolar changes in the cytoplasm. Likewise, two detailed reviews of cell death and excitotoxicity in developing and adult brain made a strong contention for a continuum of changes between necrosis and apoptosis with many intermediate forms.[11,12] The more current approach is to grade nuclear and cytoplasmic changes independently and to consider other experimental data on internucleosomal DNA cleavage, gene activation, calcium influx, and protease activation with regard to establishing evidence for apoptotic cell death in a given experimental paradigm. Any one set of criteria for a given tissue and particularly a given individual cell may fail or provide misleading evidence based on sampling biases and or timing.

Thus, current literature suggests that apoptosis and necrosis often coexist in adult tissues during injury or degeneration and that a full continuum of morphological changes is often observed between apoptosis and necrosis.[4,11,12] The observed changes must be adequately described for the particular experimental paradigm in order to define for given time points the mode of cell death and injury. The major changes to be observed include nuclear morphology and cytoplasmic morphology, organelle integrity (particularly mitochondria), and membrane integrity. An impartial inventory of these changes for a reasonable sample of the cells under consideration, of control tissue, and at reasonable time intervals is necessary for considering apoptotic mechanisms of cell death. Adequate demonstration of these features requires consistently and well fixed and prepared material. This chapter will present an overview of how to prepare and to analyze nervous system tissue by transmission electron microscopy for mechanisms of cell injury and death.

9.2 Methods of Tissue Preparation

9.2.1 Fixation

Fixation for nervous tissue is critical for preservation of membrane structure, myelin, and other organelles.[14] Good manuals exist for the detailed description of theory and methods involved in fixation, embedding, and preparation of specimens.[15,16] Proper fixation may be critical for the preservation of proteins, particularly cytosolic proteins.[17] In general, quick and even fixation must be assured through perfusion fixation via the vasculature with an appropriate fixative for that tissue. The only other alternatives for *in vivo* studies are drip fixation *in situ* or careful removal of fresh tissue and immersion fixation of small 2- to 3-mm tissue blocks dissected with the utmost of care. In addition, the fragility of nervous tissue is such that during the slow chemical process of aldehyde fixation, there must be no trauma to the tissue from rough handling or dissection. Dark neurons can be produced by postmortem changes through improper fixation or handling of brain.[18,19] Liver is also a difficult tissue for immersion fixation. For certain other tissues, fixation can be reliably performed by immersion fixation with relatively acceptable results for ultrastructure.[20] These tissues include spleen, kidney, pancreas, heart, and muscle.

Possible fixatives include paraformaldehyde, glutaraldehyde, paraformaldehyde with periodate-lysine as cross-linker, osmium tetroxide, and acrolein.[15,16] Possible buffers include phosphate-based buffers, cacodylate-based buffers, and others suitable for certain pH ranges. Variables to be considered in fixation include time length of fixation, buffer pH, temperature, osmolality and osmolarity, and volume of fixative for weight of tissue. For most studies of apoptosis, fixations with phosphate buffered paraformaldehyde at neutral pH with small concentrations of glutaraldehyde are used since tissue is also used for TUNEL staining and/or other specialized reactions that may be fixation sensitive. Representative protocols are given below, but each new experimental situation should be carefully assessed for optimal results with a trial of different fixation protocols.

9.2.2   Cell Culture

Cell culture affords the luxury of precise fixation through superfusion. Minimal periods of cell stress, hypoxia, or hypercarbia are preferable, with careful rinsing of medium and rapid fixation using cooled fixative *in situ*. Examples of suitable protocols are given in References 21 through 23, and are basically similar to fixation protocols used for *in vivo* preparations.

9.2.3   Animal Tissue

Animal tissue is best prepared by intravascular perfusion with fixative through the ascending aorta or through the left ventricle. This assures the best preservation of ultrastructure. Suitable protocols are listed below.

9.2.4   Human Tissue

Human tissue must be obtained through biopsy or autopsy procedures. Biopsies offer the most attractive means of securing defined tissue from a physiologically normal organ. Unfortunately, most biopsies are very small and result in tissue cores of 100 to 200 μm, weighing 10 to 20 mg total tissue weight. For nervous tissue, the reason underlying the biopsy must be sought, since the biopsy site may represent tumor margin, tissue resected for surgical treatment of epilepsy, or other nonnormal tissue. Perimortem delay, illness, and adrenal corticosteroid failure may well influence apoptosis.[24] Thus, prolonged periods of decline before death may result in markedly increased apoptotic counts compared to similar cases dying rapidly. In some tissues, adrenal corticosteroid failure may result in lack of apoptosis through steroid-sensitive mechanisms.[24]

9.3 Correlation with Other Histological Methods

9.3.1 Basic Histological Methods

Correlation of ultrastructural studies with light microscopic analysis is most useful for definition of cytoarchitectonic areas and for initial surveying for evidence of apoptosis. For some experiments, this involves devoting some of the experimental animals or material to biochemical analysis, some to ultrastructural analysis, and some to preparation for light microscopy.[11-13]

This can also be accomplished in the same animal by taking thick sections of 0.2 to 1 μm from blocks prepared for electron microscopy, or by selecting adjacent blocks for processing for light microscopy.[11,12] Suitable stains include Nissl stains or more general purpose stains such as hematoxylin–eosin.

Another method is to prepare thick vibratome sections of tissue of 35 to 100 μm in thickness with thinner alternating sections for histology and/or immunocytochemistry.[13] The thick vibratome sections can be flat-embedded under Mylar film, on glass slides, or under plastic coverslips to enable large areas to be analyzed for apoptosis or for precise selection of architectonic areas. These plastic-embedded sections and their adjacent sections stained for light microscopy can help direct precise dissection of small blocks 0.5 to 1 mm large for re-embedding.

9.3.2 TUNEL Methods

TUNEL methods are useful for defining distribution of cells with DNA breaks to assist in ultrastructural analysis. This method has also been used on material destined for electron microscope and provides perhaps the best and most direct way to correlate DNA cleavage with observed morphological changes.[25] The objection to a pure criterion of TUNEL staining for apoptosis is that generalized endonuclease digestion of DNA may be observed in necrotic cells.[11,12,26] Internucleosomal cleavage typical of programmed cell death should result in defined fragments identifiable by ladder pattern on electrophoresis.[11,12]

9.3.3 Immunocytochemical Methods

Proposed immunocytochemical markers for apoptosis include c-jun, proteases, and other markers.[27-29] In some experimental situations, it may be important to consider cell-specific markers for precise immunochemical identification of cells undergoing apoptosis in complex tissues such as the nervous system. An example would be glial fibrillary acidic protein for astrocytes.

9.4 Ultrastructural Analysis

9.4.1 Cell Identification

The best method is to survey areas with known, well-defined laminated cell populations such as cerebellum, hippocampus, olfactory bulb, or defined subcortical nuclei such as cranial nerve nuclei. More difficult are areas with multiple, overlapping, and intermingled neuronal populations such as cerebral cortex, thalamus, and hypothalamus, particularly where small neurons may be difficult to distinguish from glial cells. Cell identification for neurons can be facilitated by observation of synaptic structures terminating on cell (Figures 9.1 to 9.3), or by observation of typical features such as axon or dendritic structures.[13] Obviously, tightly laminated and dense cell populations such as granule cell neurons of cerebellum and hippocampus or larger projection neurons in other regions are easiest.[30]

9.4.2 Nuclear Changes

The ultrastructural quality of the material should permit observation of the nuclear membrane and nuclear pores with preservation of chromatin structure and nucleolar morphology in normal cells in the same material (Figures 9.1 to 9.4). It is of the utmost importance to have comparison to normal cells of the same category either in the same animal or in control animals. Figure 9.1 shows a large pyramidal neuron in the cerebral cortex with normal nuclear and cytoplasmic morphology. This animal was exposed to a sham intoxication in a study of delayed neuronal degeneration in carbon monoxide exposure.[13]

In Table 9.1, a list of nuclear changes is given for the spectrum between apoptosis and necrosis.[1-4,11,13,31,32] In addition, one must consider in nervous

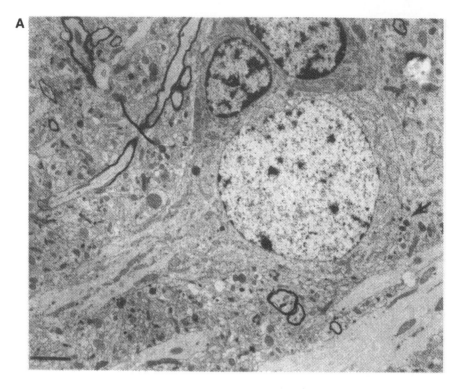

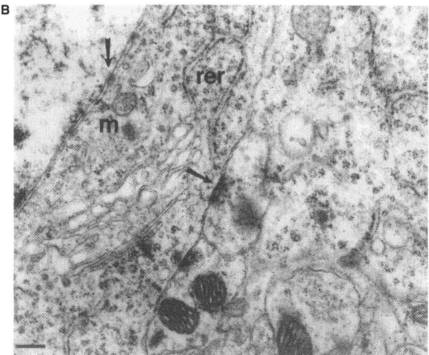

FIGURE 9.1

tissue the occurrence of cells altered during postmortem fixation such as "dark neurons" and other neuronal changes such as retrograde reaction after axonal injury.[14,18,19] The application of these criteria is illustrated in the following material from experiments on delayed neuronal degeneration after carbon monoxide exposure.

In animals exposed to carbon monoxide, there is occasional evidence of necrotic cells with swollen, electronlucent nuclei (Figure 9.2). These cells show little margination of nuclear chromatin and relatively preserved nuclear morphology. In the same animals, other neurons show early changes of chromatin margination and convolution of the nuclear membrane as illustrated for a pyramidal neuron (Figure 9.3) and for a possible nonpyramidal neuron (Figure 9.4) from the cerebral cortex. At the light microscopic level, TUNEL staining of rats exposed to sublethal carbon monoxide intoxication suggests widespread occurrence of internucleosomal DNA cleavage.[13] These ultrastructural nuclear changes are suggestive of an early stage of apoptosis when considered with the cytoplasmic changes in the same cells (see below). The interpretation of early apoptotic cells in nervous tissue is controversial. Most likely, many intermediate forms exist and many cells may show mixed characteristics of necrosis and apoptosis.[4,11,12]

For apoptotic cell death, the late changes of significant margination and formation of apoptotic bodies can also be assessed in tissue with less perfect ultrastructural preservation. This is the typical case for human tissue, particularly from the central nervous system. One can assess nuclear morphology in such samples and the relative degree of cytoplasmic swelling or condensation and preservation of organelles.[30]

More advanced changes are illustrated for apoptotic cells from autopsy samples from frontal cortex of a patient with late infantile neuronal ceroid lipofuscinosis (Figure 9.5A,B) and from cerebellum of a patient with Batten's disease or juvenile neuronal ceroid lipofuscinosis (JNCL) (Figure 9.5C,D).[30]

9.4.3 Cytoplasmic Changes

The cytoplasm of dying cells should be independently assessed for changes in membranes, organelles such as mitochondria, and for status of rough

---

FIGURE 9.1

Normal neuron. **A.** Pyramidal neuron in cerebral cortex of rat exposed to sham carbon monoxide exposure showing normal, oval nuclear morphology, cytoplasmic density with surrounding satellite cells and neuropil. Arrow shows region of plate B. Original magnification 4000×, bar represents 2 microns. **B.** Detail of pyramidal neuron in A from region marked by arrow showing normal nuclear membrane, nuclear pores, rough endoplasmic reticulum and Golgi apparatus with presence of axosomatic synapse (arrow) on plasma membrane. Nuclear pores are easily visible (oblique arrow) and mitochondria (m) and rough endoplasmic reticulum (rer) appear normal. Original magnification 25,000×, bar represents 200 nm.

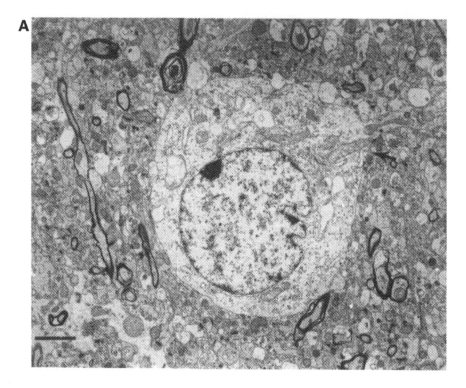

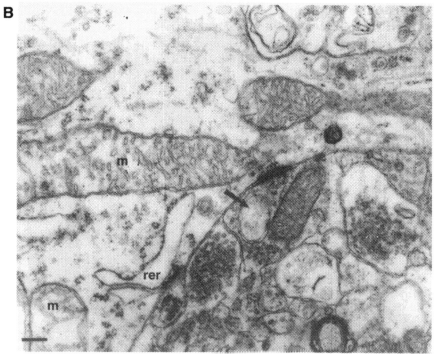

FIGURE 9.2

endoplasmic reticulum, Golgi, and lysosomal structures.[11,12] In Table 9.2, a list of cytoplasmic morphological features is listed for apoptotic cell death compared to necrotic cell death.[1-4,11,12] These features are assessed in the same illustrations (Figures 9.1 to 9.4) used above for nuclear morphology.

The cytoplasm of a normal pyramidal neuron (Figure 9.1) is contrasted with the swollen, electronlucent cytoplasm of a necrotic neuron (Figure 9.2) and the condensed cytoplasm of neurons that may be in early stages of apoptosis.[13] There is no widespread vacuolar change and all organelles and membranes are relatively intact. Of note is the occurrence of some dilatation of the rough endoplasmic reticulum (Figures 9.3B, 9.4B) compared to a normal neuron (Figure 9.1B) and to abnormal profiles in a necrotic neuron (Figure 9.2B).

9.4.4   Mitochondrial and Other Organelle Changes

In apoptotic cells, there should generally be a lack of mitochondrial abnormalities, increased lysosomal or autophagocytic profiles, or major abnormalities of the endoplasmic reticulum or Golgi apparatus.[1-4] In other words, the cytoplasm should appear relatively normal, although condensed in volume and electrondense. In necrosis, swollen abnormal mitochondria and abnormal cytoplasmic features make the cell look abnormal even at low power with swollen and electron-lucent appearance.

Mitochondria and other organelles are relatively normal appearing in cells in early stages of apoptosis (Figures 9.3B, 9.4B) compared to their somewhat swollen and electron-lucent character and disrupted cristae in the necrotic neuron (Figure 9.2B).[13]

In cells in more advanced stages of apoptosis, there is still relative preservation of cytoplasmic integrity and of mitochondrial morphology (Figure 9.5).[30] There is little indication of lysosomal or autophagocytotic activity in subacute delayed neuronal degeneration.[13]

Mitochondria may be normal in some apoptotic cells since cytochrome c release is in fact an inducer of the apoptotic program.[33] In these instances, a subpopulation of the mitochondria are biochemically abnormal and may present some morphological abnormalities. This signaling of mitochondrial injury would be contrasted with a more global event affecting energy production in all mitochondria of the cell and leading to necrosis. Such "hetero-

---

FIGURE 9.2
Swollen neuron with early signs of necrosis. **A.** Pyramidal neuron in cerebral cortex of rat exposed to carbon monoxide, showing swollen cytoplasm and disruption of cellular organelles with relatively normal appearing nucleus. There are some swollen astrocytic processes nearby. Arrow shows region of plate B. Original magnification, 4000x, bar represents 2 microns. **B.** Detail of pyramidal neuron in A from region marked by arrow showing swollen mitochondria (m) and rough endoplasmic reticulum (rer). Cytoplasm is watery and electronlucent. Presence of axosomatic synapse absolutely identifies cell as neuron (arrow). Original magnification 25,000x, bar represents 200 nm.

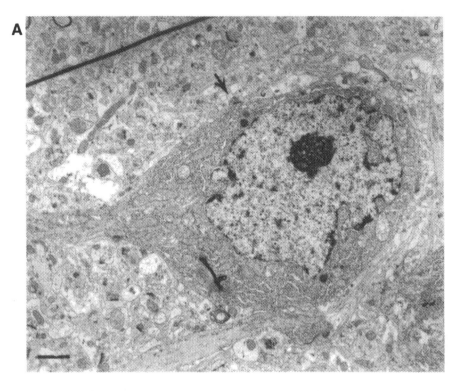

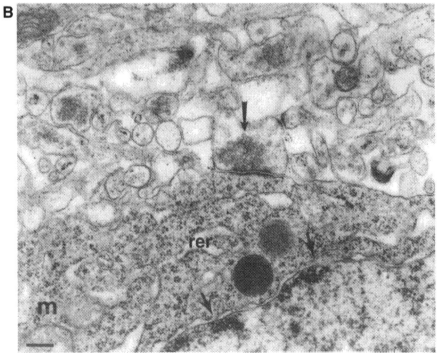

FIGURE 9.3

plasmic " injury may well be common in mitochondrial diseases accompanied by apoptosis.[34]

9.4.5 Continuum of Necrosis to Apoptosis

The occurrence of mixed nuclear and cytoplasmic morphologies supports a continuum of cell response from necrosis to apoptosis.[11,12] Most often, the combination is nuclear morphology suggestive of apoptosis with cytoplasmic morphology suggestive of necrosis. One can imagine multiple modes of cell injury in complex nervous tissue after an initial insult either due to continued presence of toxic influence (e.g., prolonged tissue levels of trimethyltin, binding of carbon monoxide to mitochondrial cytochrome oxidase), to further injury from inflammatory responses of glial cells and recruited microglial cells/macrophages, or from secondary energy failure or necrotic mechanisms supervening on a cell entering initially into apoptosis. This may be particularly true for ischemic injury. The current literature suggests that many experimental models of injury lead to cell death both by apoptosis and by necrosis with intermediate morphological forms as discussed above. This includes ischemic injury,[36-44] excitotoxic or seizure-induced injury,[45-48] and combined physical or ischemic injury.[49-51]

Thus, the ability of a cell to undergo programmed cell death may be compromised or changed by injury to neighboring cells, lack of effective heterophagocytosis, or even superimposed cell injury leading to energy failure and interruption of the steps of programmed cell death.[6]

9.4.6 Documentation and Analysis

Documentation and analysis should be based on careful survey of multiple animals and sites using electron microscopy at low magnifications. These comparisons should include similarly processed material from control animals. To document the various nuclear and cytoplasmic features, there should be electron micrographs taken at magnifications that permit visualization of nuclear, mitochondrial, endoplasmic reticulum, and plasma membranes (e.g, 15,000 to 25,000×). It is important to realize that both necrotic and apoptotic cell death and intermediate forms may affect multiple cell classes besides neurons, particularly glial cells.[52-57]

FIGURE 9.3

Condensed neuron with early signs of apoptosis. **A.** Pyramidal neuron in cerebral cortex of rat exposed to carbon monoxide, showing condensed cytoplasm, relatively normal appearing mitochondria, and abnormal nuclear shape with crenelation of nuclear membrane and margination of chromatin. Arrow indicates region of plate B. Original magnification, 5000×, bar represents 2 microns. **B.** Detail of pyramidal neuron in A with irregular nuclear profile, distention of nuclear membrane space, and loss of nuclear pores near regions of chromatin condensation (arrows). Mitochondria (m) and rough endoplasmic reticulum (rer) are relatively normal. Long arrow indicates axosomatic synapse. Original magnification 25,000×, bar represents 200 nm.

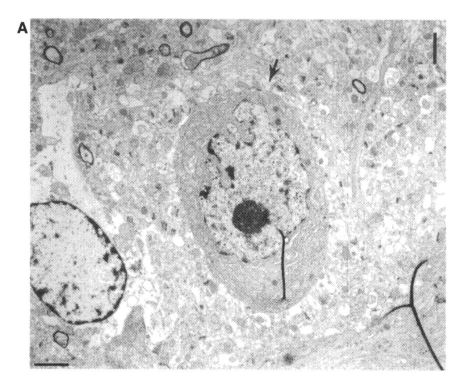

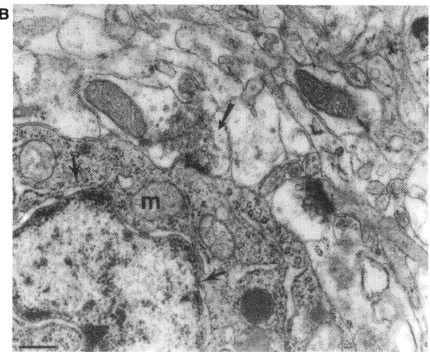

FIGURE 9.4

9.5   Protocols

9.5.1   Tissue Fixation

Materials and Equipment

- Commercial formalin (37% solution)
- Paraformaldehyde powder, reagent grade
- Purified glutaraldehyde, 25 to 50% (in sealed ampule or under nitrogen)
- 0.1 to 0.2 M sodium phosphate buffer, pH 7.4 to 7.6
- Sodium hydroxide pellets, reagent grade
- Buchner filter flask and funnel with Whatman #1 or equivalent paper

Tissue fixation is usually performed with mixed aldehydes or with paraformaldehyde solution. The chosen approach depends on tissue chosen and on potential use for ultrastructural analysis as well as issues of convenience. The following fixatives may be considered:

Routine Formalin Fixation

1. "10%" Formalin consists of 10% by volume dilution of commercial formalin in 0.1 M sodium phosphate buffer (pH 7.4 to 7.6) and is prepared by dilution of 37% commercial formalin into buffer (final concentration 3.7% by weight).
2. Fixation is from 24 h to several days, depending on size of tissue block (for example, 24 h for ca. 500 mg mouse brain).

FIGURE 9.4
Condensed neuron with early signs of apoptosis. **A.** Smaller neuron in cerebral cortex of rat exposed to carbon monoxide, showing condensed cytoplasm, normal cell organelles, and abnormal nuclear shape with crenelation of nuclear membrane and margination of chromatin. Swollen, necrotic appearing cell is in top margin of plate. Arrow indicates region of plate B. Original magnification 4000×, bar represents 2 microns. **B.** Detail of neuron in A showing nonspherical appearance of nuclear membrane (arrows), mild dilatation of endoplasmic reticulum, presence of free ribosomal rosettes, and normal appearing mitochondria (m). Presence of axosomatic synapse absolutely identifies cell as neuron (arrow). Original magnification 25,000×, bar represents 200 nm.

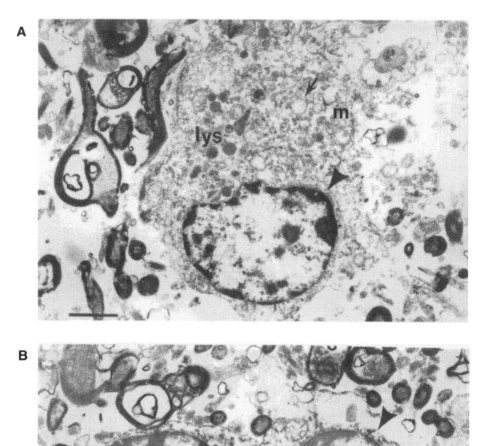

FIGURE 9.5

Apoptotic cells in human brain (courtesy of Dr. Rose Mary Boustany). **A.** Apoptotic cell profile in human frontal cortex of patient with late infantile neuronal ceroid lipofuscinosis (LINCL). Cell with marginated chromatin (arrowhead) and condensed cytoplasm represents presumed apoptotic cell. Mitochondria (m, arrow indicates one) are quite numerous, and appear relatively normal, given postmortem state. There are abundant lysosomes (lys) in this cell. Original magnification 6000×, bar represents 1 micron. **B.** Another field from patient in plate A showing two cells with chromatin margination and relatively condensed cytoplasm. Mitochondrion (m) is relatively normal sized, although electron-lucent. Original magnification 6000×, bar represents 1 micron.

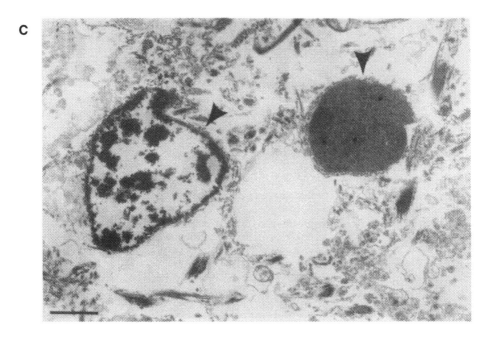

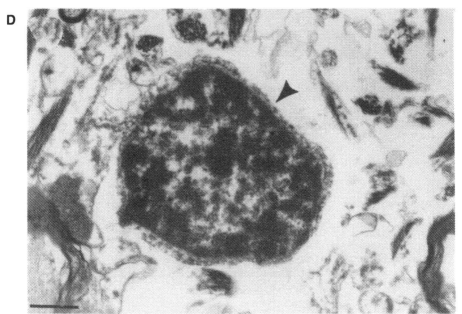

FIGURE 9.5 (continued)
C. Apoptotic cell profiles in human cerebellum of patient with Batten's disease or juvenile neuronal ceroid lipofuscinosis (JNCL). The cells are located in internal granule cell layer and are most likely granule cell neurons, but precise identification is difficult in postmortem tissue. One cell shows marginated chromatin (oblique arrowhead), while other profile is apoptotic body (fragment of nucleus, horizontal arrowhead). Original magnification 5000×, bar represents 1 micron. **D.** Another field from patient with JNCL (plate C), showing more condensed nucleus (arrowhead) representing apoptotic body consistent with late stages of apoptosis. Original magnification 10,000×, bar represents 500 nm.

TABLE 9.1

Nuclear Ultrastructure of Dying or Injured Cells

Low magnification	
Dark neuron	*Dark* nucleus and cell
Retrograde reaction	*Eccentric or displaced* euchromatic nucleus
Early apoptotic neuron	*Crenelated* heterochromatic nucleus with DNA margination
Late apoptotic neuron	*Fragmentation* into chromatin-containing apoptotic bodies
Early necrotic neuron	*Oval* electron-lucent nucleus
Late necrotic neuron	*Destruction* or karyolysis commensurate with cytoplasmic dissolution
High magnification	
Dark neuron	Condensation and collapse of normal structure — "shrunken"
Retrograde reaction	Tendency to euchromatic appearance, otherwise normal
Apoptotic neuron	Margination and condensation of chromatin, coalescence into caps or larger domains along nuclear margin; *early crenelation of nuclear membrane* with formation of protuberances; loss of pores near condensed chromatin
Necrotic neuron	Margination of chromatin in small, loose-textured aggregates; nuclear shape relatively intact or swollen; destruction of nuclear membrane, loss of pores commensurate with destruction of cytoplasmic structures

TABLE 9.2

Cytoplasmic Ultrastructure of Dying or Injured Cells

Low magnification	
Dark neuron	Dark cytoplasm with condensation
Retrograde reaction	Rearrangement of rough endoplasmic reticulum into arrays; otherwise normal organelles
Early apoptotic neuron	*Condensed, shrunken appearance with electron-dense cytoplasm;* normal appearing mitochondria and other organelles; absence of large vacuoles
Late apoptotic neuron	Phagocytosis by surrounding cells or microglial cells blebbing of cell surface to form apoptotic bodies; secondary necrosis in some cell classes
Early necrotic neuron	*Swollen, electron-lucent with abnormal organelles*
Late necrotic neuron	Destruction with cytolysis, karyolysis phagocytosis, and/or liquefaction necrosis; inflammatory response depending on extent of cell loss
High magnification	
Dark neuron	Condensed, electron-dense cytoplasm and organelles
Retrograde reaction	Normal cytoplasm with rearranged rough endoplasmic reticulum
Apoptotic neuron	*Condensation of cytosol with intact membranes and normal organelles, particularly mitochondria;* some dilatation of rough endoplasmic reticulum and Golgi apparatus, with formation of vacuoles in some experimental contexts
Necrotic neuron	*Swelling of all cytosol compartments with electron-lucency and disintegration of membrane integrity; abnormal mitochondria with matrix densities*

4% Paraformaldehyde Fixation

1. 4% Paraformaldehyde is prepared as follows (liter): 40 g of powdered paraformaldehyde (caution carcinogen) is weighed out under hood or with respiratory precautions and set aside; 500 ml of distilled or purified water is heated on stirred hot plate to 70 to 80°C and paraformaldehyde powder carefully added and stirred into slurry.

2. One to three pellets of sodium hydroxide are added by dropwise addition of 1 N sodium hydroxide solution to depolymerize paraformaldehyde. This will occur relatively suddenly, but successfully only at this temperature range. Higher temperatures will result in excessive vapor production.

3. This slightly opalescent, but clear solution is then added to 500 ml of 0.2 M sodium phosphate buffer (pH 7.4 to 7.6) to yield final solution of 4% paraformaldehyde in 0.1 M phosphate buffer. This solution should be filtered over fine grade Whatman filter paper and Buchner funnel by slow vacuum to provide clarification and removal of small precipitates and undepolymerized paraformaldehyde. For larger volumes, aldehyde mixture can be filtered before addition to buffer if that is more convenient and buffer solution is clear.

Mixed Aldehyde Fixation

1. This fixative consists of 2 to 4% paraformaldehyde (P) with 0.1% to 2% glutaraldehyde (G). Some potential combinations are 4% P to 0.1% G (light fixation), 4% P to 0.5% G (moderate fixation), or more traditional half or full strength Karnofsky formulations of 2% P to 2% G or 4% P to 4% G. The latter fixative is hyperosmolar and should not be used in most initial fixations.

2. These fixations are prepared as for 4% paraformaldehyde with addition of electron microscopy grade glutaraldehyde (25 to 50% stored in nitrogen-filled, sealed bottles or ampules) to give correct final concentration (e.g., for 1% G in 1 l of fixative, one would need 20 ml of 50% glutaraldehyde). This should be added before final filtration.

3. In some protocols, calcium is added for "membrane stabilization." Saturated calcium chloride solution is added dropwise with stirring until clouding persists. Solution is then filtered.

Fixation for Fetuses or Young Postnatal Animals

1. This fixative consists of 2.5% paraformaldehyde in 0.1 M phosphate buffer at pH 7.4 with addition of 4% sucrose (w/v). Saturated

    calcium chloride is added dropwise with stirring until clouding and solution is filtered as above.

2. Small amounts of glutaraldehyde (0.1 to 0.5%) can be added to improve ultrastructural preservation.

Other Fixative Choices

Other fixatives may be chosen for particular tissues, particularly when immunocytochemical analysis is planned and the particular antigen is fixation sensitive. Such an approach is described for a cytoplasmic, relatively soluble antigen, the various enolase isoenzymes, in Reference 17. These approaches usually involve some compromise of ultrastructural preservation, particularly in nervous tissue.

9.5.2 Tissue Preparation

Materials and Equipment

- Appropriate surgical and dissection instruments
- Anesthetic solutions (pentobarbital, ketamine, etc.) and approved euthanasia protocol
- Foam board and dissection pins/tape or other method to immobilize animal
- Fume hood
- Appropriate cannula or fine-bore needles for perfusion
- Peristaltic pump for perfusing solutions

Immersion Fixation

1. Animal is deeply anesthetized or euthanized. Appropriate tissue is gently dissected and removed. Further subdivision into small blocks 100 to 200 mg is advised with 1 to 3 mm maximum distance from external surface to deepest tissue plane (smallest blocks will fix the most adequately, but are more liable to damage during dissection and more difficult to process and to orient during cutting).

2. This method results in adequate ultrastructural preservation for tissue culture as well as a number of tissues including kidney, heart, muscle, and pancreas. Brain and liver are usually poorly served by this approach, but key is in rapidity of transfer into fixative, absence of trauma to tissue during removal, and size of tissue block. Small blocks of tissue, small organs (e.g., rodent brain at postnatal day

14 to 30), and needle biopsies may be reliably fixed in this manner. Block size should approximate 2 to 3 mm² for best results, given slow penetration rates of aldehyde fixatives. In many cases, using cooled fixative most likely decreases autolytic processes. Tissue must be handled gently without dropping or physical trauma through instrumentation.

Perfusion Fixation of Young or Adult Animals

For nervous tissue, perfusion fixation is the method of choice for assuring relatively reasonable preservation of myelin, synaptic structures, and general membrane integrity.

1. Animal is deeply anesthetized or euthanized. Reasonable anesthesia is intraperitoneal pentobarbital 100 mg/kg after appropriate restraint and/or neurolepsis with ketamine. Animal must be in deepest level of anesthesia for appropriate ablation of pain and vascular reactivity. After appropriate immobilization and confirmation of anesthetic plane by corneal reflex, breathing pattern, and lack of pain response, chest wall is rapidly opened within 5 to 15 s. Heart should still be beating. Chest wall should be excised or gently reflected without torquing great vessels or carotids, and attention given to avoiding damage to vessels. Rapid, but careful technique is sufficient for most experiments; gold standard is total control of respiration and blood pressure through artificial ventilation and vascular access.

2. Heart is exposed and pericardium removed or window opened. Right auricle or atrium is incised to provide good flow of return blood from right heart. After incision or careful insertion of needle, cannula is inserted into left ventricle and flow-guided gently into outflow tract and, if possible, into ascending aorta, taking care not to rupture intraventricular septum, heart wall, and/or aorta. Best method is to insert cannula into ascending aorta and to gently occlude around cannula with soft clamp. Cannula should be secured so as not to torque or shift heart or great vessels. Fluid build-up in lungs or through nose/mouth indicates rupture into right heart through bronchial circulation or through rupture of intraventricular septum.

3. Vasculature is rinsed with room temperature normal saline for 0.5 to 1.0 ml/gm body weight or until blood return from right heart is substantially cleared. Some add heparin and/or vasodilators such as nitrite to this rinse. This is usually not necessary in well-anesthetized animal with rapid progression through steps 1 and 2 to prevent cessation of blood flow and intravascular coagulation. For certain fixatives containing low percentages of paraformaldehyde and/or glutaraldehyde and larger animals, it is possible to

commence directly with half-strength or even full strength fixative for rinsing. Danger is clogging of capillary vasculature with chemically fixed red blood cells and/or serum proteins.

4. The usual amount of fixative is 2 to 4 ml/g body weight under pressure equivalent to 50 to 100 cm water delivered through left ventricle. For some experiments, room temperature fixative for certain fraction of volume can be followed by ice-cold fixative to minimize autolysis. Gravity flow is most reliable pressure guide but inconvenient. Volume-limited perfusion by peristaltic pump is easiest, but attention must be given to maintaining appropriately low pressures and impedance, since there is no pressure limitation. After perfusion, animal is left undisturbed for 1 to 2 h before dissection.

5. Tissue must be kept moist or hydrated at all times and dissection must be delicate particularly if performed at early time points after initiation of fixation.

6. Postfixation in ice-cold fixative after dissection is appropriate for many experiments.

7. Tissue blocks of 1 to 2 mm are prepared for processing with appropriate geometry to assure proper orientation. Alternative is to cut 60- to 100-micron sections on vibratome (allowing 2 to 4 mm of block thickness and up to 1 × 1 cm block face) which facilitates visual control of dissection and processing. Such sections can be alternated with 15- to 35-micron sections for light microscopy and immunocytochemistry.

Perfusion Fixation of Small Fetuses or Young Postnatal Animals (e.g., Rodents)

1. Animal is cooled on ice to provide anesthesia (less stressful than needle administration of anesthetic in very young animal).

2. Vasculature is rinsed using hand-held syringe and #30 needle with warm room temperature saline, room temperature fixative (10 ml), and then ice-cold fixative.

3. Brain is removed after perfusion and postfixed 4 to 6 h. Dissection and cutting as above.

9.5.3   Tissue Storage

Tissue can either be stored in imbedded blocks in plastic resin or paraffin waxes, as frozen tissue blocks for cryosection with adequate protection against dessication, or as tissue blocks in refrigerated buffer solution. The first methods allow for indefinite storage, while the keeping of unimbedded tissue blocks runs the risk of bacterial or yeast contamination. This can be minimized by addition of small percentage of fixative (e.g., 0.1% paraform-aldehyde) or antimicrobial (e.g., thymol crystal).

9.5.4 Embedding Methods

Methods will be described for processing tissue through embedding into resin. These methods can be carried out in laboratories with fume hood and vacuum oven and may be cost-effective or reasonable for large scale experiments. Further work-up with cutting of thick and thin sections on ultrami-crotome, grid collection and storage, grid coating, and examination under electron microscope are beyond scope of chapter and presume availability of fully equipped EM laboratory and trained technicians and support personnel. All reagents are EM grade and glass-distilled water should be used.

- Absolute alcohol (nondenatured), EM grade
- 0.1 M Sodium cacodylate buffer (**Danger: arsenic containing**) with 7.5% sucrose
- 4% Glutaraldehyde in 0.1 M sodium cacodylate buffer
- 1% Osmium tetroxide in S-collidine buffer
- 0.5% Uranyl acetate in veronal acetate buffer
- Propylene oxide, EM grade (**Danger: use under hood**)
- Epon resin: Luft's formula or equivalent
- Vacuum oven

Vibratome Sections

1. Place vibratome sections mounted on slides into 4% glutaralde-hyde in 0.1 M cacodylate buffer with sucrose for 10 min (tissue previously in phosphate buffer must be rinsed first to avoid pre-cipitate). For 1 liter of buffer: 21.4 g of sodium cacodylate are mixed and brought to 1000 ml distilled water. Adjust pH to 7.4 using 1 N HCl. Osmolarity is 350 mOsm and solution is filtered and kept in refrigerator. Add 4% glutaraldehyde by dilution of 25 to 50% ampules into buffer on day of use for appropriate volume.

2. Rinse tissue for 3 changes of 5 minutes each with 0.1 M sodium cacodylate buffer with 7.5% sucrose. For 1 liter of buffer: preparation is as in step 1 with addition of 75 g sucrose at onset.

3. Postfix tissue in 1% osmium tetroxide in S-collidine buffer for 10 to 30 min (shorter for vibratome sections, longer for tissue blocks). For stock OsO_4 solution: melt crystals in sealed vial under hot water and allow to run to one side of vial and recrystallize. Using acid-washed glassware, gloves, and under fume hood, score and break

vial and place in glass-stoppered bottle. Add appropriate amount of distilled water to make 2% solution. May take 2 to 4 days to dissolve in warm water bath. Store in closed container in refrigerator and only use with acid-washed one-time use pipettes to avoid organic contamination. Will blacken and discolor when no longer good from very initial faint yellowish color. For 0.2 *M* S-collidine stock buffer: 120 ml distilled water is added to 5.34 ml purified S-collidine, and then 16 ml of 1*N* HCL is added. Volume is adjusted to 200 ml with distilled water. Store in refrigerator. For working solution: add appropriate volumes of 2% OsO_4 and 0.2 *M* S-collidine buffer at 1:1 ratios. Can be stored in refrigerator until discolored.

4. Optional *en bloc* staining with uranyl acetate for greater contrast. Tissue is placed in 0.5% uranyl acetate in veronal buffer for 30 min. Stock is prepared: 2.92 g sodium acetate, 4.42 g sodium barbital, 5.10 g NaCl, and 3.75 g uranyl acetate are added to 7.26 ml of distilled water, and pH adjusted with 24 ml 1*N* HCl for pH 4.70, osmolarity 350 to 450 mOsm. Solution is kept in refrigerator.

5. Tissue is dehydrated through ascending alcohols for 3 to 5 minutes each change through following steps: 35%, 50%, 70%, 95% × 2 changes, 100% × 2 changes. Tissue is very brittle at this point and must be carefully handled or protected by solvent-resistant supports or filters.

6. Alcohol is removed with two changes of propylene oxide for 3 to 5 min each. **Very noxious solvent, use under hood.** Also will dissolve many plastics; appropriate disposable resistant plasticware must be used.

7. Tissue is placed with 1:1 mixture of epon and propylene oxide for 10 min. Epon mixture is prepared (Luft's formula) freshly for each run by combining appropriate volumes of room temperature stock solutions of mixtures A and B (see below) in 3:7 ratio with stirring with magnetic bar and addition of accelerator 2% by total volume to A+B mixture of DMP-30 catalyst. If there are excess bubbles, mixture can be degassed briefly in vacuum oven. Stock solutions of A (41 ml Epon 812 and 65 ml DDSA or dodecenyl succinic anhydride) and mixture B (82 ml Epon 812 and 70 ml MNA or methyl nadic anhydride) are prepared by stirring above proportions with stir bar and can be stored for long periods in refrigerator.

8. Flood slide with pure Epon mixture for 3 changes of 5 min each.

9. Leave enough Epon on slide to cover tissue section.

10. Put slides in 90°C oven for 30 min.

11. Strip Epon film off slide with soaking in hot water and store as thin film. Selected areas can be dissected with scissors or razor blade and re-embedded or glued on top of Epon blank blocks (stubs for ultramicrotome). Remounted tissue is re-cured for 30 min at 90°C.

12. Submit to EM laboratory for thick sections (0.2 to 0.5 µm) and thin sectioning and appropriate stains and coating.

Small Tissue Blocks (0.5 to 1.0 mm)

1. Steps are as for vibratome sections with following changes and longer times to allow for penetration of reagents:
   a. Glutaraldehyde step is 30 min.
   b. Cacodylate-sucrose rinsing step is 2 changes for 10 min.
   c. Postfixation in 1% OsO_4 is 30 min.
   d. Dehydration in ascending alcohols and propylene oxide changes are 5 to 10 min each.
   e. Infiltration with 1:1 Epon propylene oxide mixture is without accelerator for 30 min under vacuum at room temperature.
   f. Pure Epon mixture with accelerator is incubated with tissue for 30 min to 1 h under vacuum at room temperature.
   g. Blocks are placed in molding cups or block holders with appropriate identification and placed in embedding oven at 60°C for 24 to 48 h or shorter times at high temperature.

9.5.5   Ultrastructural Analysis

Appropriate record-keeping is essential to record details of dissection protocols, tissue blocks, light and electron microscopic examination of tissue, and logs of photographic negatives. It is essential to keep some thick sections (0.2 to 0.5 µm) of each EM block to facilitate orientation to anatomical landmarks in tissue. This is particularly true if blocks are further trimmed to facilitate cutting of thin sections. Alternating series of thick and thin sections can be used to step through a block for systematic analysis of larger regions or an appropriate number of blocks used to provide survey. Point-counting methods or stereological analysis may assist in determining actual incidence of described features.[58] In general, low-power 1000 to 4000× micrographs are extremely important to provide context for changes observed at higher magnifications.

Acknowledgment

We gratefully acknowledge assistance of Ms. Ruby Ange, D. Scott Burkhart, John Peterson, Larry Hawkey, Susan Reeves, and Steve Conlon. The work was conducted with support of NIH-NIA ADRC award 5P50 AG-05128, NIEHS IAA Y01-ES-40290 (DES) and numerous contributions, each less than $100, to Joseph and Kathleen Bryan Alzheimer's Disease Research Center.

References

1. Wyllie, A.H., Kerr, J.F.R., and Currie, A.R., Cell death: the significance of apoptosis, *Int. Rev. Cytol.* 68, 251, 1968.
2. Schweichel, J.-U. and Merker, H.-J., The morphology of various types of cell death in prenatal tissues, *Teratology*, 7, 253, 1972.
3. Schweichel, J.-U., Das elektronenmikroskopische Bild des Abbaues der epithelialen Scheitelleiste wahrend der Extremitatenentwicklung bei Rattenfeten, *Z. Anat. Entwickl-Gesch.*, 136, 192, 1972.
4. Clarke, P.G.H., Developmental cell death: morphological diversity and multiple mechanisms, *Anat. Embryol.*, 181, 195, 1990.
5. Robertson, A.M.G. and Thompson, J.N., Morphology of programmed cell death in the ventral nerve cord of *Caenorhabditis elegans* larvae, *J. Embryol. Exp. Morphol.*, 67, 89, 1982.
6. Driscoll, M., Cell death in *C. Elegans*: molecular insights into mechanisms conserved between nematodes and mammals, *Brain Pathol.*, 6, 411, 1996.
7. Oppenheim, R. W., Yin, Q. W., Prevette, D., and Yan, Q., Brain-derived neurotrophic factor rescues developing avian motoneurons from cell death, *Nature*, 360, 755, 1992.
8. Homma, S., Yaginuma, H., and Oppenheim, R. W., Programmed cell death during the earliest stages of spinal cord development in the chick embryo: a possible means of early phenotypic selection, *J. Comp. Neurol.*, 345, 377, 1994.
9. Lo, A. C., Houenou, L. J., and Oppenheim, R. W., Apoptosis in the nervous system: morphological features, methods, pathology, and prevention, *Arch. Histol. Cytol.*, 58, 139, 1995.
10. Chang, G. Q., Hao, Y., and Wong, F., Apoptosis: final common pathway of photoreceptor death in rd, rds, and rhodopsin mutant mice, *Neuron*, 11, 595, 1993.
11. Portera-Cailliau, C., Price, D. L., and Martin, L. J., Excitotoxic neuronal death in the immature brain is an apoptosis-necrosis morphological continuum, *J. Comp. Neurol.*, 378, 70, 1997.
12. Portera-Cailliau, C., Price, D. L., and Martin, L. J., Non-NMDA and NMDA receptor-mediated excitotoxic neuronal deaths in adult brain are morphologically distinct: further evidence for an apoptosis-necrosis continuum, *J. Comp. Neurol.*, 378, 88, 1997.

13. Piantadosi, C.A., Zhang, J., Levin, E.D., and Schmechel, D.E., Apoptosis and delayed neuronal damage after carbon monoxide poisoning in the rat, *Exp. Neurol.*, 147, 103, 1997.

14. Peters, A., Palay, S.L., and deF. Webster, H., *Fine Structure of the Nervous System*, W.B. Saunders, Philadelphia, 1994.

15. Hayat, M.A., *Principles and Techniques of Electron Microscopy*, CRC Press, Boca Raton, 1989.

16. Glauert, A.M., Fixation, dehydration, and embedding of biological specimens, Volume 3: Part I. In: *Practical Methods in Electron Microscopy*, A.M. Glauert, Ed., North-Holland, Amsterdam, 1977.

17. Schmechel, D.E., Methods of localizing cell-specific proteins in brain. In: *Neuronal and Glial Proteins: Structure, Function, and Clinical Application*, P.J. Marangos, I.C. Campbell, and R.C. Cohen, Eds., Academic Press, New York, 1988.

18. Cammermeyer, J., Is the solitary dark neuron a manifestation of postmortem trauma to the brain inadequately fixed by perfusion?, *Histochemistry*, 56, 97, 1978.

19. Cammermeyer, J., Argentophil neuronal perikarya and neurofibrils induced by postmortem trauma and hypertonic perfusates, *Acta Anat.*, 105, 9, 1979.

20. Schmechel, D.E., Burkhart, D.S., Ange, R., and Izard, M.K., Cholinergic axonal dystrophy and mitochondrial pathology in prosimian primates, *Exp. Neurol.*, 142, 111, 1996.

21. Watt, J. A., Pike, C. J., Walencewicz-Wasserman, A. J., and Cotman, C. W., Ultrastructural analysis of beta-amyloid-induced apoptosis in cultured hippocampal neurons, *Brain Res.*, 661, 147, 1994.

22. Figiel, I., and Kaczmarek, L., Cellular and molecular correlates of glutamate-evoked neuronal programmed cell death in the *in vitro* cultures of rat hippocampal dentate gyrus, *Neurochem. Int.*, 31, 229, 1997.

23. Manev, H., Kharlamov, E., Uz, T., Mason, R.P., and Cagnoli, C.M., Characterization of zinc-induced cell death in primary cultures of rat cerebellar granule cells, *Exp. Neurol.*, 146, 171, 1997.

24. Middleton, G., Reid, L.E., and Harmon, B.V., Apoptosis in the human thymus in sudden and delayed death, *Pathology*, 26, 81, 1994.

25. Migheli, A., Attanasio, A., Lee, W.-H., Bayer, S.A., and Ghetti, B., Detection of apoptosis in weaver cerebellum by electron microscopic *in situ* end-labeling of fragmented DNA, *Neurosci. Lett.*, 199, 53, 1995.

26. Nishiyama, K., Kwak, S., Takekoshi, S., Watanabe, K. and Kanazawa, I., *In situ* nick end- labeling detects necrosis of hippocampal pyramidal cells induced by kainic acid, *Neurosci. Lett.*, 212, 139, 1996.

27. Soriano, M. A., Ferrer, I., Rodriguez-Farre, E., and Planas, A. M., Apoptosis and c-Jun in the thalamus of the rat following cortical infarction, *Neuroreport*, 7, 425, 1996.

28. Hara, A., Hirose, Y., Wang, A., Yoshimi, N., Tanaka, T. and Mori, H., Localization of Bax and Bcl-2 proteins, regulators of programmed cell death, in the human central nervous system, *Virchows Arch.*, 429, 249, 1996.

29. Martinou, J. C. and Sadoul, R., ICE-like proteases execute the neuronal death program, *Curr. Opin. Neurobiol.*, 6, 609, 1996.

30. Lane, S.C., Jolly, R.D., Schmechel, D.E., Alroy, J., and Boustany, R.-M., Apoptosis is the mechanism of neurodegeneration in Batten disease, *J. Neurochem.*, 67, 677, 1996.

31. Earnshaw, W.C., Nuclear changes in apoptosis, *Curr. Opin. Cell Biol.*, 7, 337, 1995.

32. Majno, G. and Joris, I., Apoptosis, oncosis, and necrosis. An overview of cell death, *Am. J. Pathol.*, 146, 3, 1995.

33. Kluck, R.M., Bossy-Wetzel, E., Green, D.R., and Newmeyer, D.D., The release of cytochrome c from mitochondria: a primary site for Bcl-2 regulation of apoptosis, *Science*, 275, 1132, 1997.

34. Howell, N., Leber hereditary optic neuropathy: how do mitochondrial DNA mutations cause degeneration of the optic nerve?, *J. Bioenerg. Biomembr.*, 29, 165, 1997.

35. Petit, P.X., Lecoeur, H., Zorn, E., Dauguet, C., Mignotte, B., and Gougeon, M.L., Alterations in mitochondrial structure and function are early events of dexamethasone-induced thymocyte apoptosis, *J. Cell Biol.*, 130, 157, 1995.

36. Deshpande, J., Bergstedt, K., Linden, T., Kalimo, H., and Wieloch, T., Ultrastructural changes in the hippocampal CA1 region following transient cerebral ischemia: evidence against programmed cell death, *Exp. Brain. Res.*, 88, 91, 1992.

37. Nitatori, T., Sato, N., Waguri, S., Karasawa, Y., Araki, H., Shibanai, K., Kominami, E., and Uchiyama, Y., Delayed neuronal death in the CA1 pyramidal cell layer of the gerbil hippocampus following transient ischemia is apoptosis, *J. Neurosci.*, 15, 1001, 1995.

38. Li, Y., Sharov, V. G., Jiang, N., Zaloga, C., Sabbah, H. N., and Chopp, M., Ultrastructural and light microscopic evidence of apoptosis after middle cerebral artery occlusion in the rat, *Am. J. Pathol.*, 146, 1045, 1995.

39. van Lookeren Campagne, M. and Gill, R., Ultrastructural morphological changes are not characteristic of apoptotic cell death following focal cerebral ischaemia in the rat, *Neurosci. Lett.*, 213, 111, 1996.

40. Kato, H., Kanellopoulos, G. K., Matsuo, S., Wu, Y. J., Jacquin, M. F., Hsu, C. Y., Kouchoukos, N. T., and Choi, D. W., Neuronal apoptosis and necrosis following spinal cord ischemia in the rat, *Exp. Neurol.*, 148, 464, 1997.

41. Kato, H., Kanellopoulos, G. K., Matsuo, S., Wu, Y. J., Jacquin, M. F., Hsu, C. Y., Choi, D. W., and Kouchoukos, N. T., Protection of rat spinal cord from ischemia with dextrorphan and cycloheximide: effects on necrosis and apoptosis, *J. Thorac. Cardiovasc. Surg.*, 114, 609, 1997.

42. Mackey, M. E., Wu, Y., Hu, R., DeMaro, J. A., Jacquin, M. F., Kanellopoulos, G. K., Hsu, C. Y., and Kouchoukos, N. T., Cell death suggestive of apoptosis after spinal cord ischemia in rabbits, *Stroke*, 28, 2012, 1997.

43. Rosenblum, W. I., Histopathologic clues to the pathways of neuronal death following ischemia/hypoxia, *J. Neurotrauma*, 14, 313, 1997.

44. Chalmers-Redman, R. M., Fraser, A. D., Ju, W. Y., Wadia, J., Tatton, N. A., and Tatton, W. G., Mechanisms of nerve cell death: apoptosis or necrosis after cerebral ischaemia, *Int. Rev. Neurobiol.*, 40, 1, 1997.

45. Pollard, H., Charriaut-Marlangue, C., Cantagrel, S., Represa, A., Robain, O., Moreau, J., and Ben-Ari, Y., Kainate-induced apoptotic cell death in hippocampal neurons, *Neuroscience*, 63, 7, 1994.

46. Represa, A., Niquet, J., Pollard, H. and Ben-Ari, Y., Cell death, gliosis, and synaptic remodeling in the hippocampus of epileptic rats, *J. Neurobiol.*, 26, 413, 1995.

47. Sloviter, R. S., Dean, E., Sollas, A. L. and Goodman, J. H., Apoptosis and necrosis induced in different hippocampal neuron populations by repetitive perforant path stimulation in the rat, *J. Comp. Neurol.*, 366, 516, 1996.

48. Bengzon, J., Kokaia Z., Elmer, E., Nanobashvili, A., Kokaia, M., and Lindvall, O., Apoptosis and proliferation of dentate gyrus neurons after single and intermittent limbic seizures, *Proc. Natl. Acad. Sci. U.S.A.*, 94, 10432, 1997.

49. Colicos, M. A. and Dash, P. K., Apoptotic morphology of dentate gyrus granule cells following experimental cortical impact injury in rats: possible role in spatial memory deficits, *Brain Res.*, 739, 120, 1996.

50. Buchi, E. R. and Szczesny, P. J., Necrosis and apoptosis in neuroretina and pigment epithelium after diffuse photodynamic action in rats: a light and electron microscopic study, *Jpn. J. Ophthalmol.*, 40, 1, 1996.

51. Buchi, E. R., Cell death in the rat retina after a pressure-induced ischaemia-reperfusion insult: an electron microscopic study. I. Ganglion cell layer and inner nuclear layer, *Exp. Eye Res.*, 55, 605, 1992.

52. Gehrmann, J. and Banati, R.B., Microglial turnover in the injured CNS: activated microglia undergo delayed DNA fragmentation following peripheral nerve injury *J. Neuropathol. Exp. Neurol.*, 54, 680, 1995.

53. Vrdoljak, E., Bill, C. A., Stephens, L. C., van der, Kogel A. J., Ang, K. K. and Tofilon, P. J., Radiation-induced apoptosis of oligodendrocytes *in vitro*, *Int. J. Radiat. Biol.*, 62, 475, 1992.

54. Vela, J. M., Dalmau, I., Gonzalez, B. and Castellano, B., The microglial reaction in spinal cords of jimpy mice is related to apoptotic oligodendrocytes, *Brain Res.*, 712, 134, 1996.

55. Nguyen, K. B., McCombe, P. A. and Pender, M. P., Macrophage apoptosis in the central nervous system in experimental autoimmune encephalomyelitis, *J. Autoimmun.*, 7, 145, 1994.

56. Gillardon, F., Lenz, C., Kuschinsky, W., and Zimmermann, M., Evidence for apoptotic cell death in the choroid plexus following focal cerebral ischemia, *Neurosci. Lett.*, 207, 113, 1996.

57. Liu, X. Z., Xu, X. M., Hu, R., Du, C., Zhang, S. X., McDonald, J. W., Dong, H. X., Wu, Y. J., Fan, G. S., Jacquin, M. F., Hsu, C. Y., and Choi, D. W., Neuronal and glial apoptosis after traumatic spinal cord injury, *J. Neurosci.*, 17, 5395, 1997.

58. Glauert, A.M., Stereological techniques, Volume 6: Chapter 2. In: *Practical Methods in Electron Microscopy*, A.M. Glauert, Ed., North-Holland, Amsterdam, 1977.

10

Flow Cytometry in the Study of Apoptosis

Miriam J. Smyth

CONTENTS

10.1 Introduction

The aim of this chapter is to provide an introduction and guide to investigators studying apoptosis who are new to the field of flow cytometry. It should also serve as a reference for flow cytometrists who are interested in applying the methods of flow cytometry in studies of apoptosis. The chapter opens with an introduction to the topics of apoptosis and flow cytometry. An outline of the elements of apoptosis that can be measured by flow cytometry follows. The chapter goes on to detail a number of specific assays for detection of apoptosis by flow cytometry. In the case of each assay, the following information is provided: (1) an introduction to the characteristic apoptotic feature that the assay is designed to measure, (2) an introduction to the theory behind the measurement, (3) details of how to perform the assay, and (4) a discussion of the advantages and limitations of the assay. The chapter concludes with a discussion of the power and limitations of flow cytometry as a tool for studying apoptosis.

10.1.1   Introduction to Flow Cytometry

Cytometry is defined as the measurement of physical and/or chemical characteristics of cells or other biological particles. Flow cytometry refers to a technique in which these measurements are made while the cells or particles (for example, nuclei, chromosomes) are passing in a fluid stream in single file past one or more sensors in the measuring apparatus.[1] In the 25 years since it was first developed, flow cytometry has evolved into a technology capable of rapidly analyzing and separating cells, based on physical, biochemical, immunologic, or functional properties.

   The anatomy of a multilaser/multiparameter flow cytometer is shown in Figure 10.1. This illustrates the use of multiwavelength excitation and measurement of multicolor fluorescence emission and light scatter, as employed in various apoptosis assays. The details of the anatomy and use of this instrument are presented by Steinkamp.[2] For the purposes of the material presented in this chapter, the following brief summary of the operation of this instrument will suffice. Cells are introduced into the flow chamber at the rate of approximately 1000/s. The core stream containing the cells or particles to be analyzed is surrounded by a sheath solution that focuses the sample down to a narrow stream, ensuring that the particles are flowing in single file. The particles then enter the flow cell, where they impinge on the laser beam and generate the signal(s) of interest. As cells pass through the chamber, they are illuminated at a specific wavelength(s). Electrical and optical sensors measure physical, biochemical, and functional characteristics, and allow simultaneous measurement of multiple features on the same cell. Typical measurements include light scatter from cells (which correlates with diffractive, reflective, and refractive properties of external and internal features of the cell) and total or multicolor fluorescence emanating from stains bound to components of the cell.[2] The signal is detected by photomultiplier tubes. Using a dedicated computer, the data are displayed as frequency distribution histograms showing the number of cells with the characteristic of interest.

   Flow sorting represents a powerful extension of the flow cytometry method. Sorting employs electrical and/or mechanical means to divert cells

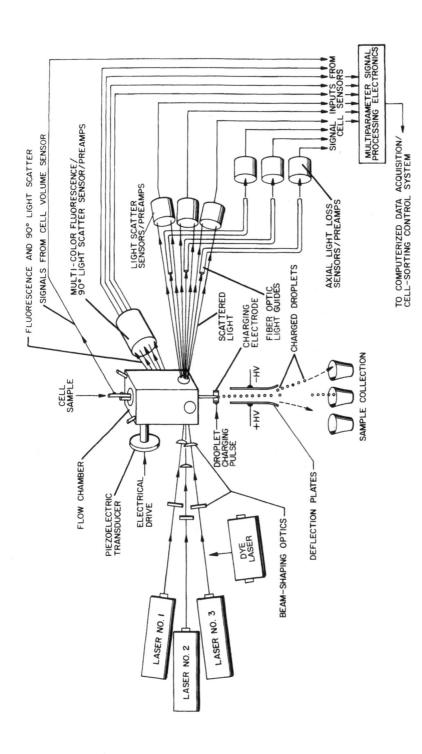

FIGURE 10.1

Multiwavelength-multiparameter flow cytometer/sorter illustrating three-laser illumination, flow chamber, cell sensors, beam-shaping optics, signal processing electronics, and droplet-charging and deflection apparatus (see text for details). (Figure courtesy of J. Steinkamp [Los Alamos National Laboratory] is adapted from Figure 9 in Reference 2.)

TABLE 10.1

Some Characteristics of Apoptosis that Can be Measured by Flow Cytometry

Characteristics of Apoptosis	Some Relevant Citations
Changes in cell volume	4–7
Changes in cell granularity	4–7
Changes in plasma membrane permeability	7–12
Alterations in cellular calcium concentration and pH	13–14
Expression of apoptosis-specific and survival-specific gene products	15–22
Apoptosis and cell cycle position	23–26
Internucleosomal DNA degradation	
Detection of hypodiploid cells	25, 27–32
Labeling DNA strand breaks	4,23,33–34
Phagocytosis of membrane-enclosed apoptotic bodies	35
Production of intracellular peroxides	36–37
Phospholipid membrane asymmetry	4,20,38

with preselected characteristics from the main fluid stream. The sorter can be programmed so that combinations of desired characteristics are used as sorting criteria.[3] This results in the isolation of subpopulations of viable or fixed cells or particles from the main population with a high degree of purity (usually >95% pure). Sorting facilitates correlation of measured properties with cell morphology and allows enrichment for functional and biochemical studies.

10.1.2   Apoptosis and Flow Cytometry

Apoptosis is characterized by a well-defined set of cellular events, including changes in cell volume, disruption of the plasma membrane, DNA fragmentation, chromatin condensation, expression of specific gene products, and generation of membrane-enclosed apoptotic bodies. There is much interest in the use of flow cytometry to monitor the morphological, biochemical, and molecular changes that occur during apoptosis. As shown in Table 10.1, many of the characteristics of apoptosis can now be measured by flow cytometry.

There are many applications of flow cytometry and cell sorting in the study of apoptosis (for reviews, see References 39 through 43). The increased use of flow cytometry in routine research and clinical applications over the past two decades has formed the basis for the development of flow cytometric assays for apoptosis, a number of which are presented below.

10.2   Description of the Methods

10.2.1   Analysis of Light-Scatter Properties

As a cell passes through the laser beam, it scatters the laser light both in the forward and perpendicular directions. The intensity of the light scattered in the forward direction (i.e., along the laser beam) provides the forward-angle light-scatter signal (FALS or FLS) and generally correlates with cell size.[44] The intensity of light scattered at a 90° angle to the laser beam is known as perpendicular light scatter (PLS), orthogonal, or side scatter, and generally correlates with granularity, refractivity, and the presence of intracellular structures that reflect the light in the perpendicular direction.[44] Most commercially available flow cytometers are capable of analyzing FALS and PLS as a routine measurement.

As a cell undergoes apoptosis, a number of changes occur that alter the ability of the cell to scatter light. The most significant of these changes are a consequence of cell shrinkage and condensation of the cytoplasm and chromatin. Cell shrinkage, an early event during apoptosis, is a result of cell dehydration, which leads to condensation of the cytoplasm and, subsequently, reduction in cell size and alterations in cell shape. Cell shrinkage leads to a decrease in the forward light-scatter signal from the cell. Condensation of the chromatin and fragmentation of the nucleus appear to lead to a transient increase in side scatter in some cell systems during apoptosis.[45,46] As the apoptotic program progresses, however, the forward and side-scatter signals decrease.[39]

Analysis of Light Scatter

Reagents

- 70% (v/v) ethanol
- PBS

Fixation

Ideally, viable cells (1×10^6 to 5×10^6 cells in 1 ml PBS) should be analyzed on the flow cytometer. If cells are to be fixed, proceed as follows:

1. Add 1×10^6 to 5×10^6 cells in 1 ml PBS to 9 ml 70% (v/v) ethanol on ice. At this point, the fixed cells can be stored at $-20°C$ for up to 1 month.
2. Centrifuge cells (5 min, 1000 rpm).
3. Remove ethanol, resuspend cells in 1 ml PBS.
4. Analyze on flow cytometer.

Instrument Set-Up

- Excitation in blue light (e.g., 488 nm line, argon ion laser).
- Use standard settings of the light-scatter detectors.

Advantages/Limitations

This assay is easy to perform and can be readily combined with other flow cytometric assays such as those investigating surface immunofluorescence, mitochondrial potential, the lysosomal proton pump, exclusion of propidium iodide (PI), and permeability to Hoechst 33342 and other dyes.[40]

This assay has two main limitations. Differences among cytometers related to scatter detectors and optical geometry make comparison between instruments inappropriate. Decreased forward light scatter alone is not a reliable measure of apoptosis[41] because a decreased FLS signal is not unique to apoptosis. For example, G_0/G_1 cells tend to produce a lower forward light scatter signal than do S and G_2 phase cells, reflecting the smaller size of G_0/G_1 cells. Reduced intensity of scattered light is also seen from cells that are mechanically broken, necrotic cells, apoptotic bodies, cell debris, and isolated cell nuclei. It is, therefore, inappropriate to rely on FLS as a sole discriminator of apoptosis, except in situations where the cell population is known to be homogeneous in size.[41] Additionally, care must be taken to use an exponential scale (logarithmic amplifiers) during analysis of light scatter, because individual cells show great heterogeneity in light scatter signals.

10.2.2   Assays of Plasma Membane Integrity

One of the features that distinguish a dead cell from a living cell is loss of the transport function of the plasma membrane and, in many cases, loss of physical integrity of the membrane. This phenomenon has served as the basis for the development of a variety of assays for measuring apoptosis. Several of these assays are described below.

10.2.2.1   *Propidium Iodide (PI) Exclusion Assay Combined with Hydrolysis of Fluorescein Diacetate (FDA)*

Cells with impaired transport function of the plasma membrane (such as cells in the late stages of apoptosis, dead cells, mechanically broken cells, and isolated nuclei) are unable to exclude charged dyes. PI, for example, crosses the plasma membrane, binds to DNA and double-stranded RNA by intercalation, and emits an intense red fluorescence signal. Unbound and in aqueous solution, this dye shows weak fluorescence. Living cells and cells in the early stages of apoptosis exclude PI. By virtue of its simplicity, the PI exclusion assay is a particularly popular flow cytometric assay of plasma membrane integrity.

As FDA is an uncharged molecule, it readily penetrates into living cells. Once inside the cell, FDA serves as a nonfluorescent substrate for esterases, which are ubiquitous enzymes present in all cell types. Subsequent hydrolysis of FDA generates fluorescein which, by virtue of its charge, remains in the living cell and is detected based on its strong green fluorescence. Therefore, living cells and cells in the early stages of apoptosis will emit green fluorescence. The combination of staining with PI and FDA allows discrimination of living and early apoptotic cells based on green fluorescence (fluorescein) and red fluorescence (PI). This assay may be combined with analysis of light scatter as an alternative method for identifying dead cells.[39]

Many of the assays using charged cationic dyes rely on the fact that a transient decline occurs in membrane transport function prior to the cells becoming totally permeable to the charged fluorochromes. This provides a window of opportunity where the rate of uptake of a number of these dyes is elevated relative to control cells—following a brief incubation with these charged fluorochromes, necrotic cells can be identified as those generating a strong fluorescence signal, while apoptotic cells will generate a less intense signal, and living cells will generate a weak signal.[47-48]

PI Exclusion Assay Combined with Hydrolysis of FDA[39]

Reagents

- FDA stock solution: dissolve 1 mg FDA (Molecular Probes Inc.) in 1 ml acetone (make fresh)
- PI stock solution: dissolve 1 mg PI (Molecular Probes Inc.) in 1 ml distilled H_2O. Can be stored at 0 to 4°C in the dark, for up to 1 month
- Hanks' buffered salt solution (HBSS)

Staining

1. Suspend approximately 1×10^6 cells in 1 ml HBSS.
2. Add 2 μl FDA stock solution.
3. Incubate cells (37°C, 15 min).
4. Add 20 μl PI stock solution.
5. Incubate (5 min, room temperature).
6. Analyze cells on flow cytometer.

Instrument Set-Up

- Excitation in blue light (e.g., 488 nm line, argon ion laser).
- Use light-scatter signal to trigger cell measurement.
- Analyze green fluorescence at 530 ± 20 nm and red fluorescence at >620 nm.

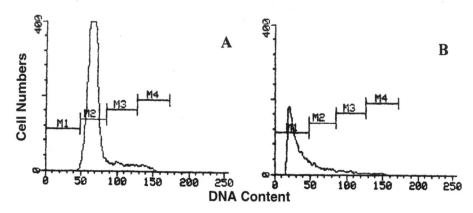

FIGURE 10.2

Induction of apoptosis in SK-N-MC cells following treatment. Control cells (A) and cells treated with retinoic acid (B) were stained with PI. Note hypodiploid peak in Panel B. (Figure courtesy of R.-M. Boustany and K. Puranam, Duke University Medical Center.)

By omitting points 2 and 3, or points 4 and 5, a simplified, single-parameter analysis of cell viability can be performed, based either on exclusion of PI alone or on hydrolysis of FDA alone, respectively.

Advantages/Limitations

While assays involving investigation of the integrity and transport function of the plasma membrane have some limitations, they have a number of advantages. They are easy to perform, inexpensive, rapid, and a powerful way to quantify apoptosis. These methods are particularly appropriate for identification of necrotic cells, cells that have suffered mechanical damage, or cells in the very late stages of apoptosis.

The main limitation relates to the fact that plasma membrane integrity is preserved during the early stages of apoptosis. This method, therefore, cannot discriminate between living cells and early apoptotic cells.

10.2.2.2 PI Exclusion Assay Followed by Counterstaining with Hoechst 33342 (HO342)

As discussed above, PI is excluded by living cells. In contrast, the bisbenz-imidazole dye HO342 crosses the plasma membrane of living and apoptotic cells and stains DNA. HO342 fluorescence is suppressed in dead cells. The latter, however, stain more intensely with PI. This assay is a modification of the method of Pollack and Ciancio.[12]

PI Exclusion Assay Followed by Counterstaining with HO342[40]

Reagents

- Fixative solution: 25% (v/v) ethanol in PBS
- PI solution: dissolve 2 mg PI in 100 ml PBS
- HO342 stock solution: dissolve 3 mg HO342 (Molecular Probes) in 10 ml distilled water
- HO342 working solution: dilute HO342 stock solution 1:4 in PBS (Ca^{2+} and Mg^{2+} free)

Fixation/Staining

1. Centrifuge 3×10^5 to 1×10^6 cells (5 min, 1000 rpm).
2. Decant medium, vortex pellet.
3. Add 100 μl PI solution, vortex, and store on ice for 30 min.
4. Add 1.9 ml fixative solution, vortex.
5. Add 50 μl HO342 working solution and vortex.

   Samples may be stored for a maximum of 3 days in this solution, at 0 to 4°C. Samples should sit for a minimum of 30 min before analysis.
6. Analyze by flow cytometry.

Instrument Set-Up

- Illuminate with UV light (351 nm line of argon ion laser, or using UG1 filter with high pressure mercury lamp).
- Measure blue fluorescence, using a combination of filters and dichroic mirrors to obtain maximum transmission at 460 ± 20 nm.
- Measure red fluorescence with long pass filter at >620 nm.

10.2.2.3 *PI Exclusion Assay Combined with Uptake of Rhodamine 123 (Rh123)*

Mitochondria lose transmembrane potential during the early stages of the apoptotic program.[49]

Rh 123 is a cationic fluorochrome that is taken up selectively by mitochondria based on their transmembrane potential.[50] Rh123 becomes concentrated in living cells and cells in the early stages of apoptosis, because of the intact plasma membrane and active mitochondria. These cells emit strong green fluorescence from the uptake of Rh 123 and exclude PI. Dead cells, in contrast, stain weakly with Rh 123 and stain strongly with PI (red).

PI Exclusion Assay Combined with Uptake of Rh 123[37]

Reagents

- PI stock solution: dissolve 1 mg PI in 1 ml distilled H_2O
- Rh 123 stock solution: dissolve 1 mg Rh 123 (Molecular Probes) in 1 ml distilled H_2O
- Tissue culture medium (or HBSS)

The Rh 123 and PI stock solutions can be stored at 0 to 4°C in the dark for up to 1 month.

Staining

1. Suspend 1×10^6 cells in 1 ml tissue culture medium (or HBSS).
2. Add 5 µl Rh 123 stock solution.
3. Incubate (5 min, 37°C).
4. Add 20 µl PI stock solution.
5. Let samples sit for 5 min, room temperature.
6. Analyze cells on flow cytometer.

Instrument Set-Up

- Excitation in blue light (e.g., 488 nm line, argon ion laser).
- Use light-scatter signal to trigger cell measurement.
- Analyze green fluorescence at 530 ± 20 nm and red fluorescence at >620 nm.

Advantages/Limitations

This method has the advantage of combining two functional assays: (1) an assay of plasma membrane integrity, and (2) an assay of mitochondrial transmembrane potential. In addition, the method also allows detection of a transient phase of cell death that occurs early in the apoptotic program when the cell becomes leaky to PI but the mitochondria remain charged, resulting in the cell staining with both PI and Rh 123.[51] In this situation, the staining with Rh123 may be more intense than in living intact cells. It has been suggested[42] that this may reflect a temporary increase in the mitochondrial transmembrane potential, concomitant with the loss of plasma membrane function. A further advantage of this method is that it can be readily combined with analysis of light scatter.

While it has been proposed that the change in mitochondrial transmembrane potential may represent the earliest identifiable biochemical change in cells undergoing apoptosis,[49] it is important to note that changes in mitochondrial transmembrane potential are not unique to apoptosis. For example, changes in mitochondrial transmembrane potential are common in cells undergoing necrosis and in cells with altered mitochondrial electron transport or proton transport (for example, following treatment with metabolic poisons, such as cyanide or azide).[41] In addition, altered transmembrane potential is not observed in all cells undergoing apoptosis (reviewed in Reference 41).

10.2.2.4   Uptake of Hoechst 33258 Combined with Exclusion of 7-Aminoactinomycin D (7-AAD)

Apoptotic cells display an increased blue HO258 fluorescence signal when compared to nonapoptotic cells. This is believed to reflect the increased plasma membrane permeability to this dye. The cells are counterstained with 7-AAD which is a charged, cationic dye that selectively labels late apoptotic and necrotic cells.

Uptake of HO258 Combined with Exclusion of 7-AAD[39]

Reagents

- HO258 stock solution: dissolve 0.1 mg HO258 (Molecular Probes) in 1 ml distilled H_2O
- PI stock solution: dissolve 0.5 mg PI in 1 ml distilled H_2O
- 7-AAD stock solution: dissolve 0.1 mg of 7-AAD (Molecular Probes) in 1 ml distilled H_2O
- HBSS solution: HBSS containing 1% (w/v) bovine serum albumin (BSA)

The H0258 and 7-AAD stock solutions can be stored in the dark at 0 to 4°C for up to 1 month.

Staining

1. Suspend approximately 1×10^6 cells in 1 ml of HBSS solution.
2. Add 10 μl HO258 stock solution.
3. Incubate cells (2 to 10 min, 37°C).
4. Cool the sample on ice.
5. Centrifuge (5 min, 1000 rpm).
6. Remove supernatant.
7. Suspend cell pellet in 1 ml HBSS solution.
8. Add 10 μl 7-AAD stock solution.
9. Keep cells on ice.
10. Analyze cells by flow cytometry.

Instrument Set-Up

- Excite HO258 with either the 351 nm or 325 nm line of the argon ion laser. Excite 7-AAD with the 488 nm line of the argon ion laser or the 543 nm line of the helium neon laser.
- Measure the blue fluorescence of HO258, using a combination of filters and dichroic mirrors to obtain maximum transmission at 460 ± 20 nm.
- Measure the fluorescence of 7-AAD with a 660 nm long pass filter.

Advantages/Limitations

This assay has two main advantages: (1) it is convenient and (2) it can be readily combined with a two color analysis of cell-surface immunofluorescence. This allows simultaneous analysis of the cell-surface phenotype of apoptotic and nonapoptotic cells by using antibodies labeled with phycoerythrin and fluorescein.[52]

A limitation of the assay is that the intensity of the HO258 fluorescence signal is dependent upon the duration of incubation with HO258. It is, therefore, necessary to determine the optimal incubation time (generally 2 to 10 min) for each cell line in order to allow discrimination of apoptotic cells.

10.2.3 Measurement of Cellular DNA Content

10.2.3.1 *Analysis of DNA Content Following Extraction of Degraded DNA from Apoptotic Cells*

In cells undergoing apoptosis, an endonuclease is activated which preferentially cleaves DNA at the internucleosomal (linker) sections, resulting in internucleosomal DNA degradation.[27,53] This generates small fragments of chromatin that can be extracted from the nucleus prior to staining the cells.

The cells are permeablized with detergents or fixed with precipitating fixatives such as alcohol or acetone. During subsequent rinsing and staining, the degraded DNA leaks from the cells. The intensity of the fluorescence signal from each cell following staining with propidium iodide (PI) or Hoechst 33258 is directly proportional to the DNA content of the cell. Therefore, apoptotic cells are detected on the basis of containing less than the diploid amount of DNA and is measured as a characteristic hypodiploid or sub-G_1 peak on a DNA histogram,[27,43,54,55] as shown in Figure 10.2. When combined with analysis of light scatter, apoptotic cells are identified as cells with lower forward light scatter signal and decreased PI fluorescence signal relative to G_1 cells.

Analysis of DNA Content Following Extraction of Degraded DNA from Apoptotic Cells[39]

Reagents

- 70% (v/v) ethanol
- PBS
- Phosphate–citric acid buffer for extraction of DNA: 192 ml 0.2 M Na_2HPO_4 + 8 ml 0.1 M citric acid; pH 7.8
- DNA staining solution (prepare fresh): 200 μg PI dissolved in 10 ml PBS + 2 mg DNase-free RNase A (if RNase is not DNase free, boil for 5 min)

Fixation/Staining

1. Add 1 × 10^6 to 5 × 10^6 cells in 1 ml PBS to 9 ml 70% (v/v) ethanol on ice. At this point, the fixed cells can be stored at –20°C for up to 1 month.
2. Centrifuge cells (5 min, 1000 rpm).
3. Remove ethanol, resuspend cells in 10 ml PBS.
4. Centrifuge cells (5 min, 1000 rpm).
5. Suspend cells in 0.5 ml PBS.
6. Add phosphate-citric acid buffer (0.2 to 1.0 ml). The volume of extraction buffer to be used depends on the extent of DNA degradation. If DNA degradation is known to be extensive, this step can

be bypassed. In cases where there is little DNA degradation, an excess of extraction buffer (1.0 ml) should be used.

7. Incubate 5 min, room temperature.
8. Centrifuge 5 min, 1000 rpm.
9. Resuspend cell pellet in 1 ml DNA staining solution.
10. Incubate cells 30 min, room temperature.
11. Analyze cells on flow cytometer.

Instrument Set-Up

- Use 488 nm laser line or blue light (BG12 filter) for excitation.
- Measure red fluorescence (>600 nm) and forward light scatter.

Advantages/Limitations

This is a simple and inexpensive assay that allows identification of apoptotic cells based on degradation of the DNA. Additionally, it allows determination of DNA ploidy and/or cell-cycle distribution of the nonapoptotic cell population. A further advantage stems from the option of using any DNA fluorochrome or instrument and the option of counterstaining other cell constituents.[39]

The main limitation of this assay is low specificity as a measure of apoptosis. The sub-G_1 peak may include not only apoptotic cells but also cells that have suffered mechanical damage, cells with altered chromatin structure (where accessibility of the fluorochrome to the DNA is impaired[56]), cells with lower DNA content (in a population of cells with heterogeneous DNA indices), cell debris, nuclear fragments, and isolated chromosomes released from mitotic cells. An additional source of error stems from analysis of cells that were treated with detergent or subjected to a hypotonic solution. These treatments lyse the cell and generate cellular components that appear in the sub-G_1 peak. If permeabilization by detergent is necessary, this source of error can be minimized by subjecting the cell to gentle permeabilization with detergent in the presence of exogenous proteins such as serum or serum albumin, as discussed in Darzynkiewicz et al.[40] The use of a logarithmic scale in DNA content distributions introduces a source of error by causing accumulation of events representing the particles mentioned above whose DNA content is less than 1% that of nonapoptotic cells. Darzynkiewicz et al. further recommend classification of apoptotic cells as those whose DNA content is no less than 10 to 20% that of normal G_1 cells.[40]

It is important to note that this method must be used on cells that are fixed in precipitating types of fixative, e.g., such as alcohols or acetone. Fixation in formaldehyde or glutaraldehyde results in cross-linking of low molecular weight DNA to other constituents and precludes its extraction. Detection of reduced DNA content following fixation with cross-linking agents is only seen in cells in the very late stages of apoptosis when DNA degradation is very advanced.

10.2.3.2   Labeling DNA Strand Breaks with BrdUTP

Activation of the apoptotic endonuclease also results in the generation of a large number of DNA strand breaks. A number of flow cytometric assays for monitoring generation of these breaks are in use. Those assays in most common use involve labeling of the 3'-OH termini in the breaks in the apoptotic cell with biotinylated or digoxygenin-conjugated nucleosides in a reaction employing exogenous terminal deoxynucleotidyl transferase (TdT) or DNA polymerase (nick translation).[34,57] Here, an alternative, recently described assay involving incorporation of BrdUTP and detection with flu-oresceinated anti-BrdUrd MoAb is presented. A comparison of the differences in fluorescence intensity of apoptotic cells following DNA strand break labeling using the various methods is presented by Li and Darzynkiewicz.[58]

Labeling DNA Strand Breaks with BrdUTP[58]

Reagents

- Fixative A: 1% (v/v) methanol-free formaldehyde (Polysciences Inc.) in HBSS, pH 7.4
- Fixative B: 70% (v/v) ethanol
- PBS
- Triton X-100
- BSA
- TdT reaction buffer (5× concentrated): 1 M potassium cacodylate (Boehringer Mannheim) + 125 mM Tris-HCl, pH 6.6, + BSA, 1.26 mg/ml.
- 10 mM Cobalt chloride ($CoCl_2$) (Boehringer Mannheim).
- TdT (Boehringer Mannheim) in storage buffer, 25 units in 1 μl.
- PI staining solution: PI (2.5 μg/ml) + DNase-free RNase A (200 μg/ml) in PBS.
- 15 mM Na_2EDTA-NaOH, pH 7.8.
- FITC-conjugated anti-BrdUrd MoAb (Becton Dickinson, San Jose, CA; clone B44).
- MoAb Staining Solution: 0.7 μg FITC-conjugated anti-BrdUrd MoAb, 0.1% (v/v) Triton X-100 + 1% BSA in PBS

Fixation/Staining

1. Fix cells in Fixative A, 15 min on ice.

2. Centrifuge (5 min, 1000 rpm).

3. Rinse with HBSS.

4. Add cell suspension to 5 ml ice-cold 70% (v/v) ethanol. At this point, the cells can be stored at –20°C for up to 4 days.

5. Rinse cells (approximately 1×10^6) twice in 5 ml PBS.

6. Centrifuge (5 min, 1000 rpm).

7. Resuspend the pellet in 50 μl of a solution containing:

    10 μl TdT reaction buffer (5×)

    0.5 μl (12.5 units) TdT in storage buffer

    5 μl 25mM CoCl$_2$

    0.25 nmoles BrdUTP (Sigma)

    Distilled H$_2$O to 50 μl

8. Incubate (60 min at 37°C).

9. Wash cells twice in 15 mM Na$_2$EDTA-NaOH, pH 7.8.

10. Incubate cells in 100 μl MoAb staining solution.

11. Counterstain with PI staining solution.

12. Incubate (30 min, room temperature in the dark).

13. Run cells on flow cytometer.

Instrument Set-Up

- Illuminate with blue light (488 nm laser line or BG12 excitation filter).

- Measure green fluorescence at 530 ± 20 nm.

- Measure red fluorescence at >610 nm.

The main advantage of this method is its specificity in identification of apoptotic cells.[39] Another advantage is that simultaneous measurement of DNA content allows determination of the cell-cycle position of both apoptotic and nonapoptotic cells. This method facilitates more sensitive and lower cost detection of DNA strand breaks than do the alternative methods.[58] This and related DNA strand break assays also facilitate detection of apoptosis in tissue sections.

A disadvantage of this assay is its complexity. A negative result may relate to a technical problem rather than absence of apoptosis. Therefore, care should be taken to utilize appropriate positive and negative controls. It is important to prefix the cells with a cross-linking agent such as formaldehyde to ensure that the degraded DNA remains in the cell despite the numerous washings that are required during cell staining.

10.3 Power and Limitations of Flow Cytometry as a Measure of Apoptosis

Use of flow cytometry ensures that precise measurements can be made at high speed on a cell-by-cell basis, thereby providing separate measurements on each individual particle within the suspension rather than averaging across the population. The ability to detect very low frequency events (i.e., detection of at least one event in 20,000) allows one to define and quantify the heterogeneity in a large cell population with high statistical precision.[2] When a multiparameter instrument is used, the instrumentation allows simultaneous measurement of a number of cellular properties. For example, a number of fluorescent dyes can be combined. Alternatively, light scatter and fluorescence signals can be collected from a given cell. The ability to sort cells for use in future experiments or assays is another strength of this technology.

Despite its many advantages, flow cytometry also has a number of limitations. The main limitation is the requirement for a sophisticated and expensive apparatus to which many investigators do not have easy access. It is also necessary to have a suspension of single particles that can flow through the instrument without compromising the smooth flow of fluid and without clogging any tubes or orifices. This requirement makes flow cytometry inappropriate for use with cell lines that are not readily dispersed into a single-cell suspension.

The following criteria should be carefully considered before deciding on the most appropriate flow cytometric apoptosis assay to use in a given set of circumstances — the cell system under investigation, the question being addressed (i.e., relationship between induction of apoptosis and cell cycle position), the nature of the inducer of cell death, and other restrictions (i.e., type of flow cytometer, whether samples need to be transported, and cost).

When apoptosis is occurring, use of one specific method alone does not guarantee detection. To identify apoptosis with a high degree of assurance, at least two apoptosis assays, each measuring different parameters, should be employed.

In recent years, evidence has accumulated showing that apoptosis can occur in the absence of one or more of the classical features of apoptosis. Therefore, caution should be exercised in interpretation of a negative result in assays of apoptosis. In addition, there is ample evidence from numerous model systems that not all assays give a positive result for apoptosis, even when apoptosis is occurring. Furthermore, as not all cells/models undergo classical apoptosis, it is always advisable to examine the morphology of the cells to confirm apoptosis in situations where conventional assays for apoptosis give conflicting results. It is specifically recommended that the cells be evaluated by light or electron microscopy, as the original description of the phenomenon of apoptosis was based on specific morphological changes.[59] An additional complication in interpretation of results is that a characteristic feature of apoptosis (i.e., DNA degradation) may occur in cells undergoing necrosis.[42]

10.4   Conclusion

In looking to the future, it is anticipated that improvements in instrumentation and staining will lead to the development of novel electro-optical techniques, improved staining capabilities, enhanced detection of fluorescence emission, and increased sorting speed. These advances will lead to exciting and important new applications of flow cytometry as a tool to monitor apoptosis.

Acknowledgments

This work was supported by The Claude D. Pepper Older Americans Independence Center (NIA 5 P60 AG11268).

References

1. Shapiro, H. M., *Practical Flow Cytometry*, 3rd ed., Wiley-Liss, New York, 1995, 542 pp.
2. Steinkamp, J. A., Flow cytometry, *Rev. Sci. Instrum.*, 55, 1375, 1984.
3. Carter, N. P. and Meyer, E. W., Introduction to the principles of flow cytometry, in *Flow Cytometry: A Practical Approach*, Ormerod, M.G., Ed., IRL Press at Oxford University Press, Surrey, U.K., 1994, Chap. 1.
4. Linnik, M. D., Hatfield, M. D., Swope, M. D., and Ahmed, N. K., Induction of programmed cell death in a dorsal root ganglia X neuroblastoma cell line, *J. Neurobiol.*, 24, 433, 1993.
5. Mirzoeva, O. K., Yaqoob, P., Knox, K. A., and Calder, P. C., Inhibition of ICE-family cysteine proteases rescues murine lymphocytes from lipoxygenase inhibitor-induced apoptosis, *FEBS Lett.*, 396, 266, 1996.

6. Carbonari, M., Cibati, M., Cherchi, M., Sbarigia, D., Pesce, A. M., Dell'Anna, L., Modica, A., and Fiorilli, M., Detection and characterization of apoptotic peripheral blood lymphocytes in human immunodeficiency virus infection and cancer chemotherapy by a novel flow immunocytometric method, *Blood*, 83, 1268, 1994.

7. Shenker, B. J., Datar, S., Mansfield, K., and Shapiro, I. M., Induction of apoptosis in human T-cells by organomercuric compounds: a flow cytometric analysis, *Toxicol. Appl. Pharmacol.*, 143, 397, 1997.

8. Sasaki, D. T., Dumas, S. E., and Engleman, E. G., Discrimination of viable and nonviable cells using propidium iodide in two color immunofluorescence, *Cytometry*, 8, 413, 1987.

9. Schmid, I., Krall, W. J., Uttenbogaart, C. H., Braun, J., and Giorgi, J. V., Dead cell discrimination with 7-amino-actinomycin D in combination with dual color immunofluorescence in single laser flow cytometry, *Cytometry*, 13, 204, 1992.

10. Telford, W. G., King, L. E., and Fraker, P. J. H., Rapid quantitation of apoptosis in pure and heterogenous cell populations using flow cytometry, *J. Immunol Methods*, 172, 1, 1994.

11. Sipe, K. J., Srisawasdi, D., Dantzer, R., Kelley, K. W., and Weyhenmeyer, J. A., An endogenous 55 kDa TNF receptor mediates cell death in a neural cell line, *Mol. Brain Res.*, 38, 222, 1996.

12. Pollack, A. and Ciancio, G., Cell cycle phase-specific analysis of viability using Hoechst 33342 and propidium iodide after ethanol preservation, in *Flow Cytometry*, Darzynkiewicz, Z. and Crissman, H. A., Eds., Academic Press, San Diego, CA, 1990, Chap. 3.

13. Novak, E. J. and Rabinovitch, P. S., Improved sensitivity in flow cytometric intracellular ionized calcium measurement using fluo-3/Fura Red fluorescence ratios, *Cytometry*, 17, 135, 1994.

14. Van Graft, M., Kraan, Y. M., Segers, I. M., Radosevic, K., De Grooth, B. G., and Greve, J., Flow cytometric measurement of [Ca^{2+}]i and pHi in conjugated natural killer cells and K562 target cells during the cytotoxic process, *Cytometry*, 14, 257, 1993.

15. Firestein, G. S., Nguyen, K., Aupperle, K. R., Yeo, M., Boyle, D. L., and Zvaifler, N. J., Apoptosis in rheumatoid arthritis: p53 overexpression in rheumatoid arthritis synovium, *Am. J. Pathol.*, 149, 2143, 1996.

16. Gomez, J., Martinez-A. C., Fernandez, B., Garcia, A., and Rebollo, A., Critical role of Ras in the proliferation and prevention of apoptosis mediated by IL-2, *J. Immunol.*, 157, 2272, 1996.

17. Hsieh, S.-C., Huang, M.–H., Tsai, C.-Y., Tsai, Y.-Y., Tsai, S.-T., Sun, K.-H., Yu, H.-S., Han, S.-H., and Yu, C.-L., The expression of genes modulating programmed cell death in normal human polymorphonuclear neutrophils, *Biochem. Biophys. Res. Commun.*, 233, 700, 1997.

18. Webb, A., Cunningham, D., Cotter, F., Clarke, P. A., di Stefano, F., Ross, P., Corbo, M., and Dziewanowska, Z., BCL-2 antisense therapy in patients with non-Hodgkin lymphoma, *Lancet*, 349, 1137, 1997.

19. Sasaki, K., Kobayashi, T., Imamura, S., Shigekura, T., Kato, R., Kawamoto, Y., Tsuji, T., and Miyama, A., Flow cytometry analysis of the Fas ligand expression of activated lymph node T-cells, *Immunol. Lett.*, 55, 11, 1997.

20. Boersma, A. W. M., Nooter, K., Burger, H., Kortland, C. J., and Stoter, G., Bax upregulation is an early event in cisplatin-induced apoptosis in human testicular germ-cell tumor cell line NT2, as quantitated by flow cytometry, *Cytometry*, 27, 275, 1997.

21. Bonfoco, E., Zhivotovsky, B., Rossi, A. D., Aguilar-Santelises, M., Orrenius, S., Lipton, S. A., and Nicotera, P., BCL-2 delay apoptosis and PARP cleavage induced by NO donors in GT1-7 cells, *NeuroReport*, 8, 273, 1996.
22. Cossarizza, A., Ortolani, C., Monti, D., and Franceschi, C., Cytometric analysis of immunosenescence, *Cytometry*, 27, 297, 1997.
23. Tonini, G. P., Mazzocco, K., di Vinci, A., Geido, E., de Bernardi, B., and Giaretti, W., Evidence of apoptosis in neuroblastoma at onset and relapse — an anlaysis of a large series of tumors, *J. Neuro-Oncol.*, 31, 209, 1997.
24. Ninomiya, Y., Adams R., Morriss-Kay, G. M., and Eto, K., Apoptotic cell death in neuronal differentiation of P19 EC cells: cell death follows reentry into S phase, *J. Cell. Physiol.*, 172, 25, 1997.
25. Matthews, C. C., and Feldman, E. L., Insulin-like growth factor 1 rescues SH-SY5Y human neuroblastoma cells from hyperosmotic induced programmed cell death, *J. Cell. Physiol.*, 166, 323, 1996.
26. Endresen, P. C., Prytz, P. S., and Aarbakke, J., A new flow cytometric method for discrimination of apoptotic cells and detection of their cell cycle specificity through staining of F-actin and DNA, *Cytometry*, 20, 162, 1995.
27. Nicoletti, J., Migliorati, G., Pagliacci, M. C., Grignani, F., and Riccardi, C., A rapid and simple method for measuring thymocyte apoptosis by propidium iodide staining and flow cytometry, *J. Immunol. Methods*, 139, 271, 1991.
28. Rozzo, C., Chiesa, V., Caridi, G., Pagnan, G., and Ponzoni, M., Induction of apoptosis in human neuroblastoma cells by abrogation of integrin-mediated cell adhesion, *Int. J. Cancer*, 70, 688, 1997.
29. Iwadate, Y., Fujimoto, S., Sueyoshi, K., Namba, H., Tagawa, M., and Yamaura, A., Prediction of drug cytotoxicity in 9L rat brain tumor by using flow cytometry with a deoxyribonucleic acid-binding dye, *Neurosurgery*, 40, 782, 1997.
30. Desole, M. S., Sciola, L., Delogu, M. R., Sircana, S., Migheli, R., and Miele, E., Role of oxidative stress in the manganese and 1-methyl-4-(2'-ethylphenyl)-1,2,3,6-tetrahydropyridine-induced apoptosis in PC12 cells, *Neurochem. Int.*, 31, 169, 1997.
31. Lane, S. C., Jolly, R. D., Schmechel, D. E., Alroy, J., and Boustany, R.-M., Apoptosis as the mechanism of neurodegeneration in Batten's disease, *J. Neurochem.*, 67, 677, 1996.
32. Ni, Y., Zhao, B., Hou, J., and Xin, W., Preventive effect of Ginkgo biloba extract on apoptosis in rat cerebellar neuronal cells induced by hydroxyl radicals, *Neurosci. Lett.*, 214, 115, 1996.
33. Olano, J. P., Wolf, D., Keherly, M., and Gelman, B. B., Quantifying apoptosis in banked human brains using flow cytometry, *J. Neuropath. Exp. Neurol.*, 55, 1164, 1996.
34. Meyaard, L., Otto, S. A., Jonker, R. R., Mijnster, M. J., Keet, I. P. M., and Miedema, F., Programmed death of T cells in HIV-1 infection, *Science*, 257, 217, 1992.
35. Hess, K. L., Babcock, G. F., Askew, D. S., and Cook-Mills, J. M., A novel flow cytometric method for quantifying phagocytosis of apoptotic cells, *Cytometry*, 27, 145, 1997.
36. Hatanaka, Y., Suzuki, K., Kawasaki, Y., Endo Y., Taniguchi, N., and Takei, N., A role of peroxides in Ca^{2+} ionophore-induced apoptosis in cultured rat cortical neurons, *Biochem. Biophys. Res. Commun.*, 227, 513, 1996.
37. Li Y., Maher, P., and Schubert, D., A role for 12-lipoxygenase in nerve cell death caused by glutathione depletion, *Neuron*, 19, 453, 1997.

38. Blankenberg, F. G., Katsikis, P. D., Storrs, R. W., Beaulieu, C., Spielman, D., Chen, J. Y., Naumovski, L., and Tait, J. F., Quantitative analysis of apoptotic cell death using proton nuclear magnetic resonance spectroscopy, *Blood*, 89, 3778, 1997.

39. Darzynkiewicz, Z. and Li, X., Measurements of cell death by flow cytometry, in *Techniques in Apoptosis*, Cotter, T.G. and Martin, S. J., Eds., Portland Press, London, U.K., 1996, Chap. 4.

40. Darzynkiewicz, Z., Juan, G., Li, X., Gorczyca, W., Murakami, T., and Traganos, F., Cytometry in cell necrobiology: analysis of apoptosis and accidental cell death (Necrosis), *Cytometry*, 27, 1, 1997.

41. Mesner, P. W., Jr. and Kaufmann, S. H., Methods utilized in the study of apoptosis, *Adv. Pharmacol.*, 41, 57, 1997.

42. Darzynkiewicz, Z., Li, X., Gong, J., Hara, S., and Traganos, F., Analysis of cell death by flow cytometry, in *Cell Growth and Apoptosis*, Studzinski, G. P., Ed., Oxford University Press, New York, 1995, Chap. 8.

43. Darzynkiewicz, Z, Li, X., and Gong, J., Assays of cell viability: discrimination of cells dying by apoptosis, *Methods Cell Biol.*, 41,15, 1994.

44. Salzman, G. C., Singham, S. B., Johnston, R. G., and Bohren, C. F., Light scattering and cytometry, in *Flow Cytometry and Sorting*, Melamed, M. R., Lindmo, T., and Mendelsohn, M. L., Eds., Wiley-Liss, New York, 1990, Chap. 5.

45. Ormerod, M. G., Cheetham, F. P. M., and Sun, X.-M., Discrimination of apoptotic thymocytes by forward light scatter, *Cytometry*, 21, 300, 1995.

46. Swat, W., Ignatowicz, L., and Kisielow, P., Detection of apoptosis of immature CD4$^+$8$^+$ thymocytes by flow cytometry, *J. Immunol. Methods*, 137, 79, 1981.

47. Belloc, F., Dumain, P., Boisseau, M. R., Jalloustre, C., Reiffers, J., Bernard, P., and Lacombe, F., A flow cytometric method using Hoechst 33342 and propidium iodide for simultaneous cell cycle analysis and apoptosis determination in unfixed cells, *Cytometry*, 17, 59, 1994.

48. Philpott, N. J., Turner, A. J. C., Scopes, J., Westby, M., Marsh, J. C. W., Gordon-Smith, E. C., Dalgerish, A. G., and Gibson, F. M., The use of 7-aminoactinomycin D in identifying apoptosis: simplicity of use and broad spectrum of application compared with other techniques, *Blood*, 87, 2244, 1996.

49. Marchetti, P., Susin, S. A., Decaudin, D., Gamen, S., Castedo, M., Hirsch, T., Zamzami, N., Naval, J., Senik, A, and Kroemer, G., Apoptosis-associated derangement of mitochondial function in cells lacking mitochondial DNA, *Cancer Res.*, 56, 2033, 1996.

50. Johnson, L. U., Walsh, M. L., and Chen, L. B., Localization of mitochondria in living cells with rhodamine 123, *Proc. Natl. Acad. Sci. U.S.A.*, 77, 990, 1980.

51. Darzynkiewicz, Z., Traganos, F., Staiano-Coico, L., Kapuscinski, J., and Melamed, M. R., Interactions of rhodamine 123 with living cells studied by flow cytometry, *Cancer Res.*, 42, 799, 1982.

52. Schmid, I., Uttenbogaart, C. H., and Giorgi, J. V., Senstive method for measuring apoptosis and cell surface phenotype in human thymocytes by flow cytometry, *Cytometry*, 15, 12, 1994.

53. Gong, J., Traganos, F., and Darzynkiewicz, Z. A selective procedure for DNA extraction from apoptotic cells applicable for gel electrophoresis and flow cytometry, *Anal. Biochem.*, 218, 314, 1994.

54. Elstein, K. H., Thomas, D. J., and Zucker, R. M., Factors affecting flow cytometric detection of apoptotic nuclei by DNA analysis, *Cytometry*, 21, 170, 1995.

55. Elstein, K. H. and Zucker, R. M., Comparison of cellular and nuclear flow cytometric techniques for discriminating apoptotic subpopulations, *Exp. Cell Res.*, 211, 322, 1994.

56. Darzynkiewicz, Z., Traganos, F., Kapuscinski, J., Staiano-Coico, L., and Melamed, M. R., Accessibility of DNA *in situ* to various fluorochromes: relationship to chromatin changes during erythroid differentiation of Friend leukemia cells, *Cytometry* 5, 355, 1984.

57. Gorczyca, W., Bruno, S., Darzynkiewicz, R. J., Gong, J., and Darzynkiewicz, Z., DNA strand breaks occurring during apoptosis: their early *in situ* detection by the terminal deoxynucleotidyl transferase and nick translation assays and prevention by serine protease inhibitors, *Int. J. Oncol.*, 1, 639, 1992.

58. Li, X. and Darzynkiewicz, Z., Labelling DNA strand breaks with BrdUTP. Detection of apoptosis and cell proliferation, *Cell Prolif.*, 28, 571, 1995.

59. Kerr, J. F. R., Wyllie, A. H., and Currie, A. R., Apoptosis: a basic biological phenomenon with wide-ranging implications in tissue kinetics, *Br. J. Cancer*, 26, 239, 1972.

11

Methods Used to Study Protease Activation During Apoptosis

Scott H. Kaufmann, Peter W. Mesner, Jr., L. Miguel Martins,
Timothy J. Kottke, and William C. Earnshaw*

CONTENTS

* Supported in part by grant R01 CA69008 from the National Cancer Institute and a "Programa Gulbenkian de Doutoramento em Biologia e Medicina" studentship to L.M.M. S.H.K. is a Scholar of the Leukemia Society of America. W.C.E. is a Principal Fellow of the Welcome Trust.

11.1   Background

A growing body of evidence suggests that proteases, particularly caspases, play important roles in the initiation and completion of apoptotic cell death (see Chapter 6 by Mesner and Kaufmann). This claim is based on a wide variety of experiments that use disparate and often complementary techniques. In the present chapter, we describe some of these techniques, examine their potential uses, and discuss their potential limitations.

11.2   Gene Deletion

Perhaps the strongest evidence in favor of a critical role for proteases in controlling the cell death process comes from studies in the nematode *Caenorhabditis elegans*. Deletion of the *ced-3* gene, which encodes a caspase, results in failure of developmental cell deaths that would ordinarily occur.[1,2] A complete description of the methodology for performing gene knockouts is beyond the scope of the present chapter. However, a brief discussion of this approach — particularly the limitations of this approach — appears to be in order.

At least 10 human caspases have been identified to date (see Chapter 6). For several of these enzymes, alternatively spliced variants have also been described. Although studies reported to date indicate that different caspases have different substrate preferences,[3-5] these same studies also indicate considerable overlap in ability to cleave individual substrates. Accordingly, the demonstration of a developmental phenotype in a caspase knockout mouse is informative; but the lack of a phenotype cannot rule out a role for a deleted protein in developmentally regulated cell death because of the possibility that closely related enzymes compensate for the loss of a particular caspase. The observation that caspase-3$^{-/-}$ mice have supernumerary neurons and generally die *in utero* with neurological abnormalities[6] indicates that caspase-3 plays an important role in developmentally regulated neuronal apoptosis that is not usually or efficiently subsumed by other caspases in this tissue at critical times during development. On the other hand, the ability of lymphoid cells from caspase-3$^{-/-}$ mice to undergo apoptosis[6] indicates one of two possibilities: (a) caspase-3 might not play a role in lymphoid apoptosis; or (b) other caspases might subsume the critical functions of caspase-3 in its absence. This possibility of redundant functions must be kept in mind as ongoing studies examine the role of other caspases and caspase-activating polypeptides in programmed cell death (PCD).

In addition, it must be remembered that gene knockout experiments by themselves are designed only to elucidate the role of the deleted protein in developmentally regulated cell deaths. Only by subjecting the knockout mice to additional stresses does it become possible to determine the role of the deleted protein in other types of cell death. This is elegantly illustrated by the recent demonstration that deletion of caspase-1, which has no effect on nervous system development,[7,8] results in diminished infarct size after cerebral artery occlusion.[9] Accordingly, it appears that caspase-1 contributes to postischemic cell death in the central nervous system. This result, however, also illustrates yet another limitation of gene knockout studies: they do not indicate how the deleted polypeptide contributes to the observed phenotype. In the case of caspase-1 and cerebral infarction, it has been proposed that the contribution might be indirect, with caspase-1 contributing to secretion of the inflammatory cytokine interleukin-1β, which in turn contributes to cerebral edema and further cellular damage.[9]

11.3   Overexpression Studies

As an alternative approach, forced overexpression has also been utilized to explore the role of various caspases in apoptosis. Studies have indicated that overexpression of active caspases results in apoptosis of the transfected cells (e.g., References 10 to 16) and mutation of the active site cysteine, which abolishes the potential protease activity of the translated gene product, prevents this induction of apoptosis by the transfected cDNA.[10] Although these results have often been interpreted as providing evidence that caspases play an active role in apoptosis, more cautious interpretation appears to be required. Two separate studies have indicated that loading of cells with nonspecific proteases such as trypsin, chymotrypsin, or proteinase K also results in apoptotic cell death.[17,18] Accordingly, the occurrence of apoptosis after transfection of a particular cDNA can be taken as evidence for toxicity of the gene product, but not as evidence that the gene product plays a direct role in apoptosis.

11.4   Cleavage of Candidate Substrates

If the transfection experiments can be misleading, then which approaches can be utilized to examine the role of proteases in apoptosis? Three complementary techniques have been widely utilized: (1) examination of the cleavage of candidate substrates in intact cells; (2) assessment of the cleavage of small fluorogenic or chromogenic substrates after cell disruption; and (3) evaluation of the effect of divers protease inhibitors on various aspects of the apoptotic process. When thoughtfully performed, these techniques can provide evidence for the role of proteases in apoptosis. As illustrated below, however, each of these techniques also has important limitations that must be kept in mind.

At first glance, evaluating the cleavage of cellular polypeptides during apoptosis appears to be straightforward: cells are treated with a proapoptotic stimulus, lysed in a suitable buffer, subjected to SDS-PAGE, transferred to a solid support, and probed with antibodies using standard techniques that are widely available.[19] Utilizing this approach, a rapidly growing number of polypeptides have been shown to be cleaved during apoptosis (Table 11.1). A number of potential limitations, however, must be kept in mind. First, proteolysis can also occur after rupture of lysosomal membranes during necrosis. Accordingly, the occurrence of proteolysis — even proteolysis of the polypeptides listed in Table 11.1 — cannot necessarily be taken as evidence that cells have died an apoptotic death. Second, cleavage of a particular polypeptide does not identify the protease responsible for the cleavage. This is perhaps best illustrated in the case of poly(ADP-ribose) polymerase (PARP), a polypeptide that is cleaved during apoptotic cell death in a wide variety of cell types.[20,21] PARP is cleaved at asparate 216[22] by catalytic amounts of caspase-3 and caspase-7.[13,23,24] Thus, cleavage of PARP in intact cells has often been assumed to represent activation of either or both of these pro- teases. However, PARP or PARP subfragments have also been shown to be cleaved by caspases -1, -4, -6, -8, -9, and -10 at high enzyme:substrate ratios.[16,24-27] Because little is known about the concentration and subcellular distribution of active forms of the caspases in apoptotic cells, the possibility that PARP cleavage could reflect the action of caspases -1, -4, -6, -8, -9 or -10 under some circumstances cannot be excluded at present. Thus, the cleavage of PARP does not by itself demonstrate that a "caspase-3-like" protease has been activated.

It is also important to consider the possibility that substrates might be cleaved by proteases that are activated downstream of caspases. Although lamin B_1 is cleaved relatively efficiently by caspase-6[28] and somewhat less efficiently by caspase-3,[29] this polypeptide can also be cleaved *in vitro* by a calcium-dependent serine protease.[30] Even though it remains to be demonstrated that this latter protease is responsible for lamin B_1 cleavage to the fragments observed in most types of apoptotic cells, cleavage of lamin B_1 by itself can no longer be taken as definitive evidence of caspase-6 activation. In short, cleavage of candidate substrates demonstrates that proteases are activated during apoptosis but does not ipso facto demonstrate the identity of the protease(s) in apoptosis.

In addition, the demonstration that a particular substrate is cleaved does not establish the role of that cleavage in the apoptotic process. As outlined in a recent review by Villa et al.,[31] the cleavage of a particular intracellular polypeptide during apoptosis might reflect any of three different possibili- ties: (1) cleavage might disable the polypeptide from performing its normal function and thereby facilitate subsequent apoptotic events; (2) cleavage might generate fragments with novel biological activities that enable subse- quent apoptotic events; and (3) cleavage might reflect the presence of pro- teolytic activity within the cell without having any direct consequence on apoptosis, i.e., a bystander effect. Once a cleavage event is demonstrated, further studies are required to distinguish between these possibilities. Two examples illustrate some of the approaches that have been utilized.

Lamins, the major structural polypeptides of the nuclear envelope, are degraded early during the course of apoptosis.[20,32-34] Sequencing of the cleaved fragment of lamin A demonstrated that this cleavage occurs within the coiled–coiled domain that is involved in the protein–protein interactions of these intermediate filament polypeptides,[28] raising the possibility that cleavage might disrupt the structural integrity of the lamin polymer. In additional experiments, it was shown that inhibition of lamin cleavage abolished nuclear fragmentation during apoptosis *in vitro*[35,36] and slowed nuclear fragmentation *in vivo*,[37] suggesting that the cleavage of the lamins facilitated nuclear fragmentation.

Cleavages of other substrates appear to generate novel enzyme activities that play an active role in further apoptotic events. For example, caspases generate constitutively active fragments of protein kinase $C\delta$[38] and PITSLRE kinase[39,40] during apoptosis. To examine the potential role of these fragments in subsequent apoptotic events, cDNAs encoding the constitutively active kinase domains have been transfected into cells. These experiments indicate that the forced expression of the constitutively active kinase fragments results in apoptosis in the absence of any other proapoptotic stimulus.[39,41] As indicated in the preceding section, these types of overexpression experiments can be difficult to interpret. It is possible that the transfected protein is simply toxic to cells without having any direct role in the apoptotic process. Additional experiments in cells lacking the active kinase as a result of treatment with inhibitors, transfection with dominant negative mutants, or exposure to antisense constructs have been required to support the conclusion that these kinases play essential roles in apoptotic events.

As illustrated by these examples, the demonstration that a candidate substrate has been cleaved merely indicates that one or more proteases become active during apoptosis. This observation by itself is limited, because it does not necessarily indicate the identity of the protease or the role of the cleavage in the apoptotic process. In addition to these limitations, two other potential disadvantages of this approach need to be kept in mind. First, it relies on the availability of antibodies, which are often costly and sometimes unavailable from commercial sources. Second, it is relatively labor-intensive and is not amenable to high through-put screening, e.g., in the setting in which one is looking for inhibitors of apoptosis in neurons.

11.5   Detection of Protease Activity Using Small Substrates

Sequencing of the sites at which protease substrates are cleaved[22,28] has facilitated the development of rapid and convenient fluorogenic and chromogenic

TABLE 11.1
Partial List of Polypeptides Cleaved during Programmed Cell Death*

Polypeptides	Cleavage site	Enzyme responsible	Proposed effect of cleavage	References
Polypeptides Whose Cleavage Inhibits Function				
PARP	DEVD/G	caspase-3, -7, and/or -9	Separates DNA binding and catalytic domains, inhibiting synthesis of poly(ADP-ribose)	16,20–23, 78,79
DNA-PKcs	DEVD/N	caspase-3 and ?	Inactivation	80–83
Rb retinoblastoma protein	DEAD/G	?	Inhibition of cell cycle progression	84,85
Focal adhesion kinase	DQTD/S	caspase-3 and 7	Detachment from basement membrane	86
Replication Factor C	DEVD/G	caspase-3	Inhibition of DNA replication	87
MDM2	DVPD/G	caspase-3, -6, or -7	Enhancement of P53 function	88
Polypeptides Whose Cleavage Activates Enzymatic Activity				
Protein kinase C delta	DMQD/M	caspase-3 and ?	Constitutively active kinase domain	38,41
Protein kinase C theta	DEVD/K	caspase-3 and ?	Constitutively active kinase domain	89
PITSLRE kinase	YVPD/S	caspase-3 and caspase-1	Constitutively active kinase domain	39,40
p21-activated kinase 2	SHVD/G	?	Constitutively active kinase domain	90
Gelsolin	DQTD/G	caspase-3	Calcium-insensitive actin filament cleavage	91
ICAD endonuclease inhibitor	DEPD/S	caspase-3	Liberates active CAD endonuclease	92

Polypeptides Whose Cleavage Alters Structural Integrity or Association/Dissociation Properties of Structural Proteins

Lamin A	VEID/N	caspase-6	Nuclear lamina disassembly	28,32
Lamin B	DSVD/S	caspase-6	Nuclear lamina disassembly	20,32,35,93
Gas-2	SRVD/G	?	Cytoskeleton rearrangement	94
Fodrin	?	caspase-3 and ?	Plasma membrane blebbing	95,96

Polypeptides Whose Cleavage Does Not Have Demonstrated Biological Effects to Date

Huntingtin	DSVD/L	caspase-3 and ?	No known effect	97
70 kDa protein of U1snRNA	DGPD/G	caspase-3 and ?	Inhibition of RNA processing?	81,98
HnRNP proteins C1 and C2	?	caspase-3 and ?	Inhibition of RNA processing?	99
Steroid response element binding proteins	DEPD/S	caspase-3 and caspase-7	Nonphysiological cleavage	100–103
D4-GDP dissociation inhibitor	DELD/S	caspase-3 and ?	No demonstrated effect	104

Polypeptides Whose Cleavage Has Been Reported during Apoptosis but Might not Be Caspase-Mediated

Topoisomerase I	Unknown	Unknown	Unknown	20,93,105
Topoisomerase II	Unknown	Unknown	Unknown	20,106,107
APC	Unknown	Unknown	Unknown	108
Plasminogen activator inhibitor-2	Unknown	Unknown	Unknown	109
NuMA	Unknown	Unknown	Nuclear shape changes	93,110,111

* For updated list of caspase substrates and potential substrates, see Ref. 112.

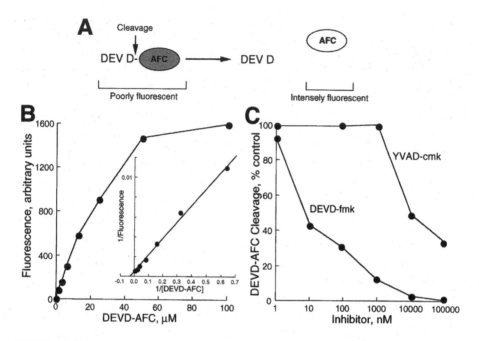

FIGURE 11.1

Detection of caspase activity using fluorogenic substrates. **A,** Schematic showing release of AFC from tetrapeptide-coupled substrate as a consequence of caspase action. **B,** Examination of product release as a function of substrate concentration. Cytosol from apoptotic HL-60 leukemia cells[73] was incubated for 2 h at 37°C with the indicated final concentration of DEVD-AFC, diluted, and assayed for fluorescence as described in Section 11.5.1. **Inset,** Lineweaver-Burke plot of the data in panel B. **C,** Inhibition of DEVD-AFC cleavage activity in HL-60 cytosol by preincubation with DEVD-fluoromethylketone or YVAD-chloromethylketone for 5 min before addition of the substrate DEVD-AFC.[42]

assays for the proteases that are activated during PCD. A protocol currently in use in our laboratories[42] is described below. Results obtained using this protocol are illustrated in Figure 11.1.

11.5.1 Protocol for Fluorogenic Assay

1. **Prepare Cytosol or Other Subcellular Fractions** (modified from Reference 42). After the proapoptotic treatment, all steps are performed at 4°C. Cells are released by trypsinization (if adherent), sedimented at 200 × g for 10 min, washed twice in calcium-/magnesium-free phosphate-buffered saline, and resuspended in a small volume of buffer A (25 mM HEPES [pH 7.5 at 4°C], 5 mM $MgCl_2$, 5 mM EDTA, 1 mM EGTA supplemented immediately before use with 1 mM PMSF, 1 mM DTT, 10 µg/ml pepstatin A, and 10 µg/ml leupeptin) (see notes 1 and 2). After a 20 min incubation, cells are lysed with 20 to 30 strokes in a tight-fitting Dounce homogenizer.

Following removal of nuclei by sedimentation (800 × g for 10 min or 16,000 × g for 3 min), the supernatant is sedimented at 280,000 × g_{max} for 60 min in a Beckman TL-100 ultracentrifuge. The supernatant (cytosol) is then frozen in 50-μl aliquots at –70°C (see note 3). Other subcellular fractions can be isolated from the homogenate using standard techniques.[43]

2. **Assay the Samples for Ability to Cleave Fluorogenic or Chromogenic Substrates**. Aliquots of cytosol or other subcellular fractions containing 50 μg of cytosolic or nuclear protein (estimated by the bicinchoninic acid method[44]) in 50 μl buffer A are thawed on ice and diluted with 225 μl of freshly prepared buffer B (25 mM HEPES [pH 7.5], 0.1% [w/v] CHAPS, 10 mM DTT, 100 U/ml aprotinin, 1 mM PMSF) containing 100 μM substrate and incubated at 37°C (see notes 4 and 5). Reactions are terminated by addition of 1.225 ml ice-cold buffer B. Reagent blanks containing 50 μl of buffer A and 225 μl of buffer B are incubated at 37°C for 2 h, then diluted with 1.225 ml ice-cold buffer B. Fluorescence is measured in a fluorometer, using an excitation wavelength of 360 nm and emission wavelength of 475 nm (see note number 6). Standards containing varying amounts of the liberated fluorophore (e.g., 0 to 1500 pmoles of 7-amino-4-trifluoromethylcoumarin) can be utilized to determine the absolute amount of fluorochrome released. Applications of this assay are depicted in Figure 11.1B and 11.1C.

11.5.2   Notes on the Procedure

1. The procedure as outlined was designed for the examination of caspase activity in various subcellular fractions. It is also possible to assay caspase activity after disrupting cells by incubation in buffer B. Because this buffer contains a zwitterionic detergent, membranous organelles are disrupted; and it is not necessary to perform ultracentrifugation after removal of the nuclei by sedimentation at 800 × g for 10 min or 16,000 × g for 3 min.

2. Volumes utilized might need to be adjusted, depending on the availability of adequate cell numbers. As a starting point, we typically add 1 ml buffer A to 10^8 cells. The resulting cytosolic extracts typically have a protein concentration of 4 to 8 mg/ml.

3. Control experiments from our laboratory indicate that asp-glu-val-asp-AFC (DEVD-AFC) cleavage activity is stable for at least 3 months at –70°C.

4. The choice of substrate concentration is extremely important. If the activity in two samples is being compared and the substrate concentration is below the K_m for the enzyme, changes in amount of product released might reflect either an alteration in v_{max} (e.g.,

altered number of enzyme molecules) or a change in K_m (altered affinity for the substrate). Although either of these outcomes is potentially interesting, experiments demonstrating changes in amount of product released are typically interpreted as showing that the number of active enzyme molecules (v_{max}) has been altered. This interpretation is valid only if the activity has been assayed under conditions where the substrate is saturating. A procedure for estimating the K_m is depicted in Figure 11.1B.

5. Length of incubation. If suitable equipment is available, continuous monitoring of fluorochrome release can be utilized to examine kinetics of product release and/or kinetics of enzyme inhibition.[45,46] If this equipment is not available, the assay can be run as an endpoint assay, with cold buffer B being added at a fixed time point to stop the reaction. When this latter approach is utilized, it is important to perform control experiments to verify that the release of substrate is a linear function of time. Common causes of non-linearity include "substrate exhaustion" if the ratio of enzyme/substrate is too high, as well as lability of enzymes under the reaction conditions. Utilizing the conditions described above, we have found that the release of AFC from DEVD-AFC is linear for at least 2 h.

6. Although peptide derivatives of 7-aminomethylcoumarin (AMC) and 7-amino-4-trifluoromethylcoumarin (AFC) are commonly utilized for these assays, peptide derivatives containing the two fluorophores AMC and dinitrophenol (dnp) have also been described.[47] The basis for using these latter substrates is the observation that the substrates exhibit low fluorescence due to quenching of the closely spaced fluorophores, whereas the two products each exhibit higher fluorescence because the fluorophores are no longer close enough to interact. Alternatively, the same assay can be adapted to a colorimetric format by using a chromogenic substrate, e.g., peptide linked to *p*-nitroanaline (pNA), although the sensitivity is diminished somewhat compared to the fluorimetric assay.

11.5.3 Strengths and Limitations of the Assay

This type of assay can be readily adapted to a microtiter plate format and utilized to examine various small molecules or macromolecules as potential inhibitors of apoptotic protease activity. For such an assay, one merely adds the test compounds or diluent to replicate aliquots of an extract containing the active protease, incubates for a period of time, and then adds substrate to initiate the reaction described above. An example of this approach is illustrated in Figure 11.1C. The ability to rapidly screen a large number of compounds (or column fractions) for inhibitory activity is clearly a strength of this approach.

When applied to the study of protease activation (see below), one simply adds potential activators to various aliquots of cytosol from control (nonapoptotic cells), incubates for the desired length of time, and adds substrate to the reaction mixture. Once again, this technique has the potential to rapidly screen a large number of compounds or column fractions.

Despite this versatility, an important limitation must also be kept in mind. Because of overlapping substrate specificities, any of a variety of proteases present in cellular extracts might be responsible for cleaving commercial substrates. For example, DEVD-AFC, a substrate based on the cleavage site in PARP,[22] is efficiently cleaved by caspases -3, -6, and -7, but also cleaved by caspases -8 and -10.[5,13,23,24,48,49] Likewise, VEID-AFC, which is efficiently cleaved by caspase-6, is also cleaved by recombinant caspase-3, albeit less efficiently[3] (T.J.K. and S.H.K., unpublished observations). Similar overlap in substrate specificities is observed with other synthetic caspase substrates as well.[3] Accordingly, the demonstration that a particular fluorogenic or chromogenic substrate is cleaved in a cellular extract does not indicate the identity of the protease or proteases responsible for the cleavage. In the most striking example published to date, the threonine-dependent protease activity of the proteosome was shown to cleave YVAD-AMC,[50] although the K_m (700 μM) is much higher than the K_m of caspase-1 for the same substrate.[50,51]

11.6 Protease Inhibitors

As an alternative approach, the potential role of proteases in PCD has been examined using a variety of low molecular weight protease inhibitors. Cells are typically preincubated in the presence of these low molecular weight inhibitors, then subjected to proapoptotic conditions in their continued presence. As illustrated in Chapter 6 (Mesner and Kaufmann), the observation that neuronal PCD is prevented by certain protease inhibitors — notably caspase inhibitors — has provided strong support for the role of caspases in neuronal cell death. Although a complete review of protease inhibitors is beyond the scope of this chapter, a few comments might help place the results obtained with inhibitors in perspective.

As outlined in Reference 51, several different approaches have been utilized to inhibit caspases. Although these enzymes, like other cysteine proteases, are sensitive to nonspecific sulfhydryl alkylating agents such as *N*-ethylmaleimide and iodoacetamide,[52] peptide derivatives containing a carboxyl-terminal aspartate have generally been employed in an attempt to improve potency and specificity of the inhibitors. When these peptides are derivatized to aldehydes or nitriles, the resulting compounds are reversible inhibitors. In contrast, when the same peptides are derivatized to halome-thylketones or acyloxymetylketones, the resulting compounds are irreversible inhibitors. (For a review of the chemistry involved, see Reference 51) In applying these inhibitors using the approach described above, several potential limitations must be kept in mind.

Even with the so-called specific inhibitors, it is usually not clear which enzymes have been inhibited to produce the observed effect. For example, early studies of PCD in rat thymocytes demonstrated that tosyl-L-lysine chloromethylketone (TLCK) and tosyl-L-phenylalanine chloromethylketone (TPCK) inhibited PCD induced by the topoisomerase II poison tenipo-side.[53-55] Although TLCK and TPCK were originally synthesized as mechanism-based inhibitors of the serine proteases trypsin and chymotrypsin, respectively, both of these agents inhibit sulfhydryl-dependent proteases such as cathepsin B[56] as well as multiple caspases.[13,35] Accordingly, the observation that TPCK or TLCK inhibits teniposide-induced apoptosis cannot be taken as evidence that a chymotrypsin-like or trypsin-like enzyme is involved in the process.

This lack of specificity is not unique to the chloromethylketones. The aldehyde-based inhibitors N-acetyl-leu-leu-norleucinal and N-acetyl-leu-leu-methional, which were originally utilized as inhibitors of cysteine-dependent proteases calpain I and II, have been more recently been shown to inhibit the threonine-dependent protease activity of the proteosome as well.[57] In fact, the inhibition of apoptosis in neuronal cells by these compounds, but not the calpain inhibitor calpastatin, has been cited as evidence to support the potential role of the proteosome in neuronal apoptosis.[58]

The preceding examples indicate that some of the currently available protease inhibitors inhibit members of multiple mechanistically distinct protease families. Accordingly, inhibition of PCD by these inhibitors does not necessarily indicate the family of proteases involved in the process. Further experience is required to determine whether the newer agents that are currently viewed as "specific" inhibitors will in fact be specific for a single class of proteases like the caspases.

Even if currently available caspase inhibitors are eventually shown to preferentially target caspases, the observation that a particular inhibitor slows or abolishes PCD does not indicate which individual protease has been targeted. For example, YVAD-aldehyde will inhibit caspase-1 with a K_i of 3 nM *in vitro*; yet 2700 nM YVAD-aldehyde is required to inhibit production of interleukin-1β in intact cells.[51] If it were not known that total inhibition of caspase-1 activity (e.g., gene knockout) abolishes interleukin-1β production,[7,8] the requirement for such high concentrations of YVAD-aldehyde to inhibit interleukin-1β production might be viewed as evidence that a less sensitive protease plays a more important role in the process than caspase-1 does. In the case of DEVD-aldehyde, there is likewise a disparity between the nanomolar K_i of the compound against caspase-3 *in vitro*[59] and the 10 to 100 μM concentrations required to inhibit PCD in intact cells.[23,60] Although these disparities have been attributed to poor penetration of the inhibitors into intact cells, measurements of intracellular inhibitor concentrations that would be required to confirm this explanation have not been reported. If limited cell penetration is responsible for the poor potency of low molecular weight caspase inhibitors in intact cells, then coupling of the peptide portions of these inhibitors to permeability-enhancing peptides (e.g., the hydrophobic region of the signal peptide of Kaposi fibroblast growth factor or the antennapedia penetration sequence that carries impermeant molecules directly into the cytoplasm of cells[61]) would be expected to result in more effective inhibitors. Results of studies with these new cell-permeable inhibitors remain to be reported.

Even if the potency of these inhibitors in intact cells is improved, however, it must be kept in mind that the DEVD-aldehyde has been shown to inhibit caspase-1 (K_i 15 nM) and caspase-4 (K_i 135 nM) about as well as the caspase-3-like caspase-7 (K_i 35 nM).[4] Accordingly, inhibition of a cellular process by DEVD-aldehyde cannot be taken as evidence of involvement of caspase-3 or even a caspase-3 like enzyme. Because of overlapping substrate specificity, similar limitations might apply to the interpretation of results obtained with other caspase inhibitors as well.

Two final points make the interpretation of inhibitor data even more problematic. First, a recent study utilizing purified caspases *in vitro* demonstrated that the specificity of irreversible caspase inhibitors for certain family members can be lost when the inhibitors are utilized at high concentrations and/or for prolonged periods of time. In particular, the inhibitors Z-VAD-[(2,6-dichlorobenzoyl)oxy]methyl ketone and Ac-YVAD-nitrile were shown to have K_i values 10- to 200-fold lower for caspases -1 and -4 than for caspases -3 and-7. Nonetheless, activities of caspases -3 and -7 were also irreversibly inhibited by prolonged (i.e., 1 h) incubation with these inhibitors.[4] Because the inhibition is irreversible, these types of inhibitors do not have to have a high affinity to eventually inhibit the activity. This property must be kept in mind in interpreting experiments utilizing prolonged incubations with irreversible inhibitors. Second, it must be remembered that so-called "protease inhibitors" might also inhibit other classes of enzymes. TPCK, which inhibits chymotrypsin-like serine proteases, sulfhydryl-dependent cathepsins, and at least some caspases (see above), can also enhance the phosphorylation of a 42-kDa polypeptide in cells treated with a proapoptotic stimulus, possibly by inhibiting a phosphatase.[62] These observations serve as a reminder that enzymes other than proteases have reactive nucleophiles at their active sites. Accordingly, the previously cited observation that TPCK inhibits teniposide-induced PCD might reflect some alteration in signal transduction rather than (or in addition to) inhibition of proteases. Only by applying multiple structurally and/or mechanistically distinct inhibitors of an enzyme under study and showing that each has the same effect is it possible to guard against these types of unsuspected cross-reactivities. Unfortunately, multiple mechanistically distinct *cell permeable* inhibitors of the proteases implicated in PCD are not currently available. Until they are, inhibitor data will have to be interpreted cautiously.

11.7   Detection of Active Caspases by Affinity Labeling

The mapping of caspase cleavage sites in PARP, the lamins and other substrates has also enabled the design of caspase-specific affinity labeling
reagents. Each of these reagents consists of a small peptide chain linked to
a detector group (e.g., biotin) and a moiety that irreversibly binds to the
active site, e.g., chloromethyl, fluoromethyl, or acyloxymethyl ketone
(Figure 11.2A). In principle, the peptide binds to the substrate binding pocket
of the caspase, the reactive group forms a covalent bond with the active site
cysteine (Cys[285] for the case of caspase-1[63]) as established by Thornberry
et al.,[64] and the covalently modified subunit of the enzyme can then be
detected or purified using a reagent that reacts with the detector group (e.g.,
streptavidin). Results obtained using this approach are illustrated in
Figures 11.2B and 11.2C.

This approach is feasible because caspases can tolerate the insertion of
detector groups on substrate-like peptides. As first shown by Sleath et al. for
caspase-1, these enzymes exhibit an absolute requirement for aspartate in
the P_1 position, the amino acid immediately on the amino side of the scissile
bond.[65] The P_2 residue, in contrast, has little effect on substrate recognition
by caspases[45] because this sidechain is pointed away from the enzyme.[63] The
observation that lysine was tolerated in the P_2 position[45] opened the door to
a new family of labeling reagents that contain biotin on the ε-amino group
of lysine. The first member of this family, tyr-val-(ε-biotin)lys-asp-
[(2,6-dimethylbenzoyl)oxy]methylketone (YVK[biotin]D-aomk) was shown
by Thornberry et al. to be a potent inhibitor of caspase-1 that could be readily
detected with avidin-based reagents.[64] The reagent described below, N-(N^α-
benzyloxycarbonylglutamyl-N^ε-biotinyllysyl)aspartic acid [(2,6-dimethyl-
benzoyl)oxy]methyl ketone (zEK[biotin]D-aomk),[42] was synthesized based
on the EVD cleavage site from PARP (EVD⇓G)[22] and the EID cleavage site
from lamin A (EID⇓N).[28] A protocol currently used in our laboratory for
detection of active caspases in cytosol preparations using zEK(biotin)D-
aomk is described below. This protocol is based on the original labeling
technique developed for caspase-1[64] as modified by us for use with the
apoptotic caspases[42] and then further adapted by Faleiro et al.[66]

11.7.1   Labeling and Detecting Enzymatically Active Caspases

1. **Affinity Label the Active Sites of Activated Caspases.**   All steps
   prior to incubation with the affinity labeling agent should be performed at 4°C. Cells are sedimented at 300 × g for 5 min and
   washed twice with KPM buffer (50 m*M* KCl, 50 m*M* PIPES, 10 m*M*
   EGTA, 2 m*M* MgCl$_2$, pH 7). They are then gently resuspended in
   150 μl of KPM buffer supplemented with 0.1 m*M* PMSF, 20 m*M*

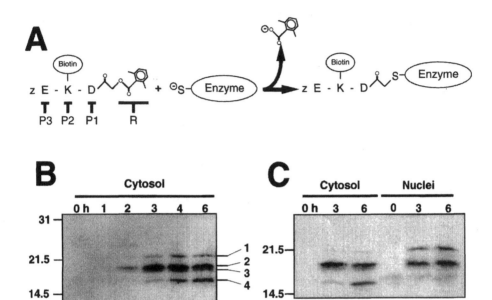

Figure 11.2
Labeling of active caspases by acyloxymethylketones. **A**, Schematic showing the reaction mechanism of *N*-(*N*$^\alpha$-benzyloxycarbonylglutamyl-N$^\varepsilon$-biotinyllysyl)aspartic acid [(2,6-dimethylbenzoxyl)oxy]methyl ketone, otherwise known as zEK(bio)D-aomk, the affinity-labeling reagent currently used in our lab.[42] Enzyme inactivation occurs through the displacement of the carboxylate leaving group (R) by the cysteine nucleophile, resulting in formation of a thiomethylketone covalent bond. P_3 to P_1 correspond to amino acid residues at the amino-terminal side of the scissile bond according to the nomenclature introduced by Schecter and Berger. Note the absolute requirement of aspartic acid at the P_1 position, a feature common to all caspases. The design of this affinity label[42] was based on the mapping of the apoptotic cleavage sites in PARP[22] and lamin A,[28] together with the indication from previous studies that substitutions (including lysine) are well tolerated in the P_2 position.[45] Although the leaving group R is bulky, ICE and other caspase family members are able to accommodate bulky residues in the P_1' position.[45] **B**, **C**, results obtained when zEK(bio)D-aomk is reacted with cytosol or nuclei from HL-60 leukemia cells treated with the chemotherapeutic agent etoposide for the indicated length of time.[42] Each labeled band corresponds to the labeled large subunit of an active caspase.

cytochalasin B, and 1 μg/ml of the following protease inhibitors: chymostatin, pepstatin, leupeptin, and antipain. After transfer to a 1.5-ml microcentrifuge tube, cells are sedimented at 800 × g. zEK(biotin)D-aomk (see note 1) is diluted to 2 μ*M* in MDB buffer (50 m*M* NaCl, 2 m*M* MgCl$_2$, 5 m*M* EGTA, 10 m*M* HEPES, 1 m*M* DTT, pH 7); and equal volumes of the 2 μ*M* zEK(bio)D-aomk solution and cell suspension are mixed. Cells are then lysed by three cycles of freezing and thawing. For each cycle, cells are frozen in liquid nitrogen and thawed at room temp. The lysates are then incubated at room temperature for 1 h and spun for 2 h at 157,000 × g_{max} in a Beckman Optima TLX ultracentrifuge. The clear lysates are then subjected to one of several different protein separation techniques and analyzed. (See notes 2 and 3.)

2. **Analyze the Labeled Polypeptides.** Prior to separation by SDS-PAGE, samples are diluted with 0.5 volume of 3× sample buffer (0.15 M Tris-HCl [pH 6.8], 45% [w/v] sucrose, 6 mM EDTA, 9% [w/v] SDS, 0.03% [w/v] bromophenol blue, 10% [v/v] β-mercaptoethanol) and boiled for 3 min. Separation is performed on a 16% SDS-polyacrylamide gel using the technique described by Laemmli.[67] In our experience, optimal resolution of the caspases is obtained using 0.75 mm thick 16 × 18 cm gels, with separation at 15 mA constant current for 6 h. Following separation, polypeptides are electrophoretically transferred to nitrocellulose (e.g., 2 h at 55V in a Bio-Rad Trans Blot apparatus using transfer buffer that consists of 20% [v/v] methanol, 0.1% [w/v] SDS, 25 mM Tris and 188 mM glycine). Unoccupied protein binding sites are blocked with 5% nonfat dry milk in PBS-T buffer (calcium/magnesium-free phosphate buffered saline containing 0.1% [v/v] Tween-20). Blots are then incubated with peroxidase-coupled streptavidin for 3 h 30 min at room temperature in PBS-T, washed once for 15 min and four times for 5 min each with PBS-T to remove unbound peroxidase-coupled streptavidin, and detected using Amersham ECL enhanced chemiluminescent detection as described by the supplier. (See notes 4-6.)

11.7.2   Notes on the Procedure

1. zEK(biotin)D-aomk is conveniently prepared as a 25 mM stock in DMSO and stored in small aliquots at –80°C. Although these affinity labeling agents are generally considered to be relatively unstable, in our experience they can be stored dissolved in DMSO for relatively long periods of time at –80°C. For example, zEK(biotin)D-aomk has retained its initial reactivity after storage for 2 years under these conditions.

2. The described procedure is designed for the detection of caspases in cytosol. As illustrated in Figure 11.2C, it is also possible to prepare various subcellular fractions and then react them with zEK(biotin)D-aomk. In this case, the affinity labeling agent is added to the purified subcellular fraction at a concentration of 1 µM. After a suitable incubation time, subsequent protein separation and detection procedures are performed as described.

3. It is possible to competitively inhibit binding of the affinity label with other caspase inhibitors. For example, preincubation of cell lysates with certain peptide chloromethyl ketones or fluoromethyl ketones completely abolishes subsequent labeling with zEK(bio) D-aomk.[42] This type of competition experiment serves several purposes. First, cell lysates contain many polypeptides that are intrinsically biotinylated. These are unaffected by preincubation of the lysates with Ac-tyr-val-ala-asp-chloromethylketone (YVAD-cmk), whereas labeling of caspases can be completely suppressed under the same conditions.[42] Second, the same approach can sometimes be utilized to identify the labeled caspase responsible for a particular biological function. For example, the Sp1 serpin from rabbitpox virus and underivatized peptides corresponding to the lamin A cleavage site — two different types of reversible caspase inhibitors — were shown to inhibit lamin cleavage *in vitro* and simultaneously abolish covalent modification of one particular caspase by affinity labeling.[36] In performing this latter type of experiment, it is important to optimize the labeling conditions so that the time of exposure to the affinity label is minimized (e.g., 1 min).[42] Because reversible inhibitors are, by definition, dissociating and reassociating with the caspase active site (thereby providing the affinity label an opportunity to covalently label the enzyme), longer labeling times will allow covalent modification of a caspase in question, even in the presence of a potent reversible inhibitor.

4. As an alternative, it is possible to purify the derivatized caspases using immobilized avidin according to procedures described by Nicholson et al.[23] or Faliero et al.[66]

5. Where levels of active caspases are very low, it is possible to achieve approximately 25-fold greater sensitivity in enhanced chemiluminescence (at 10-fold greater reagent cost) using SuperSigal Ultra™ from Pierce (Rockford, IL).

6. It is essential to optimize the detection conditions to achieve maximum signal and minimum background. High backgrounds can be a serious problem with this assay, particularly when caspase concentrations in the cell lysates are low. The most common cause for high backgrounds in our experience is the choice of blocking agent. Although milk is traditionally discouraged when using avidin/biotin detection systems, we have found powdered milk to be far superior to biotin-free blocking agents. However, we have also found that powdered milk from different sources results in different levels of background. Accordingly, powdered milk from several different sources might have to be compared to arrive at an optimal signal to noise ratio.

11.7.3   Strengths and Limitations of the Assay

This assay is extremely sensitive. When combined with the powerful new chemiluminescent substrates, this assay can detect pg levels of active proteases. In addition, when utilized in conjunction with appropriate standards, this assay can be utilized to identify the individual proteases that are active during the course of PCD.[42] Moreover, it is possible to purify the affinity-labeled caspases for further study.[23,66] Finally, because only two inhibitor molecules bind to each caspase, the labeling can be used to determine the concentration of particular labeled active caspases.

Despite its strengths, this technique also has several potential limitations. By far the most serious of these is the difficulty in procuring reagents that label caspases with high affinity and do not give unacceptable background labeling of nonapoptotic cell lysates. We have tested multiple affinity labeling reagents from several commercial sources. Unfortunately, all of these proved to give unacceptable background labeling. One possibility is that this non-specific binding proceeds through derivitization of other thiol nucleophiles present in the samples. Nonspecific reactivity is known to be a particular problem with chloromethylketones, which react with a variety of activated thiols[68] as typified by the promiscuity of the chloromethyl ketones TLCK and TPCK (See Section 11.6). Accordingly, we recommend using peptidyl acyloxymethyl ketones in preference to the more readily available chloromethyl ketones. It should be noted, however, that the chemical synthesis of acyloxymethyl ketones is very tricky, a factor that undoubtedly contributes to their expense and to the variable quality of commercial reagents. Availability, therefore, is a key consideration in determining whether to employ these compounds in particular study. Hopefully, the presence of a potentially lucrative market will stimulate suppliers to improve the quality and availability of these products.

11.8   *In Vitro* Activation Assays

The realization that caspases play a role in PCD raises questions about how these enzymes are activated and how the activation process is regulated. This question became amenable to experimental investigation with the report by Liu et al. that caspase-3 could be activated by treatment of cytosol from nonapoptotic cells with cytochrome c and deoxyadenosine triphosphate.[69] Subsequent studies have indicated that during apoptosis cytochrome c is released from mitochondria to cytosol prior to caspase activation, suggesting that cytochrome c might be a physiological participant in caspase activation.[69-74] The role of deoxyadenosine triphosphate remains to be more fully established. Nonetheless, these observations provide the framework for current methods to study the activation of caspases under cell-free conditions,

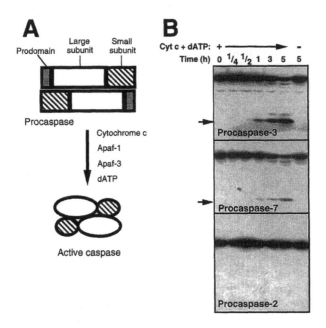

FIGURE 11.3

Caspase activation *in vitro* upon addition of cytochrome c to cytosol from nonapoptotic cells.
A, Schematic representation of caspase activation *in vitro*. The constitutively expressed protease-activating factor Apaf-1 was recently shown to share a region of limited amino acid similarity with the *C. elegans* protein ced-4.[76] The protease-activating factor Apaf-3 was recently identified as procaspase-9.[77] 2'-Deoxyadenosine triphosphate (dATP) is required for this reaction *in vitro*,[29,77] although the involvement of this nucleotide *in vivo* remains to be established. **B**, Example of protease activation *in vitro*. Cytosol prepared from K562 leukemia cells as described[73] was incubated in the absence or presence of 500 µg/ml cytochrome c and dATP for the indicated length of time, then subjected to SDS-PAGE followed by immunoblotting with reagents that recognize the indicated caspase. Note that caspases -3 and -7 are proteolytically activated (arrowheads) under these conditions, whereas caspase-2 is not. Similar results have been observed with cytochrome c concentrations as low as 1 to 10 µg/ml in subsequent experiments.

where various activators or inhibitors can be added without concern about their ability to cross membranes and penetrate into various cellular compartments. A current protocol for performance of this type of assay is described below and illustrated in Figure 11.3.

11.8.1   Protocol for *In Vitro* Activation Assay

1. **Prepare Cytosol or other Subcellular Fractions.** All steps are performed at 4°C. Cells are released by trypsinization (if adherent), sedimented at 200 × g for 10 min, washed twice in RPMI 1640 medium containing 10 mM HEPES (pH 7.4). Cytosol is prepared as described above (Section 11.5.1). In brief, cells are swelled in buffer A and disrupted by 20 to 50 strokes in a pre-chilled Dounce homogenizer with a tight-fitting pestle; nuclei are sedimented at

$800 \times g$ for 10 min; and the postnuclear supernatant is sedimented at $50,000 \times g_{max}$ for 60 min in a Beckman TL-100 ultracentrifuge. The resulting cytosolic extract is then frozen in 100-µl aliquots at –70°C. (See notes 1,2.)

2. **Assemble *in Vitro* Activation Reactions.** Reaction mixtures are assembled on ice in the wells of a pre-chilled 96-well microtiter plate. A typical 40-µl reaction contains 50 µg of cytosolic protein, additives (inhibitors, activators, etc.) prepared in buffer A, and freshly prepared buffer A as diluent to adjust final volume to 40 µl. Reactions are assembled by adding the appropriate volume of diluent to each well, followed by the addition of any additives, and finally by the addition of freshly thawed cytosolic extract. Reactions are initiated by placing the microtiter plate in a humidified 30°C incubator. (See note 3.)

3. **Analyze *in vitro* activation reactions.** Following incubation at 30°C for the desired length of time, reactions are terminated by placing the microtiter plate on ice. Individual reaction mixtures are then transferred to labeled, pre-chilled microcentrifuge tubes and assayed by one or more of the following methods:

   - **Immunoblotting.** Samples are mixed with 1/3 volume of 4x Laemmli sample buffer, heated to 65°C for 15 min, and subjected to SDS-PAGE essentially as described above (Section 11.7.1). Following electrophoretic separation, polypeptides are transferred to PVDF for blotting with anticaspase antibodies. Control samples containing unincubated cytosol should be included in order to determine the extent of procaspase cleavage (i.e., activation).

   - **Assay for Ability to Cleave Fluorogenic or Chromogenic Substrates.** As an alternative, the cold-terminated reaction mixture can be added to buffered substrate; and the fluorogenic or chromogenic assay of caspase activity can be performed as described in Section 11.5.

11.8.2   Notes on the Procedure

1. When stored at –70°C, cytosolic extracts remain stable and activatable by cytochrome c for at least 9 months.

2. Dounce homogenization is not sufficient to lyse some adherent cell lines (e.g., A549 non-small cell lung cancer cells or human foreskin fibroblasts). For these cell lines, use of a nitrogen cavitation chamber or repeated passage through a 22 gauge might be required to prepare cytosol.

3. Control experiments from our laboratory indicate that the proforms of caspases -2, -3, and -7 are stable for up to 6 h at 30 to 37°C or up to 72 h at 4°C in the absence of an activator.

11.8.3   Strengths and Limitations of the Assay

As with other cell-free systems utilized to study physiological processes, the *in vitro* caspase activation assay described above offers two major advantages over intact cells. First, it allows the testing of compounds that would penetrate intact cells poorly because of size or charge. With its microtiter plate format, the present assay should be amenable to relatively rapid screening for agents that facilitate or inhibit caspase activation. Second, cell-free systems are ideally suited to biochemical fractionation/reconstitution schemes, thus providing a powerful approach for identifying critical biochemical components of the system.[69,75]

Although the cell-free activation of caspases can potentially provide important new information regarding the control of protease activation, information obtained from experiments utilizing this approach must be evaluated carefully. It appears that the ability to activate caspases in cytosol is highly dependent on the manner in which the cytosol is prepared (P. Mesner, unpublished observation). This observation suggests that critical variables in the caspase activation process remain to be determined. Furthermore, a cell-free system by its very nature involves disruption and/or removal of organelles and cellular membranes. Compartmentalization of important functions is a basic tenant of cell biology. The loss of compartmentalization during preparation of cell-free extracts results in the dissipation of pH, concentration, and electrochemical gradients, leading to the possibility of nonphysiological molecular interactions. Accordingly, it is important that molecular interactions identified using cell-free systems of caspase activation be confirmed in intact cells.

References

1. Ellis RE, Yuan JY, Horvitz HR: Mechanisms and functions of cell death. *Annu. Rev. Cell Biol.* 7:663, 1991.
2. Xue D, Shaham S, Horvitz HR: The *Caenorhabditis Elegans* cell-death protein CED-3 is a cysteine protease with substrate specificities similar to those of the human CPP32 protease. *Genes Dev.* 10:1073, 1996.
3. Talanian RV, Quinlan C, Trautz S, Hackett MC, Mankovich JA, Banach D, Ghayur T, Brady KD, Wong WW: Substrate specificities of caspase family proteases. *J. Biol. Chem.* 272:9677, 1997.
4. Margolin N, Raybuck SA, Wilson KP, Chen W, Fox T, Gu Y, Livingston DJ: Substrate and inhibitor specificity of interleukin-1β-converting enzyme and related caspases. *J. Biol. Chem.* 272:7223, 1997.
5. Thornberry NA, Rano TA, Peterson EP, Rasper DM, Timkey T, Garcia-Calvo M, Houtzager VM, Nordstrom PA, Roy S, Vaillancourt JP, Chapman KT, Nicholson DW: A combinatorial approach defines specificities of members of the caspase family and granzyme B. Functional relationships established for key mediators of apoptosis. *J. Biol. Chem.* 272:17907, 1997.

6. Kuida K, Zheng TS, Na S, Kuan C, Yang D, Karasuyama H, Rakic P, Flavell RA: Decreased apoptosis in the brain and premature lethality in CPP32-deficient mice. *Nature* 384:368, 1996.

7. Kuida K, Lippke JA, Ku G, Harding MW, Livingston DJ, Su MS, Flavell RA: Altered cytokine export and apoptosis in mice deficient in interleukin-1 beta converting enzyme. *Science* 267:2000, 1995.

8. Li P, Allen H, Banerjee S, Franklin S, Herzog L, Johnston C, McDowell J, Paskind M, Rodman L, Salfeld J: Mice deficient in IL-1 beta-converting enzyme are defective in production of mature IL-1 beta and resistant to endotoxic shock. *Cell* 80:401, 1995.

9. Hara H, Friedlander RM, Gagliardini V, Ayata C, Fink K, Huang Z, Shimizu-Sasamata M, Yuan J, Moskowitz MA: Inhibition of interleukin-1 Beta converting enzyme family proteases reduces ischemic and excitotoxic Neuronal Damage. *Proc. Natl. Acad. Sci., U.S.A.* 94:2007, 1997.

10. Wang L, Miura M, Bergeron L, Zhu H, Yuan J: Ich-1: an ICE/ced-3-related gene, encodes both positive and negative regulators of programmed cell death. *Cell* 78:739, 1994.

11. Fernandes-Alnemri T, Litwack G, Alnemri ES: CPP32, A novel human apoptotic protein with homology to *Caenorhabditis Elegans* cell death protein Ced-3 and mammalian interleukin-1 beta-converting enzyme. *J. Biol. Chem.* 269:30761, 1994.

12. Kumar S, Kinoshita M, Noda M, Copeland NG, Jenkins NA: Induction of apoptosis by the mouse Nedd2 gene, which encodes a protein similar to the product of the *Caenorhabditis Elegans* cell death gene Ced-3 and the mammalian IL-1 beta-converting enyzme. *Genes Dev.* 8:1613, 1994.

13. Fernandes-Alnemri T, Takahashi A, Armstrong R, Krebs J, Fritz L, Tomaselli KJ, Wang L, Yu Z, Croce CM, Salveson G, Earnshaw WC, Litwack G, Alnemri ES: Mch3, a novel human apoptotic cysteine protease highly related to CPP32. *Cancer Res.* 55:6045, 1995.

14. Munday NA, Vaillancourt JP, Ali A, Casano FJ, Miller DK, Molineaux SM, Yamin TT, Yu VL, Nicholson DW: Molecular cloning and pro-apoptotic activity of ICErelII and ICErelIII, members of the ICE/CED-3 Family of cysteine proteases. *J. Biol. Chem.* 270:15870, 1995.

15. Lippke JA, Gu Y, Sarnecki C, Caron PR, Su MS: Identification and characterization of CPP32/Mch2 Homolog 1, a novel cysteine protease similar to CPP32. *J. Biol. Chem.* 271:1825, 1996.

16. Duan H, Orth K, Chinnaiyan AM, Poirier GG, Froelich CJ, He WW, Dixit VM: ICE-LAP6, A novel member of the ICE-Ced-3 gene family, is activated by the cytotoxic T cell protease granzyme B. *J. Biol. Chem.* 271:16720, 1996.

17. Williams MS, Henkart PA: Apoptotic cell death induced by intracellular proteolysis. *J. Immunol.* 153:4247, 1994.

18. Sarin A, Williams MS, Alexander-Miller MA, Berzofsky JA, Zacharchuk CM, Henkart PA: Target cell lysis by CTL granule exocytosis is independent of ICE/Ced-3 family proteases. *Immunity* 6:209, 1997.

19. Kaufmann SH, Kellner U: Erasure of western blots after autoradiographic or chemiluminescent detection, in Pound J (Ed): *Methods in Molecular Biology*, Vol. 80, Humana Press, 1998, p.223.

20. Kaufmann SH: Induction of Endonucleolytic DNA cleavage in human acute myelogenous leukemia cells by etoposide, camptothecin, and other cytotoxic anticancer drugs: a cautionary note. *Cancer Res.* 49:5870, 1989.

21. Kaufmann SH, Desnoyers S, Ottaviano Y, Davidson NE, Poirier GG: Specific proteolytic fragmentation of poly(ADP-ribose) polymerase: an early marker of chemotherapy-induced apoptosis. *Cancer Res.* 53:3976, 1993.

22. Lazebnik YA, Kaufmann SH, Desnoyers S, Poirier GG, Earnshaw WC: Cleavage of Poly(ADP-ribose)polymerase by a proteinase with properties like ICE. *Nature* 371:346, 1994.

23. Nicholson DW, Ali A, Thornberry NA, Vaillancourt JP, Ding CK, Gallant M, Gareau Y, Griffin PR, Labelle M, Lazebnik YA: Identification and inhibition of the ICE/CED-3 protease necessary for mammalian apoptosis. *Nature* 376:37, 1995.

24. Fernandes-Alnemri T, Litwack G, Alnemri ES: Mch2, A new member of the apoptotic Ced-3/ICE cysteine protease gene family. *Cancer Res.* 55:2737, 1995.

25. Gu Y, Sarnecki C, Aldape RA, Livingston DJ, Su MS: Cleavage of poly(ADP-ribose) polymerase by interleukin-1 beta converting enzyme and its homologs TX and Nedd-2. *J. Biol. Chem.* 270:18715, 1995.

26. Vincenz C, Dixit VM: Fas-associated death domain protein interleukin-1 beta-converting enzyme 2 (FLICE2), and ICE/Ced-3 homologue, is proximally involved in CD95- and p55-mediated death signaling. *J. Biol. Chem.* 272:6578, 1997.

27. Muzio M, Chinnaiyan AM, Kischkel FC, O'Rourke K, Shevchenko A, Ni J, Scaffidi C, Bretz JD, Zhang M, Gentz R, Mann M, Krammer PH, Peter ME, Dixit VM: FLICE, A novel FADD-homologous ICE/CED-3-like protease, is recruited to the CD95 (Fas/APO-1) death-inducing signaling complex. *Cell* 85:817, 1996.

28. Takahashi A, Alnemri ES, Lazebnik YA, Fernandes-Alnemri T, Litwack G, Moir RD, Goldman RD, Poirier GG, Kaufmann SH, Earnshaw WC: Cleavage of lamin A by Mch2α but not CPP32: multiple ICE-related proteases with distinct substrate recognition properties are active in apoptosis. *Proc. Natl. Acad. Sci., U.S.A.* 93:8395, 1996.

29. Liu X, Zou H, Slaughter C, Wang X: DFF, a heterodimeric protein that functions downstream of caspase-3 to trigger DNA fragmentation during apoptosis. *Cell* 89:175, 1997.

30. McConkey DJ: Calcium-dependent, interleukin 1-converting enzyme inhibitor-insensitive degradation of lamin B1 and DNA fragmentation in isolated thymocyte nuclei. *J. Biol. Chem.* 271:22398, 1996.

31. Villa P, Kaufmann SH, Earnshaw WC: Caspases and caspase inhibitors. *trends in Biochem. Sci.* 22:388, 1997.

32. Lazebnik YA, Cole S, Cooke CA, Nelson WG, Earnshaw WC: Nuclear events of apoptosis *in vitro* in cell-free mitotic extracts: a model system for analysis of the active phase of apoptosis. *J. Cell Biol.* 123:7, 1993.

33. Oberhammer FA, Hochegger K, Froschl G, Tiefenbacher R, Pavelka M: Chromatin condensation during apoptosis is accompanied by degradation of lamin A+B, without enhanced activation of cdc2 kinase. *J. Cell Biol.* 126:827, 1994.

34. Neamati N, Fernandez A, Wright S, Kiefer J, McConkey DJ: Degradation of lamin B_1 precedes oligonucleosomal DNA fragmentation in apoptotic thymocytes and isolated thymocyte nuclei. *J. Immunol.* 154:3788, 1995.

35. Lazebnik YA, Takahaski A, Moir R, Goldman R, Poirier GG, Kaufmann SH, Earnshaw WC: Studies of the lamin proteinase reveal multiple parallel biochemical pathways during apoptotic execution. *Proc. Natl. Acad. Sci., U.S.A.* 92:9042, 1995.

36. Takahashi A, Musy P-Y, Martins LM, Poirier GG, Moyers RW, Earnshaw WC: CrmA/SPI-2 Inhibition of an endogenous ICE-related protease responsible for lamin a cleavage and apoptotic nuclear fragmentation. *J. Biol. Chem.* 271:32487, 1996.

37. Rao L, Perez D, White E: Lamin proteolysis facilitates nuclear events during apoptosis. *J. Cell Biol.* 135:1441, 1996.

38. Emoto Y, Manome Y, Meinhardt G, Kisaki H, Kharbanda S, Robertson M, Ghayur T, Wong WW, Kamen R, Weichselbaum R, Kufe D: Proteolytic activation of protein kinase C δ by an ICE-like protease in apoptotic cells. *EMBO J* 14:6148, 1995.

39. Lahti JM, Xiang J, Heath LS, Campana D, Kidd VJ: PITSLRE protein kinase activity is associated with apoptosis. *Mol. Cell. Biol.* 15:1, 1995.

40. Beyaert R, Kidd VJ, Cornelis S, Van de Craen M, Denecker G, Lahti JM, Gururajan R, Vandenabeele P, Fiers W: Cleavage of PITSLRE kinases by ICE/CASP-1 and CPP32/CASP-3 during apoptosis induced by tumor necrosis factor. *J. Biol. Chem.* 272:11694, 1997.

41. Ghayur T, Hugunin M, Talanian RV, Ratnofsky S, Quinlan C, Emoto Y, Pandey P, Datta R, Huang Y, Kharbanda S, Allen H, Kamen R, Wong W, Kufe D: Proteolytic activation of protein kinase C δ by an ICE/Ced-3-like protease induces characteristics of apoptosis. *J. Exp. Med.* 184:2399, 1996.

42. Martins LM, Kottke TJ, Mesner P, Basi GS, Sinha S, Frigon N, Jr., Tatar E, Tung JS, Bryant K, Takahashi A, Svingen PA, Madden BJ, McCormick DJ, Earnshaw WC, Kaufmann SH: Activation of multiple interleukin-1β converting enzyme homologues in cytosol and nuclei of HL-60 human leukemia cell lines during etoposide-induced apoptosis. *J. Biol. Chem.* 272:7421, 1997.

43. *Subcellular Fractionation: A Practical Approach,* IRL Press, 1997.

44. Smith PK, Krohn RI, Hermanson GT, Mallia AK, Gartner FH, Provenzano MD, Fujimoto EK, Goeke NM, Olson BJ, Klenk DC: Measurement of protein using bicinchoninic acid. *Anal. Biochem.* 150:76, 1985.

45. Thornberry NA, Bull HG, Calaycay JR, Chapman KT, Howard AD, Kostura MJ, Miller DK, Molineaux SM, Weidner JR, Aunins J: A novel heterodimeric cysteine protease is required for interleukin-1 beta processing in monocytes. *Nature* 356:768, 1992.

46. Deveraux QL, Takahashi R, Salvesen GS, Reed JC: X-linked IAP is a direct inhibitor of cell-death proteases. *Nature* 388:300, 1997.

47. Enari M, Talanian RV, Wong WW, Nagata S: Sequential Activation of ICE-like and CPP32-like proteases during Fas-mediated apoptosis. *Nature* 380:723, 1996.

48. Fernandes-Alnemri T, Armstrong RC, Krebs J, Srinivasula SM, Wang L, Bullrich F, Fritz LC, Trapani JA, Tomaselli KJ, Litwack G, Alnemri ES: *In Vitro* Activation of CPP32 and Mch3 by Mch4, a novel human apoptotic cysteine protease containing two FADD-like domains. *Proc. Natl. Acad. Sci., U.S.A.* 93:7464, 1996.

49. Srinivasula SM, Ahmad M, Fernandes-Alnemri T, Litwack G, Alnemri ES: Molecular ordering of the Fas-apoptotic pathway: The Fas/APO-1 protease Mch5 is a CrmA-inhibitable protease that activates multiple Ced-3/ICE-like cysteine proteases. *Proc. Natl. Acad. Sci., U.S.A.* 93:14486, 1996.

50. Kobayashi T, Shinozaki A, Momoi T, Arahata K, Tsukahara T: Identification of an interleukin-1β converting enzyme-like activity that increases upon treatment of P19 Cells with retinoic acid as the proteasome. *J. Biochem.* 120:699, 1996.

51. Thornberry NA, Molineaux SM: Interleukin-1β Converting enzyme: a novel cysteine protease required for IL-1β production and implicated in programmed cell death. *Protein Sci.* 4:3, 1995.

52. Black RA, Kronheim SR, Sleath PR: Activation of interleukin-1β by a Co-induced protease. *FEBS Lett.* 247:386, 1989.

53. Bruno S, Lassota P, Giaretti W, Darzynkiewicz Z: Apoptosis of rat thymocytes triggered by prednisolone, camptothecin, or teniposide is selective to G_0 cells and is prevented by inhibitors of proteases. *Oncol. Res.* 4:29, 1992.

54. Bruno S, Del Bino G, Lassota P, Giaretti W, Darzynkiewicz Z: Inhibitors of proteases prevent endonucleolysis accompanying apoptotic death of HL-60 leukemic cells and normal thymocytes. *Leukemia* 6:1113, 1992.

55. Weaver VM, Lach B, Walker PR, Silorska M: Role of proteolysis in apoptosis: involvement of serine proteases in internucleosomal DNA fragmentation in immature thymocytes. *Biochem. Cell Biol.* 71:488, 1993.

56. Barrett AJ: Human Cathepsin B1: Purification and some properties of the enzyme. *Biochem. J.* 131:809, 1973.

57. Rock KL, Gramm C, Rothstein L, Clark K, Stein R, Dick L, Hwang D, Goldberg AL: Inhibitors of the proteasome block the degradation of most cell proteins and the generation of peptides presented on MHC Class I molecules. *Cell* 78:761, 1994.

58. Sadoul R, Fernandez PA, Quiquerez AL, Martinou I, Maki M, Schroter M, Becherer JD, Irmler M, Tschopp J, Martinou JC: Involvement of the proteasome in the programmed cell death of NGF-deprived sympathetic neurons. *EMBO J* 15:3845, 1996.

59. Thornberry NA, Rosen A, Nicholson DW: Control of apoptosis by proteases. *Adv. Pharmacol.* 41:155, 1997.

60. Dubrez L, Savoy I, Hamman A, Solary E: Pivotal role of a DEVD-sensitive step in etoposide-induced and Fas-mediated apoptotic pathways. *EMBO J* 15:5504, 1996.

61. Prochiantz A: Getting hydrophilic compounds into cells: lessons from homeopeptides. *Curr. Opin. Neurobiol.* 6:629, 1996.

62. Lumelsky NL, Schwartz BS: Protease inhibitors induced specific changes in protein tyrosine phosphorylation that correlate with inhibition of apoptosis in myeloid cells. *Cancer Res.* 56:3909, 1996.

63. Wilson J, Thomson JB, Kim E: Structure and mechanism of interleukin-1β converting enzyme. *Nature* 370:270, 1994.

64. Thornberry NA, Peterson EP, Zhao JJ, Howard AD, Griffin PR, Chapman KT: Inactivation of interleukin-1 beta converting enzyme by peptide (acyloxy)methyl ketones. *Biochem.* 33:3934, 1994.

65. Sleath PR, Hendrickson RC, Kronheim SR, March CJ, Black RA: Substrate Specificity of the protease that processes human interleukin-1β. *J. Biol. Chem.* 265:14526, 1990.

66. Faleiro L, Kobayashi R, Fearnhead H, Lazebnik Y: Multiple species of CPP32 and Mch2 are the major active caspases present in apoptotic cells. *EMBO J* 16:2271, 1997.

67. Laemmli UK: Cleavage of structural proteins during the assembly of the head of bacteriophage T4. *Nature* 227:680, 1970.

68. Shaw E: Cysteinyl proteinases and their selective inactivation. *Adv. Enzymol. Relat. Areas Mol. Biol.* 63:271, 1990.

69. Liu X, Kim CN, Yang J, Jemmerson R, Wang X: Induction of apoptotic program in cell-free extracts: requirement for dATP and cytochrome C. *Cell* 86:147, 1996.

70. Krippner A, Matsuno-Yagi A, Gottlieb RA, Babior BM: Loss of function of cytochrome C in jurkat cells undergoing Fas-mediated apoptosis. *J. Biol. Chem.* 271:21629, 1996.

71. Yang J, Liu X, Bhalla K, Kim CN, Ibrado AM, Cai J, Peng T-I, Jones DP, Wang X: Prevention of Apoptosis by Bcl-2: Release of cytochrome c from mitochondria blocked. *Science* 275:1129, 1997.

72. Kluck RM, Bossy-Wetzel E, Green DR, Newmeyer DD: The release of cytochrome c from mitochondria: a primary site for Bcl-2 regulation of apoptosis. *Science* 275:1132, 1997.

73. Martins LM, Mesner PW, Kottke TJ, Basi GS, Sinha S, Tung JS, Svingen PA, Madden BJ, Takahashi A, McCormick DJ, Earnshaw WC, Kaufmann SH: Comparison of caspase activation and subcellular localization in HL-60 and K562 cells undergoing etoposide-induced apoptosis. *Blood* 90:4283, 1997.

74. Kharbanda S, Pandey P, Schofield L, Israels S, Roncinske R, Yoshida K, Bharti A, Yuan ZM, Saxena S, Weichselbaum R, Nalin C, Kufe D: Role for Bcl-xL as an inhibitor of cytosolic cytochrome C accumulation in DNA damage-induced apoptosis. *Proc. Natl. Acad. Sci., U.S.A.* 94:6939, 1997.

75. Zou H, Henzel WJ, Liu X, Lutschg A, Wang X: Apaf-1, a human protein homologous to *C. elegans* CED-4, participates in cytochrome c-dependent activation of caspase-3. *Cell* 90:405, 1997.

76. Zhou Q, Snipas S, Orth K, Muzio M, Dixit VM, Salvesen GS: Target protease specificity of the viral serpin CrmA. analysis of five caspases. *J. Biol. Chem.* 272:7797, 1997.

77. Li P, Nijhawan D, Budihardjo I, Srinivasula SM, Ahmad M, Alnemri ES, Wang X: Cytochrome C and dATP-dependent formation of Apaf-1/caspase-9 complex initiates an apoptotic protease cascade. *Cell* 91:479, 1997.

78. Tewari M, Quan LT, O'Rourke K, Desnoyers S, Zeng Z, Beidler DR, Poirier GG, Salvesen GS, Dixit VM: Yama/CPP32 beta, a mammalian homolog of CED-3, is a CrmA-inhibitable protease that cleaves the death substrate poly(ADP-ribose) polymerase. *Cell* 81:801, 1995.

79. Rosenthal DS, Ding R, Simbulan-Rosenthal CM, Vaillancourt JP, Nicholson DW, Smulson M: Intact cell evidence for the early synthesis, and subsequent late apopain-mediated suppression, of poly(ADP-Ribose) during apoptosis. *Exp. Cell Res.* 232:313, 1997.

80. Song Q, Lees-Miller SP, Kumar S, Zhang Z, Chan DW, Smith GC, Jackson SP, Alnemri ES, Litwack G, Khanna KK, Lavin MF: DNA-dependent protein kinase catalytic subunit: a target for an ICE-like protease in apoptosis. *EMBO J* 15:3238, 1996.

81. Casciola-Rosen LA, Nicholson DW, Chong T, Rowan KR, Thornberry NA, Miller DK, Rosen A: Apopain/CPP32 cleaves proteins that are essential for cellular repair: a fundamental principle of apoptotic death. *J. Exp. Med.* 183:1957, 1996.

82. Han Z, Malik N, Carter T, Reeves WH, Wyche JH, Hendrickson EA: DNA-dependent protein kinase is a target for a CPP32-like apoptotic protease. *J. Biol. Chem.* 271:25035, 1996.

83. Henkels KM, Turchi JJ: Induction of apoptosis in cisplatin-sensitive and -resistant human ovarian cancer cell lines. *Cancer Res.* 57:4488, 1997.

84. Tan X, Martin SJ, Green DR, Wang JYJ: Degradation of retinoblastoma protein in tumor necrosis factor- and CD95-induced cell death. *J. Biol. Chem.* 272:9613, 1997.

85. Janicke RU, Walker PA, Lin XY, Porter AG: Specific cleavage of the retinoblastoma protein by an ICE-like protease in apoptosis. *EMBO J* 15:6969, 1996.

86. Crouch DH, Fincham VJ, Frame MC: Targeted proteolysis of the focal adhesion kinase pp125 FAK during c-MYC-induced apoptosis is suppressed by integrin signaling. *Oncogene* 12:2689, 1996.

87. Ubeda M, Habener JF: The large subunit of the DNA replication complex C (DSEB/RF-C140) cleaved and inactivated by caspase-3 (CPP32/YAMA) during fas-induced apoptosis. *J. Biol. Chem.* 272:19562, 1997.

88. Erhardt P, Tomaselli KJ, Cooper GM: Identification of the MDM2 oncoprotein as a substrate for CPP32-like apoptotic proteases. *J. Biol. Chem.* 272:15049, 1997.

89. Datta R, Kojima H, Yoshida K, Kufe D: Caspase-3-mediated cleavage of protein kinase C theta in induction of apoptosis. *J. Biol. Chem.* 272:20317, 1997.

90. Rudel T, Bokoch GM: Membrane and morphological changes in apoptotic cells regulated by caspase-mediated activation of PAK2. *Science* 276:1571, 1997.

91. Kothakota S, Azuma T, Reinhard C, Klippel A, Tang J, Chu K, McGarry TJ, Kirschner MW, Koths K, Kwiatkowski DJ, Williams LT: Caspase-3-generated fragment of gelsolin: effector of morphological change in apoptosis. *Science* 278:294, 1997.

92. Sakahira H, Enari M, Nagata S: Cleavage of CAD inhibitor in CAD activation and DNA degradation during apoptosis. *Nature* 391:96, 1998.

93. Casiano CA, Martin SJ, Green DR, Tan EM: Selective cleavage of nuclear autoantigens during CD95 (Fas/APO-1)-mediated T cell apoptosis. *J. Exp. Med.* 184:765, 1996.

94. Brancolini C, Benedetti M, Schneider C: Microfilament reorganization during apoptosis: the role of Gas2, a possible substrate for ICE-like proteases. *EMBO J* 14:5179, 1995.

95. Martin SJ, O'Brien GA, Nishioka WK, McGahon AJ, Mahboubi A, Saido TC, Green DR: Proteolysis of fodrin (non-erythroid spectrin) during apoptosis. *J. Biol. Chem.* 270:6425, 1995.

96. Cryns VL, Bergeron L, Zhu H, Li H, Yuan J: Specific cleavage of alpha-fodrin during Fas- and tumor necrosis factor-induced apoptosis is mediated by an interleukin-1beta-converting enzyme/Ced-3 protease distinct from the poly(ADP-ribose) polymerase protease. *J. Biol. Chem.* 271:31277, 1996.

97. Goldberg YP, Nicholson DW, Rasper DM, Kalchman MA, Koide HB, Graham RK, Bromm M, Kazemi-Esfarjani P, Thornberry NA, Vaillancourt JP, Hayden MR: Cleavage of Huntingtin by apopain, a proapoptotic cysteine protease, is modulated by the polyglutamine tract. *Nat. Genet.* 13:442, 1996.

98. Casciola-Rosen LA, Miller DK, Anhalt GJ, Rosen A: Specific cleavage of the 70 kDa protein component of the U1 small nuclear ribonucleoprotein is a characteristic biochemical feature of apoptotic cell death. *J. Biol. Chem.* 269:30757, 1994.

99. Waterhouse N, Kumar S, Song Q, Strike P, Sparrow L, Dreyfuss G, Alnemri ES, Litwack G, Lavin M, Watters D: Heteronuclear ribonucleoproteins C1 and C2, components of the spliceosome, are specific targets of interleukin 1β-converting enzyme-like proteases in apoptosis. *J. Biol. Chem.* 271:29335, 1996.

100. Wang X, Pai JT, Wiedenfeld EA, Medina JC, Slaughter CA, Goldstein JL, Brown MS: Purification of an interleukin-1 beta converting enzyme-related cysteine protease that cleaves sterol regulatory element-binding proteins between the leucine zipper and transmembrane domains. *J. Biol. Chem.* 270:18044, 1995.

101. Pai JT, Brown MS, Goldstein JL: Purification and cDNA Cloning of a second apoptosis-related cysteine protease that cleaves and activates sterol regulatory element binding proteins. *Proc. Natl. Acad. Sci., U.S.A.* 93:5437, 1996.

102. Wang X, Zelenski NG, Yang J, Sakai J, Brown MS, Goldstein JL: Cleavage of sterol regulatory element binding proteins (SREBPs) by CPP32 during apoptosis. *EMBO J* 15:1012, 1996.

103. Sakai J, Duncan EA, Rawson RB, Hua X, Brown MS, Goldstein JL: Sterol-regulated release of SREBP-2 from cell membranes requires two sequential cleavages, one within a transmembrane segment. *Cell* 85:1037, 1996.

104. Na S, Chuang TS, Cunningham A, Turi TG, Hanke JH, Bokoch GM, Danley DE: D4-GDI, a substrate of CPP32, Is proteolyzed during Fas-induced apoptosis. *J. Biol. Chem.* 271:11209, 1996.

105. Voelkel-Johnson C, Entingh AJ, Wold WS, Gooding LR, Laster SM: Activation of intracellular proteases is an early event in TNF-induced apoptosis. *J. Immunol.* 154:1707, 1995.

106. Nakajima T: Degradation of topoisomerase II alpha precedes nuclei degeneration during adenovirus E1A-induced apoptosis and is mediated by the activation of the ubiquitin dependent proteolysis system. *Nippon Rinsho* 54:1828, 1996.

107. Beere HM, Chresta CM, Hickman JA: Selective inhibition of topoisomerase II by ICRF-193 does not support a role for topoisomerase II activity in the fragmentation of chromatin during apoptosis of human leukemia cells. *Mol. Pharmacol.* 49:842, 1996.

108. Browne S, Williams A, Hague A, Butt A, Parakeva C: Loss of APC Protein expressed in human colonic epithelial cells and the appearance of a specific low-molecular-weight form is associated with apoptosis *in vitro. Int. J. Cancer* 59:56, 1994.

109. Jensen PH, Cressey LI, Gjertsen BT, Madsen P, Mellgren G, Hokland P, Gliemann J, Doskeland SO, Lanotte M, Vintermyr OK: Cleaved Intracellular plasminogen activator inhibitor 1 in Human myeloleukaemia cells is a marker of apoptosis. *Br. J. Cancer* 70:834, 1994.

110. Hsu HL, Yek NH: Dynamic Changes of NuMA During the cell cycle and possible appearance of a truncated form of NuMA during apoptosis. *J. Cell Sci.* 109:277, 1996.

111. Gueth-Hallonet C, Weber K, Osborn M: Cleavage of the nuclear matrix protein NuMA during apoptosis. *Exp. Cell Res.* 233:21, 1997.

112. Earnshaw WC, Martins LM, Kaufmann SH: Mammalian Caspases: structure, activation, substrates and functions during apoptosis. *Ann. Rev. Biochem.* 68: in press, 1999.

12

Qualitative and Quantitative Methods for the Measurement of Ceramide

Tim R. Bilderback, Kam M. Hoffmann, and Rick T. Dobrowsky

CONTENTS

12.1   Introduction

Ceramide has recently emerged as an important signaling molecule in response to cellular stress and as a mediator of apoptosis.[1] A critical aspect to the advancement of the role of ceramide as a signaling molecule has been the development of sensitive methods to accurately measure changes in ceramide levels. These methods have provided an important tool to the repertoire of analytic techniques used by investigators interested in the biologic role of ceramide. This chapter will provide a discussion of and protocols for several methods used to measure both qualitative and quantitative changes in ceramide levels. The methods assume some basic understanding of procedures involved in analytic biochemistry, i.e., chromatography, quantitative transfers, use of internal standards, and handling of radioisotopes, and are sufficiently detailed to provide newcomers to the field reliable and reproducible procedures for the measurement of ceramide.

12.2   Qualitative and Quantitative Methods
for Measuring Ceramide

Metabolic radiolabeling of ceramide is one of the easiest and most manageable methods for measuring ceramide levels, especially in large numbers of samples. This procedure typically uses [³H]palmitate or [¹⁴C]serine to label the fatty acyl chain or sphingolipid backbone, respectively, of ceramide and its precursor sphingomyelin (Figure 12.1). This method gives no measure of mass and therefore can only give qualitative analysis of ceramide changes. Nonetheless, this approach is very useful in screening assays to determine if particular experimental treatments may impact upon ceramide signaling.

   A few considerations about this approach to measuring ceramide are warranted. Previous studies have demonstrated that the various cellular pools of sphingomyelin, the direct precursor of ceramide in response to many agonists, are metabolically labeled at different rates.[2] Therefore, it is important to incubate the cells with the radioactive precursor for a sufficient duration to ensure that all the cellular pools of sphingomyelin are labeled to metabolic equilibrium. Failing to do so may abrogate the ceramide signal in response to treatment because the signal-sensitive pool of sphingomyelin did not sufficiently incorporate the label. This duration is directly related to the doubling time of the cells. In general, incubating the cells with radiolabel for three population doublings should be sufficient to bring all the cellular pools of sphingomyelin to metabolic equilibrium with the radiolabel.

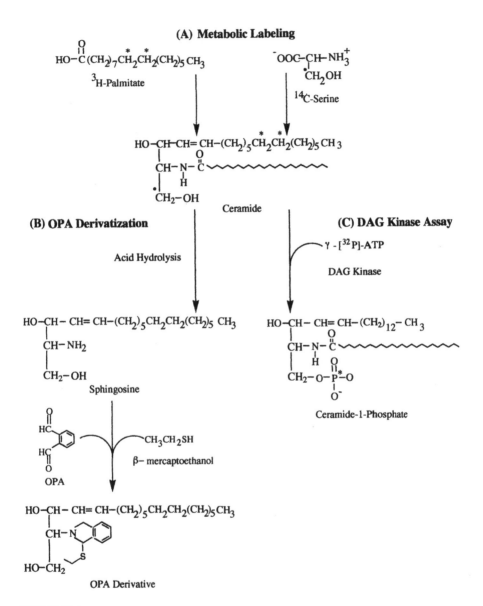

FIGURE 12.1

Schematic of methods for ceramide analysis. (A) Metabolic labeling. The position of [³H]palmitate and [¹⁴C]serine after metabolic incorporation into ceramide is indicated by the ˙ and • , respectively. Palmitate may also undergo N-acylation. (B) OPA derivatization. Acid hydrolysis releases the fatty acid from ceramide. The free long-chain base (sphingosine) is then reacted with o-phthalaldehyde (OPA) in the presence of β-mercaptoethanol, forming a fluorescent OPA derivative of sphingosine. (C) DAG kinase assay. DAG kinase transfers ³²PO₄ from γ-[³²P]ATP to ceramide-producing ceramide-1-[³²P]phosphate. The position of the ³²P is indicated by ˙.

An additional consideration when metabolically labeling ceramide is whether or not the radiolabeled precursor is equally utilized by the different metabolic pathways that produce ceramide. For example, some *in vivo* studies have erringly used a short incubation of cells with lysophosphatidylcholine containing [14C]palmitate in the *sn*-1 position of glycerol to label ceramide.[3] Since the acylation of sphingomyelin and ceramide does not directly involve donation of a fatty acyl group from lysophosphatidylcholine,[4] the amount of label incorporated into these molecules would be expected to be rather low and would undoubtedly require a very long incubation to reach equilibrium. In the absence of more convincing evidence on the biochemical mechanisms justifying this strategy,[3] lysophosphatidylcholine is an extremely poor choice as a precursor. Further, when metabolically labeling ceramide, it is important to consider the effect of availability of the precursor in the medium. Thus, when using radiolabeled serine as a precursor, incorporation will be greatest in serine-deficient media.

The first quantitative method to be discussed for determining ceramide levels employs the strategy of derivatizing the sphingoid base of ceramide. Since this method involves the isolation and degradation of ceramide to sphingosine, an internal standard is added to a lipid extract. The internal standard is chemically similar to natural ceramide and is expected to undergo the same amount of loss during the analytic workup, but has a distinct chromatographic migration enabling separation from endogenous ceramide. By calculating the ratio of the analyte signal (ceramide) to that of the internal standard, the amount of ceramide in a biologic sample can be determined by extrapolation from a standard curve constructed using the internal standard and an authentic ceramide standard.

The derivatization of the sphingoid base of ceramide is accomplished with *o*-phthalaldehyde (OPA).[5] OPA derivatizes the primary amine of the sphingoid base, forming a fluorescent derivative which can be separated by high performance liquid chromatography (HPLC) and detected fluorometrically. This procedure has low pmol sensitivity and permits a true mass measurement. The major drawback to OPA derivatization is that there is a need to purify ceramide and perform acid hydrolysis before derivatization. This increases the difficulty of handling a large number of samples, but is readily accomplished with a little practice. Additionally, this procedure requires a specialized detector which may not be readily available with many HPLC systems.

The second quantitative procedure involves the use of the enzyme diacylglycerol kinase (DAG kinase) to enzymatically convert ceramide to ceramide phosphate.[6-8] Although DAG kinase preferentially phosphorylates diacylglycerol to phosphatidic acid, under the correct conditions, the enzyme will also quantitatively phosphorylate ceramide. In the presence of [γ-32P]ATP the radiolabeled ceramide phosphate is easily detected and can be quantitated by scintillation spectrometry or by densitometry using a phosphoimager. Similar to OPA derivatization, the sensitivity of the DAG kinase assay is in the low pmol range. Further, crude lipid extracts may be used for analysis, making this method reasonable for handling a large number of samples.

The ensuing sections will describe the procedures involved for metabolic radiolabeling, formation of OPA derivatives of the long chain bases of ceramide, and labeling of ceramide with $[\gamma\text{-}^{32}P]$ATP by DAG kinase. These assays are easily adaptable for numerous cell types and in some cases for tissue samples.

12.3   Protocols for the Qualitative and Quantitative Determination of Ceramide

12.3.1   Metabolic Labeling

12.3.1.1   *Materials*

[^{3}H]Palmitate (30 to 60 Ci/mmol) is from American Radiolabeled Chemical. [^{14}C]Serine (50 to 62 mCi/mmol) is from Amersham. Type III ceramide for lipid standards is from Sigma. Thin-layer chromatography Silica Gel 60 plates are from Merck.

12.3.1.2   *Cell Culture*

The critical consideration is the initial cell density. Cells should be seeded so that at the required density for experimental treatment, the cells will have incubated for 48 to 72 h or three doubling times in radiolabeled medium. This time may need to be increased if the doubling time for the cells is particularly long. We generally incubate fibroblast and PC12 cell lines for 72 h in the presence of the radiolabel. Freshly split cells are labeled with [^{3}H]palmitate or [^{14}C]serine in complete or serine-deficient medium which contains 0.5 to 1 µCi/ml of the radioisotope. Following the incubation, the medium should be removed, the cells washed with fresh medium or phosphate-buffered saline, and then placed in fresh medium lacking the radiolabel for a period of 2 to 6 h. Previous studies have demonstrated that simple medium changes dramatically increases the level of sphingolipid metabolites.[9] The levels of these metabolites typically return to baseline levels after 2 to 6 h, although this may need to be determined empirically for any given cell type.

12.3.1.3   *Lipid Isolation*

Following experimental treatment, the cells may be harvested by adding 1 ml of any standard lysis buffer (detergent is not absolutely necessary) and scraping the cells from the plate. Nonadherent cells may be pelleted by centrifugation before this treatment. An aliquot of the lysate is reserved for analysis of total protein. Alternatively, cells may be scraped directly into 2 ml of ice-cold methanol. After adding one 1 ml of $CHCl_3$ proceed with the lipid extraction described below.

Lipids are extracted using a modification of the method of Bligh and Dyer.[10] Briefly, to a 13 × 100-mm screw-cap test tube containing 3 ml of $CHCl_3$: CH_3OH (1:2, v/v) add 0.8 ml of the lysate and mix the solution. If the cells were extracted directly into methanol, a little less water (about 0.6 to 0.7 ml) is necessary, since some liquid is always recovered with the cells. At this point the solution should be a monophase. If the solution is not forming a monophase, add a few drops of methanol to break the emulsion and achieve a monophase. After 5 min, phase separation is achieved by the addition of 1 ml of $CHCl_3$ and 1 ml of 1% perchloric acid or water. Mix the samples well by vigorous shaking and centrifuge for 5 to 10 min at room temperature. Centrifugation in a table-top clinical centrifuge at 2000 to 3000 rpm is sufficient. After centrifugation, the upper aqueous layer is aspirated and an aliquot of the organic phase is transferred to a fresh tube and dried under nitrogen gas, in a speed vac, or in an oven at 45°C .

12.3.1.4 Alkaline Methanolysis and Thin-Layer Chromatography

Using [³H]palmitate, it is especially advisable to base hydrolyze the samples prior to chromatography, since many glycerophospholipids will also incorporate the palmitate label and interfere with the chromatography of ceramide. To accomplish this, resuspend the dried lipid samples in 1 ml of $CHCL_3$ and add 0.1 ml of $2N$ KOH dissolved in CH_3OH. Incubate the tubes for 1 h at 37°C. During this incubation a white precipitate of glycerophosphate will form due to the hydrolysis of the acyl groups. Neutralize the samples with 0.1 ml of $2N$ HCL in CH_3OH and add 0.3 ml of CH_3OH. Complete a Folch extraction by the addition of 0.2 ml of water.[11] Vortex and centrifuge the sample. Carefully aspirate the upper aqueous layer and quantitatively transfer an aliquot of the 1 ml organic phase to a fresh tube. Add 5 μg of cold ceramide standard and evaporate the solvent as described above. The lipid residue is dissolved in 40 μl of $CHCl_3$ and an aliquot of this sample is spotted on a 20 × 20-cm TLC plate about 2 cm from the bottom of the plate. The plate may be developed in a solvent system consisting of $CHCl_3$:CH_3OH:NH_4OH (200:25:2.5 [v/v]).[12]

Here are a few cautionary notes on the practice of TLC that many newcomers to lipidology overlook. Be sure to use a fully equilibrated TLC chamber. This is accomplished by placing a sheet of Whatman paper inside the tank to serve as a wick and allow the solvent to fully saturate the paper. This usually takes several hours and should be prepared before the samples are spotted on the TLC plate. Do not individually add the solvent components directly to the TLC tank, especially in the presence of the wick. Mix all the components of the solvent together in a flask before pouring the solvent into the TLC chamber. Use only enough solvent so that when the TLC plate is placed in the chamber, the solvent covers about the bottom 1 cm of the plate.

Be sure that the solvent will not cover the area of the plate spotted with your sample when the plate is placed in the chamber. Once the plate has been inserted into the chamber, do not open the chamber until the solvent has migrated about 2 to 4 cm below the top of the plate. Two TLC plates may be developed at the same time in one chamber, but it is advisable to place both plates in the chamber together, ensuring that they do not touch and that they are both sufficiently in the solvent. Once the solvent front has reached the desired position, remove the plate from the chamber, rapidly mark the position of the solvent front with a pencil, and place the plate in a fume hood to dry.

Following chromatography, the isolated lipids may be visualized using iodine vapors. In a fume hood, place some crystalline iodine in a clean TLC chamber and allow the vapors to accumulate for several hours. Be careful when opening the tank as the iodine vapors are very toxic. Do not open the tank with your face near the opening, even with the fume hood on. Place the plate in the tank and after 2 to 5 min, depending upon the amount of lipid, iodine positive spots can be visualized as yellow bands. Remove the plate from the chamber, mark the bands with a pencil, and allow the iodine to dissipate from the plate in the fume hood.

Alternatively, radioactive bands may be visualized by fluorography. Following chromatography and after the solvent has evaporated, spray the plate evenly with En^{3}Hance (New England Nuclear). It is not necessary to saturate the silica gel with the spray. To help orient the film after development, mark the plate with 3 drops of radioactive dye (add 1 μl of [^{14}C]serine stock solution to 200 μl of bromophenol blue or India ink). Wrap the plate in a single layer of plastic wrap and expose it to film for 2 to 3 days at –80°C. After developing the film, place it on top of the TLC plate and line up the radioactive dye spots on the film with those on the plate. Firmly holding the film in place, mark the radioactive ceramide bands with a pencil.

To determine the amount of radioactivity in each spot, spray the silica gel with a water mister. Thoroughly saturate the gel with water without causing an excess of water run off. This procedure minimizes the generation of radioactive silica dust when scraping the silica from the plate. Using a sharp razor, scrape each band outlined by pencil onto a piece of weigh paper and transfer the moist silica to a scintillation vial containing 8 ml of scintillation fluid. Count the radioactive content of each sample in a scintillation spectrometer using a channel appropriate for the radioisotope. Alternatively, a densitometer may be used to quantitate the radioactive bands on the film. For accurate quantitation, be sure that the intensity of the bands is in the linear range of the densitometer.

12.3.1.5   *Protein and Phospholipid Measurement*

To normalize for variations in the number of cells extracted, the data from scintillation counting or densitometry may be normalized to protein or phospholipid content. Protein levels are measured using any standard protein assay. Using the specific activity of the starting isotope the mols of labeled lipid can be calculated and normalized to mg of protein. Alternatively, an aliquot of the organic extract may be saved prior to base hydrolysis and used to determine the total phospholipid content. Although several procedures are available, we prefer the method of Ames and Dubin.[13]

Duplicate aliquots of sample from the organic layer of the initial lipid extraction as well as standards of NaH_2PO_4 (0 to 80 nmols) are aliquoted into 13×100-mm test tubes. Be sure not to use aliquots of the organic extract following base hydrolysis, as most of the phospholipids have been lost. Following addition of 100 μl of ashing buffer (10% $Mg(NO_3)_2$ in ethanol [w/v]), the samples are dried down at 80°C in an oven. The samples are ashed in a strong flame just until the generation of the brown gas is complete. Ashing too long will char the sample, producing a black residue which will interfere with the spectrophotometery. After the tubes have cooled, add 0.3 ml of 0.5 N HCl and boil for 15 min to hydrolyze pyrophosphates. To each sample add 0.6 ml of 0.42% acid ammonium molybdate stock (4.2 g ammonium molybdate in 1 l of 1 N H_2SO_4) and 0.1 ml ascorbic acid (10% w/v, made fresh in water). Mix the tubes thoroughly and incubate for either 30 min at 45°C or 60 min at 37°C. After incubation, the OD_{820} is read to determine nmoles of phospholipids. A standard curve is constructed and the nmols of phosphate in the samples are extrapolated from the curve.

12.3.2   OPA Derivatization of Long-Chain Sphingoid Bases

12.3.2.1   Ceramide Isolation and Acid Hydrolysis

The cells are recovered as described above, aliquots are saved for protein determination and the lipids are extracted using the method of Bligh and Dyer.[10] Transfer an aliquot of the crude extract to a fresh tube and evaporate the solvent. Resuspend the residue in 40 μl of $CHCL_3$ and quantitatively apply the samples to a TLC plate. The lipids are resolved by developing the plate in $CHCL_3$:CH_3OH:triethylamine:2-propyl alcohol:0.25% potassium chloride (30:9:25:18:6).[14] The location of ceramide is determined by comparison to a ceramide standard using the iodine vapors to detect the lipid and 5 nmol of N-acetyl-C20-sphinganine are applied directly to the ceramide spot of each sample. The TLC plate is misted with water and the ceramide spots are scraped off and transferred to a fresh 13×100-mm screw-cap tube. Ceramide is eluted from the silica gel using 2×1 ml of $CHCL_3$:CH_3OH (2:1) followed by 1 ml of methanol. Add the solvent to the silica gel, vortex the samples vigorously, and sediment the silica by brief centrifugation. These eluates are combined in a fresh tube and evaporated.

Since the OPA derivatization requires a free amine group, the ceramide sample must be acid hydrolyzed prior to derivatization. This is accomplished by adding 1 ml of 0.5 M HCl in methanol and incubating the sample at 65°C for 15 h[14] in a 13×100-mm screw-cap test tube. Be sure to tightly cap the tube with a Teflon-lined screw cap. Do not use a rubber-lined screw cap.

After cooling, the sample is then neutralized with 1 ml of 1M KOH in methanol, and 1 ml of $CHCL_3$ is added. Phase separation is achieved by the addition of 1 ml of 1 M NaCl and 1 ml of $CHCL_3$; the mixture is vortexed and centrifuged in a bench top centrifuge for 5 min. The aqueous layer is aspirated and the organic phase is washed several times with water containing 50 µl of 1 N NH_4OH per 15 ml of water. The free long chain bases in the organic layer are then dried under nitrogen.

12.3.2.2   Formation of the o-Phthalaldehyde Derivative of Long-Chain Bases

The ceramide mass is determined using the *o*-phthalaldehyde derivatization of the long-chain base.[5] The long-chain base is dissolved in methanol (50 µl) and mixed with 50 µl of freshly prepared OPA reagent. To prepare the OPA reagent mix 99 ml 3% (w/v) boric acid in water, adjusted to pH 10.5 with KOH, 1 ml of ethanol containing 50 mg of *o*-phthalaldehyde and 50 µl of β-mercaptoethanol. Allow the solution to sit in the dark for 10 min before use. The OPA reagent is incubated with the sample for 5 min at 25°C and 500 µl of methanol:5 mM potassium phosphate (pH 7.0) (90:10, v/v) is added. The samples are then cleared by centrifuging for 1 minute in a microcentrifuge.

12.3.2.3   HPLC Analysis

The OPA derivatives are injected onto a C18 (5 mm) column and eluted with CH_3OH:5 mM potassium phosphate, pH 7.0 (90:10, v/v). An excitation wavelength of 340 nm and emission wavelength of 455 nm are used for detection of the *o*-phthalaldehyde derivatives. Under these conditions, the internal standard C-20 sphinganine will have a longer retention time than sphingosine. The amount of ceramide recovered is calculated by integrating the peak areas for the sphingosine-OPA and C-20 sphingosine OPA derivative and taking the ratio of sample to the internal standard. Optimally, quantitation of the sphingosine bases is accomplished by the construction of a standard curve using known amounts of ceramide and the internal standard, which have been subjected to the same analytic workup as the samples. The standards should be spiked into lipid extracts prepared from untreated cells so that the standards and samples are recovered from the same biologic matrix. The standard curve is linear in the region from 20 to 250 pmol. The amount of ceramide per sample should be normalized to mg of protein extracted.

12.3.3   Diacylglycerol Kinase Assay

12.3.3.1   Materials

Type III ceramide standard is purchased from Sigma. [γ-^{32}P] ATP is purchased from New England Nuclear. *n*-Octyl-β-D-gluco-pyranoside (β-octylglucoside) is purchased from Calbiochem (catalog #494459). L-α-Dioleoylphosphati-dylglycerol is obtained from Avanti Polar Lipids. *E. coli* DAG kinase may be purchased from Boehringer Mannheim.

12.3.3.2 Lipid Extraction

Lipids are extracted from cell samples by the modified method of Bligh and Dyer as described above. Since the DAG kinase assay is very sensitive,[6-8] we typically use 0.5 ml or less of the 2 ml organic layer for determination of ceramide from lipid extracts prepared from whole cell lysates. Aliquots of the remaining organic phase can be used for the determination of total phospholipid phosphate.

12.3.3.3 Diacylglycerol Kinase Assay Reagents

2× Buffer

100 mM imidazole (pH 6.6), 100 mM LiCl, 25 mM MgCl$_2$, 2 mM EGTA (pH 6.6)

Dilution Buffer

10 mM imidazole (pH 6.6), 1 mM diethylenetriaminepentaacetic acid (DTPA)

Mixed Micelles

Mixed micelles are made by drying 0.97 ml of 20 mg/ml L-α-dioleoyl-phos-phatidylglycerol (DOPG) under nitrogen. To the dried DOPG, add 1 ml of 7.5% β-octylglucoside. Vortex and sonicate the mixture until the DOPG is completely dissolved. Alternatively, store the samples at 4°C overnight to hydrate the lipid and vortex vigorously the next day. The lipid will easily go into solution with minimal or no sonication. We typically prepare a 10 ml batch of mixed micelles and store the solution in 1 ml aliquots at –20°C.

The indicated grade of β-octylglucoside can be used without further puri-fication. Less pure grades of detergent may give spurious results and will require recrystallization from acetone at –20°C.[6] Recover the crystals by filtration through a chilled fine sintered glass funnel, wash the crystals with 200 to 500 ml of ice-cold ethyl ether, and dry in a vacuum dessicator.

12.3.3.4 DAG Kinase Assay Method

Transfer an aliquot of the organic layer of the lipid extracts from the samples to new 13 × 100-mm screw cap-test tubes. Similarly, aliquot ceramide type III standards (0 to 640 pmol) in duplicate to prepare the external standard curve. Evaporate the solvent by one of the methods described above. The lipid residue is then resuspended in 20 μl of the mixed micelles by vortexing, and 70 μl of the reaction mixture is added to each sample. The appropriate amount of reaction mixture may be easily prepared by following Table 12.1.

TABLE 12.1

Preparation of DAG Kinase Reaction Mixture

Solution	Volume/Assay	Number of Assays	Total Volume
2× Buffer	50 μl		
0.1*M* DTT	2 μl		
0.5 mg/ml DGK membrane	10 μl		
Dilution buffer	8 μl		

* The indicated volume of reagent for each assay is indicated in column two. Enter the number of assays to be performed in column 3 and multiply column 2 by column 3 to get the total volume of reagent needed. Mix all the reagents thoroughly and add 70 μl per assay.

An ATP mixture is made (2 m*M* ATP (in dilution buffer) containing 4 μCi [γ-^{32}P] ATP per assay) and 10 μl is added to each sample to initiate the reaction. The reaction is allowed to proceed for 30 min at room temperature and is stopped by the addition of 3 ml of CHCL$_3$:CH$_3$OH (1:2). Next, 0.7 ml of water is added and the tubes are mixed and allowed to sit for at least 5 min. Phase partitioning is achieved by the addition of 1 ml of CHCl$_3$ and 1 ml of 1% perchloric acid. The samples are centrifuged for 5 min at 2000 × g and the upper phase is discarded as radioactive waste. The lower phase may be washed again if desired. If an emulsion forms upon washing, add a few drops of methanol. Transfer 1.5 ml of the lower phase to a fresh tube and evaporate the solvent. When dry, the samples are resuspended in 40 μl CHCl$_3$ and 20 μl is immediately spotted on a TLC plate. The lipids are resolved by development in chloroform:acetone:methanol:acetic:acid:water (10:4:3:2:1). Mark the plate with radioactive ink and expose to film overnight at –80°C. No En^{3}Hance spraying is necessary. The film is developed and the ceramide phosphate spots are located by comparison to the standards. The radioactive products are scraped from the plate and the radioactivity quantitated as detailed in Section 12.3.1.4 Alternatively, the ceramide may be quantitated using a phosphoimager. The amount of ceramide per sample is extrapolated from the standard curve and normalized to either total phosphate or protein. Alternatively, the mass of ceramide can be calculated from the specific activity of the ATP after correcting for the purity of the cold ATP stock solution.[7] The m*M* extinction coefficient for ATP at 259 nm is 15.4.

12.4 Conclusions

The three assays which have been described are all useful analytical methods for determining ceramide levels. No one assay is best for all circumstances. It may be that none of these methods are appropriate for the investigation proposed or that a combination of methods is most appropriate. For studies of the metabolic movement of ceramide, it is likely that incorporation of a radiolabeled or fluorescent precursor would be the most appropriate method. For information about the structure of the long-chain base of ceramide, OPA derivatization would be the most valuable of the methods discussed. To rapidly obtain the mass of both ceramide and diacylglycerol present, the DAG kinase assay would be the most helpful. For more detailed information about ceramide structure, techniques such as gas chromatography/mass spectrometry analysis may be best. We have described some of the more easily applied methods for measuring ceramide, but other methods are available and may be more suited to the aims of your research.

Acknowledgments

This work was supported by grant MCB 9513596 from the National Science Foundation, a Career Development Award from the Juvenile Diabetes Foundation International, and by the Higuchi Biosciences Center at the University of Kansas to RTD. TRB is supported by a fellowship from the American Heart Association, Kansas Affiliate.

References

1. Hannun, Y.A. Functions of ceramide in coordinating cellular responses to stress. *Science*, 274, 1855, 1996.
2. Linardic, C.M. and Hannun, Y.A. Identification of a distinct pool of sphingomyelin involved in the sphingomyelin cycle. *J. Biol. Chem.*, 269, 23530, 1994.
3. Schutze, S., Potthoff, K., Machleidt, T., Berkovic, D., Weigmann, K., and Kronke, M. TNF Activates NF-κB by phosphatidylcholine-specific phospholipase C-induced "acidic" sphingomyelin breakdown. *Cell*, 71, 765, 1992.
4. Merrill, A.H.J. and Jones, D.D. An update of the enzymology and regulation of sphingomyelin metabolism. *Biochim. Biophys. Acta*, 1044, 1, 1990.
5. Merrill, A.H., Wang, E., Mullins, R.E., Jamison, W.C.L., Nimkar, S. and Liotta, D.C. Quantitation of free sphingosine in liver by high-performance liquid chromatography. *Anal. Biochem.*, 171,373, 1988.
6. Priess, J., Loomis, C.R., Bishop, W.R., Stein, R., Niedel, and Bell, R.M. Quantitative measurement of *sn*-1,2-diacylglycerols present in platelets, hepatocytes, and *ras*- and *sis*-transformed normal rat kidney cells. *J. Biol. Chem.*, 261, 8597, 1986.
7. Priess, J. Loomis, C. R., Bell, R.M., and Niedel, J.E. Quantitative measurement of sn-1,2-diacylglycerols. *Methods Enzymol.* 141, 294, 1987.
8. Van Veldhoven, P.P., Bishop, W.R., and Bell, R.M. Enzymatic quantification of sphingosine in the picomole range in cultured cells. *Anal. Biochem.*, 183,177,1989.

9. Smith, E.R. and Merrill, A.H.J. Differential roles of *de novo* sphingolipid bio-synthesis and turnover in the burst of free sphingosine and sphinganine, and their 1-phosphates and N-acyl-derivatives, that occurs upon changing the medium of cells in culture. *J. Biol. Chem.*, 270, 18749, 1995.

10. Bligh, E.G. and Dyer W.J. A rapid method of total lipid extraction and purification. *Can. J. Biochem. Physiol.*, 37, 911, 1959.

11. Folch, J., Lees, M., Sloane, and Stanley, G.H. A simple method for the isolation and purification of total lipides from animal tissues. *J. Biol. Chem.*, 226, 497, 1957.

12. Liu, P. and Anderson, R.G.W. Compartmentalized production of ceramide at the cell surface. *J. Biol. Chem.*, 270, 27179, 1995

13. Ames, B.N. and Dubin, D.T. The role of polyamines in the neutralization of bacteriophage deoxyribonucleic acid. *J. Biol. Chem.*, 235, 769, 1960.

14. Nikolova-Karakashian, M., Morgan, E.T., Alexander, C., Liotta, D.C., and Merrill, A.H.J. Bimodal regulation of ceramidase by interleukin-1β. *J. Biol. Chem.*, 272, 18718, 1997.

13

In Vivo Neuronal Targeting of Genes

Andrea Amalfitano

CONTENTS

13.1   Introduction

The editors of this book have set out to compile "state of art" literature concerning aspects of the neurobiology of apoptosis. This coincides with the rapid knowledge currently being generated regarding the underlying genetic causes of multiple neurogenetic conditions. Many of these neurogenetic disease processes can be broadly classified as either dominant or recessive conditions. Recessive conditions are usually due to the lack of expression of specific genetic information, while dominant conditions may be caused by abnormal production of a normal (or abnormal) protein product.

> *Understanding the neurobiological role of apoptosis in each of these types of conditions will have direct impact upon potential therapeutic modalities. The impact of apoptosis on these conditions can be explored by the in vivo neuronal targeting of genes active in the apoptotic pathways of neuronal cells.*

This simple statement belies an enormous complexity of technologic issues that must be addressed. To overcome each of these obstacles will require the most sophisticated and exotic applications of molecular biology currently available. Therefore, this chapter has been organized in a manner that addresses some of the parameters that must be considered by any *in vivo*, neuronal cell-targeting strategy. Specifically, we will review some of the more successful methods that can be utilized to facilitate delivery and expression of genes active in the apoptotic pathways of neuronal cells *in vivo*. It is hoped that this summary will allow an investigator to intelligently decide whether a particular method of exploration will be relevant to the specific hypothesis he or she is attempting to prove or disprove. Before discussion of some of these methods, a brief consideration of caveats regarding gene expression in general is presented below.

13.2   Efficient Expression of Neuron-Targeted Transgenes

13.2.1   Polyadenylation and Splicing Elements

While a number of genes active in apoptosis can be hypothesized to be useful in the potential investigation and/or treatment of neurogenetic diseases (see Chapter 3), the potential of each respective gene will only be realized after full evaluation *in vivo*. This will require that the specific gene sequence (via utilization of a transgene cassette) be expressed in a suitable manner within the neuronal cells being investigated (targeted). Expression of the gene requires not only the specific cDNA sequence, but also specific RNA transcription enhancing signals necessary for correct and high level transcription. For example, a polyadenylation signal should be present downstream of the gene to permit correct processing and stabilization of the resultant transgene-derived RNA transcript. Usually, well defined polyadenylation sequences derived from DNA viruses such as simian virus 40 (SV-40) are incorporated into the 3' end of the transgene construct. The SV–40 derived polyadenylation signal is functionally encompassed within a 200-base pair fragment of DNA. The processing and stability of the RNA transcript can be further enhanced by the inclusion of intronic elements flanking the specific gene-coding region. Intronic elements allow splicing of the primary RNA transcript and further enhance the stability of a given primary RNA transcript, resulting in increased expression within a targeted cell. Several intronic elements have been characterized and utilized, including a small intronic sequence derived from the major late promoter transcript of the human adenovirus.[1] If size constraints are not an issue, inclusion of the native introns and polyadenylation signals originally present in the genomic location of the normal gene will in many instances provide for optimal RNA processing and stability.

13.2.2 Enhancer/Promoter Elements

13.2.2.1 Virus-Derived Enhancer/Promoter Elements

While each of the elements presented above serve to augment transgene transcription, high level expression of the transgene is critically dependent upon the transcriptional enhancer/promoter element utilized to initiate RNA transcription from the respective transgene cassette. High level transgene expression is the key to maximally impacting upon the neurobiologic process being investigated. Transcriptional enhancer/promoter elements derived from human cytomegalovirus (CMV) or Rous sarcoma virus (RSV) are routinely utilized to facilitate high level transcription of transgenes in multiple cell and tissue types, including cells of a neuronal lineage.

13.2.2.2 Neuronal Cell-Specific Enhancer/Promoter Elements

Enhancer/promoter elements that are known to direct gene expression in a neuronal cell type specific fashion have also been utilized, though the levels of RNA expression derived from these elements is (in general) quantitatively less than that obtained with the use of viral enhancer/promoter elements. In many instances, however, it may be desirable to restrict expression of a transgene (i.e; potentially toxic transgenes such as those that induce apoptosis) to a specific desired cell type, even if decreased levels of overall gene expression will be the result. This level of control can be facilitated by the use of neuronal tissue-specific enhancer/promoter elements. A number of these elements have been defined and utilized for *in vivo* studies. For example, neuron-specific enhancer/promoter elements include those derived from the tyrosine hydroxylase gene, the dopamine β-hydroxylase gene, the synapsin I and II genes, the transferrin gene, the HGPRT gene, the dystrophin gene, and many others.[2-13] Each of these elements has tissue-specific limitations and/or benefits that can only be adequately assessed in the context of the specific question being investigated; therefore the reader is referred to the original sources to explore these important issues in the necessary detail.[2-13]

13.2.2.3   Inducible Enhancer/Promoter Elements

Expression of a transgene during a critical period of time is another level of expressional control that should be considered in any neuronal targeting strategy. The temporal regulation of transcription during defined periods of time can be achieved via the use of inducible (and potentially tissue-specific) enhancer/promoter elements. Utilization of these types of systems can allow the investigator not to only restrict expression of a transgene to a specific cell type (such as neuronal cells) *in vivo,* but also to exogenously control the temporal expression of the transgene cassette during any point of neuronal development. Several systems have been recently described that allow for both tissue specific *and* temporal control of transgene expression *in vivo.*[14-18] By far, the most successfully utilized system for use in living animal systems has been the tetracycline-responsive transactivating system established by Dr. H. Bujard and colleagues.[19-21] These investigators have devised a unique "on-off switch" that allows for the temporal control of any gene, both *in vitro* and *in vivo.*

This binary gene expression system is composed of two parts, the target gene of interest, and a tetracycline-responsive transactivator protein. Briefly, *E. coli* has evolved an extremely sensitive method for restricting expression of the tetracycline-resistance genes (*TetR*) encoded by the TN10-encoded transposable element. The *TetR* genes are flanked on their 5′ end by a bacterial promoter known as the tetracycline-operator sequence (tet-op). Tet-op is physically bound by the tet-repressor protein, effectively preventing transcription of the *TetR* genes. In the presence of extremely low levels of tetracycline, the tet-repressor protein is prevented from binding the tet-op sequence, and transcription of the *TetR* gene ensues. This exquisitely sensitive gene operon has evolved to specifically allow the bacteria to express the tetracycline resistance genes only when exposed to tetracycline compounds in their growth medium.

As depicted in Figure 13.1, this system has been modified to enable temporal (tetracycline-responsive) control of gene expression in mammalian cells. The gene of interest is now placed under the transcriptional control of the tet-op sequence (transcriptionally silent in mammalian cells, since the tet-op sequence is a bacterial promoter). The tet-repressor protein has been molecularly fused to the highly acidic domain of the herpes-derived VP-16 protein (a potent mammalian transactivating protein) to form the tetracyline-responsive transactivator, tTA. If tTA is present, a tet-op–linked gene will be expressed in mammalian cells, but if tetracycline is also present, the tTA will be prohibited from binding the tet-op sequence, preventing transcription initiation. Therefore, by merely adding or removing tetracycline from the medium (or from the drinking water of transgenic animals; see below) one can exogenously control the expression of any tet-op–linked gene that is transcriptionally dependent upon the tTA. The level of transcription control afforded by these types of systems could potentially be utilized to explore the impact of genes active in the apoptotic pathways of neuronal cells, *in vivo, as described below.*

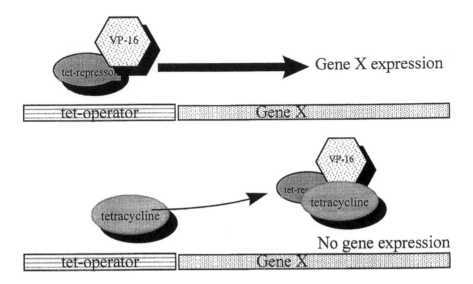

FIGURE 13.1
Schematic representation of the tetracycline-responsive mammalian transactivation system. Note that in the presence of tetracycline, tet operator-linked gene expression is *prevented*.

13.3   Specific, *In Vivo* Targeting Strategies

In this section we will review methods of introducing genetic material *in vivo*, to allow expression of relevant gene products in neuronal cells. While there are a number of potential ways of achieving this goal, we have chosen to focus attention on two very successful methods, those that utilize the benefits of transgenic mouse technology, and those utilizing vectors for somatic cell gene transfer.

13.3.1   Transgenic Animals (Mice)

Cell culture systems allow the investigator a convenient system for assessing gene activities in a controlled setting. Specific questions regarding the role and interactions of a specific gene in the biology of various tissue culture cell lines can be explored via these systems. Eventually, however, evaluation of the gene activity in the context of the living animal is an outgrowth of initial tissue culture evaluations. One of the most successful methods of achieving high level expression of a gene in the living animal is via the use of transgenic mouse technology. Transgenic technology has been refined to such a level of technical expertise that it is now routinely being made generally available to researchers either via "transgenic core labs" in academic institutions, or *via* commercial vendors. The production of a transgenic animal however, needs to be carefully assessed before dedicating ones time and resources to such explorations.

In general, a transgene cassette is constructed via the ligation, subcloning, and eventual release of a transgene that is potentially capable of allowing adequate expression of the gene of interest, in the appropriate tissues. Usually, the transgene encompasses an enhancer/promoter element, the coding region of the respective gene, and other RNA processing signals such as polyadenylation signals, and/or splicing signals. Once correctly assembled, transient transfection of the plasmid into appropriate tissue culture cells to functionally assess the resultant protein product is recommended when possible. Because of the multiple subcloning steps required to generate a typical transgene cassette and the possibility of a point mutation inadvertently being introduced into the cassette (prohibiting adequate expression of a functional protein), functional analyses are highly recommended. If one cannot adequately assess function in tissue culture systems, actual DNA sequence analysis of the transgene may reassure the investigator that the transgene can express a potentially functional protein. The transgene portion of the bacterial plasmid is then released (bacterial plasmid sequences are toxic to developing embryos if included), purified, and prepared for microinjection into fertilized one-cell embryos. The quality of the DNA is of utmost importance, and requires meticulous purification, since any contamination with impurities will result in a significantly diminished capability to eventually isolate the desired transgenic animal.

Prior to microinjection of the transgene, female mice are superovulated via hormonal treatment, and then mated with the appropriate male mouse (choice of mouse strain varies from lab to lab) and fertilized eggs are harvested the next day. Via micromanipulatory technology, each individual embryo nucleus is microinjected with ~1 picoliter of an appropriate buffer containing the transgene DNA at a concentration of 1 to 5 µg/ml. The injected eggs are then pooled and surgically implanted into the uterus of a pseudopregnant female mouse. Delivery of pups derived from the transfer should occur approximately 3 weeks after surgical implantation. We have superficially described this extremely labor-intensive, technically challenging technique, but it should be clear that those individuals/groups who repeatedly and successfully do this type of work have invested great efforts to allow for favorable outcomes to be routinely achieved. Timing of egg harvesting, the process of storing, identifying and confirming injection of the eggs, pooling of the viable eggs after microinjection, surgical adeptness, successful egg transfer, and final implantation and delivery of the eggs in appropriately preconditioned pseudopregnant female mice are just some of the hurdles that are always faced in this procedure. However, because of the incredible usefulness of transgenic animals for progress in biomedical research, dedicated transgenic core labs at universities and biotechnology companies have increased in number, all to improve the likelihood that an individual researcher can isolate an appropriate transgenic animal to explore previously unexplorable biological questions.

Once the pups are born, one must identify those individuals that actually contain the transgene sequences integrated into their genomes. This can be done by isolating a small tail snip from each individual animal (after appropriate tagging), extracting DNA from the tail tissue, and analyzing via the polymerase chain reaction or Southern blot analysis for the presence of the transgene sequences. Rapid processing of multiple, crude tail derived DNA preparations for these analyses has been described.[22] The individuals found to contain the transgene sequences in their tail DNAs are typically referred to as the F_0 generation. Since each F_0 represents an individual derived from a single microinjected embryo, an F_0 offspring also represents a unique integration site for the transgene. Once all other potential problems have been overcome, the site of integration of the transgene will be the single most important factor regarding the level of transgene expression achieved in a given strain of transgenic animals. If the transgene integrates into a portion of the mouse genome that is transcriptionally silent, significant transcription from the transgene may be unachievable. Additionally, the achievement of tissue-specific expression via the use of various promoter/enhancer elements will also be significantly impacted upon by where in the genome the transgene has integrated. Finally, though less significant, the actual copy number of the transgene at the integration site may also affect overall transgene expression levels.

Once identified, each F_0 that has the transgene present is mated, and positively transgenic offspring derived from this mating are referred to as the F_1 generation of animals. Animals that are transgene-positive in the F1 generation must by default have the transgene DNA present in every cell throughout their body, since they are all derived from a single cell gamete from the F_0 parent. Note that this is very likely not the situation in the Fo generation, because integration of the transgene DNA may have not occurred until the embryo had undergone several rounds of cell division. Therefore, each Fo must be considered a potentially mosaic/chimeric animal. Indeed, because the gonads of F_0 individuals may not contain the transgene DNA, 1/3 to 1/2 of all F_0 transgenic animals will never transmit the transgene to their progeny, a further stumbling block on the road to isolating a transgenic line of animals.

Once F1 individuals are isolated, a large colony of transgenic animals heterozygous for the transgene can be generated from a single male transgenic F1 mouse. Once expanded, the colony can be maintained, while individual animals are removed and analyzed. Usually the first investigations are directed to determine which of the various F1 transgenic strains actually express the transgene of interest. Some strains may not express the transgene at all (due to the integration site and/or copy number of the transgene) or the expression may not be at a level that can be detected, or biologically relevant. Furthermore, expression may be present, but not in the desired tissue-specific spectrum. Finally, after much work and effort (6 months to a year) the investigator may eventually end up with a line or two of transgenic animals expressing significant levels of the transgene in a manner that impacts upon a particular cellular process or biological phenotype.

13.3.1.1 Examples of Neuronal Cell-Targeted Transgenic Animals

Once a strain of transgenic animals is isolated that is expressing the desired protein product, the transgene can be bred into various mouse backgrounds, to ascertain what impact the expression of the transgene has on the phenotype of the resultant animal. For example, Bcl-2 is an apoptosis regulatory protein, and transgenic mice that specifically expressed Bcl-2 in their neurons *via* neuron-specific promoters have been described.[23-26] Bcl-2 overexpression in the transgenic animals provided protection of their neurons from neurotrophic factor deprivation and/or axotomy, while underexpression of Bcl-2 in "knockout" animals predisposed their cerebellar neurons to be more susceptible to hypokalemic or nerve growth factor deprivation cell death.[23-27] There are quite a number of genetic animal models of neurological disease (such as for Parkinson's disease, spinal cerebellar atrophy-type I, Machado-Joseph disease, spinal-bulbar muscular atrophy, Huntington chorea, Alzheimer diseases, familial amyotrophic lateral sclerosis, and others) whose pathologic processes could also be explored by neuron-targeted transgenic expression of neuro-active apoptotic genes.[28-40]

13.3.1.2 Limitations of Transgenic Animals

With the ever-expanding variety of mouse models of neuropathology becoming available (either via isolation of spontaneous mutants, or by deliberate elimination of gene activities via gene knockout technologies), the expression of proteins active in the apoptosis pathways of abnormal neuronal cells becomes an interesting target for investigation. Obviously, the respective transgene must be expressed to high enough levels in the correct tissues to allow such investigation to be productive, but another caveat must also be considered, namely, the temporal expression of the transgene relative to a specific disease process. For example, it may be desirable to allow expression of a transgene only during critical periods of CNS development, either because expression of the transgene may be toxic to the developing transgenic animal, or in order to mimic human aspects of temporally progressive neurodegenerative processes.[39,40] In order to overcome these difficulties, one might want to consider a "genetic on–off switch" for use in transgenic animals. The tetracycline-responsive transactivating system previously described is a method that has been utilized in such a manner.

For example, it has been demonstrated that tTA-dependent transgene expression in animals can be exogenously regulated by the addition or removal of tetracycline from the drinking water of the animals.[41] We have utilized this method to study aspects of the the X-linked recessive condition Duchenne muscular dystrophy, in the *mdx* mouse, the murine equivalent of the human condition, by demonstrating that muscle-specific expression of transgenes via the tetracycline responsive transactivating system is achievable.[42] Recently, a group of investigators has demonstrated that the tetracycline-responsive trans-activation system can be utilized to express calcium–calmodulin-dependent kinase (CaMKII) specifically in the forebrains of transgenic animals.[43] The temporal expression of CaMKII was regulated by the addition or removal of tetracycline derivatives from the drinking water of the animals. As a result, the ability of the animals to store memory patterns was made dependent on the temporally regulated expression of the CaMKII transgene.[43]

In summary, the tools of molecular biology and transgenic technology are now available for an investigator to scientifically explore the role of genes active in neuoroapoptosis, *in vivo*, in a manner that could not have been approached only 10 years ago.

13.4   Gene-Transfer Vectors

13.4.1   *Ex Vivo* vs. *In Vivo* Gene Transfer

Ex vivo approaches to gene therapy are based on the following paradigm. Cells (i.e., neuronal cell precursors or fetal cells) have a specific gene inserted into them, and then the cells themselves serve as the vector by physically being introduced into the targeted region of the nervous system. While this form of gene transfer has shown promise in some settings, (i.e., for the potential treatment of Parkinson's disease), we will limit our discussion to the transfer of genetic material directly to cells already residing in the central nervous system, or, in other words, an *in vivo* gene transfer approach. Though each is unique, both approaches face the same initial dilemma, that is the stable, long-term introduction of genetic information into a neuronal cell *via* the utilization of a variety of potential vectors.

13.4.2   Vectors for Somatic Gene Transfer

13.4.2.1   *Nonviral-Mediated Gene Transfer*

Nonviral methods of gene transfer are a common method of gene transfer, and in the appropriate settings can allow the investigator to ask very sophisticated questions regarding gene functions/interactions at the cellular level. In many instances, this simply involves the physical mixing of naked DNA molecules with chemical substances to facilitate DNA transport from the extracellular to the intracellular (nuclear) compartments of cells. Many times this involves a polycationic chemical compound that *via* physical interaction sequesters the negative charges of the naked DNA, resulting in a complex that can facilitate transit of the DNA through the cellular milieu. There are several examples of successful gene transfer into neuronal cells utilizing these types of techniques, including a biolistics approach of hurling DNA (bound to gold beads) into cells after a controlled explosive force.[44-46] Although these methods work satisfactorily in tissue culture systems, the results are less compelling *in vivo*. It is clear that as the technology moves forward, nonviral methods of gene transfer into the CNS will become more efficient. However, for purposes of this discussion, we will move forward to those virally mediated vectoring systems that are presently widely available and which have been repeatedly demonstrated to allow for highly efficient gene transfer into the CNS.

13.4.2.2   Viral-Mediated Gene Transfer

Viruses have evolved to become extremely efficient vehicles for the transfer of their genomes into a host cell. Transduction efficiencies of tissues exposed to viruses can approach 100%, if the appropriate amount of virus is utilized. Therefore, a great deal of experimentation has been underway to modify viruses for transfer of transgenes into cells *in vitro* and *in vivo*. However, the benefit of extremely efficient transgene delivery must always be weighed against the many shortcomings of viral-mediated gene transfer technologies. These shortcomings are multiple, and their significance must be measured in the context of the question that is being addressed by the use of the vector. In this portion of our discussion, we will concentrate on some of the viral vectoring systems that have demonstrated great promise in their potential to transfer genetic information into the neuronal tissues of living animals. We hope that this discussion will allow an investigator to begin to understand the benefits and limitations of each system, in order to determine which may be the most appropriate for the specific hypothesis to be tested.

13.4.2.3   Herpes Simplex Virus Vectors

The herpes simplex virus is a neurotropic DNA virus that is composed of 150 kb of DNA organized into right- and left-hand ends. As a consequence of the normal neurotropism of the wild-type virus, HSV vectors are being intensively studied as vectors for transfer of genes into the nervous system, *in vivo*. Upon infection of neurons, the virus can either replicate and produce infectious virions, or it can enter a latent cycle in which most of the HSV genes are transcriptionally silent, allowing the virus to remain in neurons for extended periods of time. Although the HSV genome contains the genetic information to express well over 70 unique proteins, nearly half of these proteins could potentially be deleted without harming the viruses' ability to be grown to relatively high titers in tissue culture cells. Therefore, the combined attributes of natural neural tropism, large carrying capacity, and long-term persistence *in vivo* make the potential utilization of HSV vectors for *in vivo* neural cell specific gene delivery very attractive.

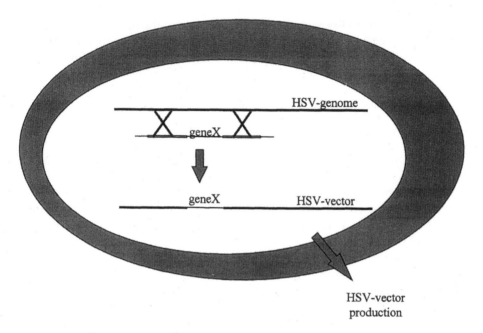

FIGURE 13.2
Schematic representation of HSV vector production in human cells. Note that nonrecombinant HSV viruses will also be produced at some frequency; therefore plaque purification of recombinant HSV vectors is required. Once the recombinant HSV vector is isolated, it can be propagated by simple reinfection of the human cell line.

Two types of HSV-based systems have been demonstrated to allow for at least transient expression of transgenes after HSV-mediated delivery *in vivo*. The first utilizes an approach that is very similar to other viral-based systems (Figure 13.2). The genomic DNA derived from a wild-type HSV is cointroduced into a VERO cell line along with a shuttling plasmid. The shuttling plasmid usually contains two critical elements: a specific transgene expression cassette that is flanked by regions of homology with the HSV genome. Recombination between the homologous regions of the two DNA molecules results in the generation of a recombinant HSV genome that contains a transgene cassette that has simultaneously replaced portions of the HSV genome.[47] Usually the transgene inserts in a manner such that HSV genes not required for growth in tissue culture are simultaneously deleted. The recombinant HSV vector, however, retains its neurotropic capabilities. Many regions of the HSV genome have been deleted in this manner, and each of the resultant vectors can be shown to direct transient (3 to 4 days) expression of transgenes in neurons *in vivo*. For example, neurons previously infected with a recombinant HSV vector expressing Bcl-2 were found to be protected from ischemic cell death.[48] Unfortunately, transient expression from HSV vectors limits their usefulness, a characteristic that is independent of the type of promoter utilized.[49,50]

Another form of HSV vecto rare referred to as HSV amplicons. These vectors are devoid of any HSV genes, except for important packaging elements required for HSV genome encapsidation into the mature viral particle. HSV amplicons are therefore *dependent* on contamination with a helper HSV virus (usually a mutant or recombinant HSV vector) for propagation in tissue culture. These types of HSV vector have also been demonstrated to allow for transgene delivery and expression in the nervous system of living animals, and in some reports for extended periods of time.[51,52]

13.4.2.3.1 Limitations of HSV-Based Vectors

Transient durations of transgene expression either by HSV mutant or amplicon based vectors has been a key limitation of HSV vectors *in vivo*. While there are some reports suggesting that HSV amplicon-based vectors may be capable of sustained transgene expression *in vivo*, these results are obscured by the presence of HSV helper virus contamination, possibly resulting in low level replication of the amplicon *in vivo*.[51,52] It is also clear that in many instances the HSV genome persists in neuronal cells that have been transduced with a respective transgene, suggesting that promoter shut-down is the cause of the transient nature of trangene expression in many experimental protocols.[50] The utilization of various enhancer/promoter elements (including viral derived or neuronal cell-specific) has not improved the situation.[49,50] Longer term expression from HSV vectors is being explored via the use of the HSV derived "latency-activated promoter (LAP)," a promoter that is persistently active during the chronic latent phase of HSV infection.[53] Recently, the deletion of multiple genes (ICP4, ICP0, ICP27, and ICP22 via the utilization of packaging cell lines supplying some of these gene activities *in trans*) in a mutant HSV vector has been reported to reduce cytopathic effects of the vector after *in vivo* use, along with sustained expression of transgenes.[49,54] These new generations of vector suggest that HSV-mediated gene transfer is an important technological tool that can potentially be utilized by investigators studying the *in vivo* impact of genes active in the apoptotic pathways of neurons.

13.4.2.4 Adenovirus Vectors

The human adenovirus (Ad) is a double-stranded DNA virus composed of 35,995 bp of DNA. Ads can infect a great variety of tissues, including muscle, liver, kidney, and brain. One extremely beneficial aspect of Ad biology is its ability to infect, deliver, and express transgenes into cells that are terminally differentiated and nondividing, a critical aspect when considering gene transfer into neuronal tissues *in vivo*. First generation Ad vectors are produced as demonstrated in Figure 13.3. Briefly, a shuttle plasmid containing the transgene of interest is cotransfected into an appropriate packaging cell line along with restriction enzyme-digested full-length Ad DNA. A recombination event between the shuttle plasmid and the full length virion DNA will result in the production and isolation of a recombinant Ad vector capable

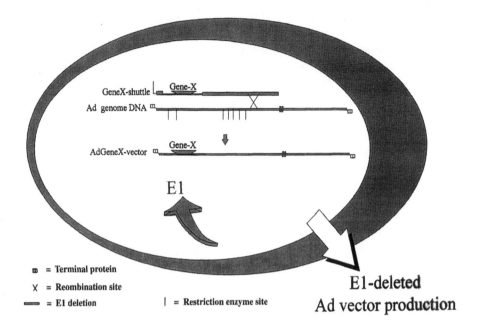

GeneX-shuttle Gene-X

Ad genome DNA

AdGeneX-vector Gene-X

E1

□ = Terminal protein

Χ = Reombination site

━━ = E1 deletion | = Restriction enzyme site

**E1-deleted
Ad vector production**

FIGURE 13.3
Schematic representation of E1 deleted Ad vector production. The 293 cell line supplies (*trans-complements*) the critical E1 gene activities deleted in the recombinant vector. Note that non-recombinant Ad viruses will also be produced at some frequency, but the restriction enzyme digestion of the cotransfected Ad vector genome decreases this possibility. Once the recombinant vector is isolated, it can be propagated by simple reinfection of the 293 cells.

of delivering the respective transgene to the various cell types described earlier, including postmitotic neurons. Once identified and clonally purified, the Ad vector can be serially propagated by ever repeated infection of increasing numbers of the packaging cells. Stocks of vector can approach concentrations of 10^{13} particles/ml after cesium chloride purification and concentration. The relative ease of production of Ad vectors is a critical benefit, since efficient, *in vivo* tissue transduction with any vector requires delivery of large amounts of the vector. The utilization of other virally based vectoring systems can be limited by this critically important characteristic, i.e., AAV-based vectors.

In vivo studies have repeatedly demonstrated the efficacy of Ad vectors to deliver and sustain expression of transgenes in neuronal tissues, *in vivo*. A number of early studies first demonstrated that neuronal cells could be transduced by Ad vectors carrying indicator (nontherapeutic) transgenes efficiently *in vivo*.[55-57] Subsequently, transduction of potentially therapeutic transgenes into animal models of neurogenetic disease have also recently shown evidence of significant impact upon the neuropathological disease process being investigated.[58,59] For example, an Ad vector transducing a glial cell line-derived neurotrophic growth factor (GDNF)-encoding transgene was stereotactically injected into the striatum of rats. Subsequently, the rats were lesioned to simulate the deficit of Parkinson's disease. This study demonstrated that Ad vector-derived GDNF significantly protected the neurons in the lesioned rats.[58] The administration of Ad vectors encoding genes active in the apoptotic pathways of neuronal cells could also be investigated in such a manner, in the previously described animal models of neuronal disease.

13.4.2.4.1 Limitations of Ad Vectors

Despite the relative ease of generating high titer stocks of recombinant Ad vectors, there are many short comings to the use of first generation Ad vectors that must also be considered. The most critical of these shortcomings is the fact that first generation Ad vectors are not completely replication defective. Therefore, Ad genes still resident in the vector genome have the potential for both expression and allowing vector genome replication, despite the presence of an E1 deletion.[60] This results in a number of problems, including interference with normal cellular processes as well as induction of a potent immune response *in vivo*.[61] The immunogenic potential *in vivo* appears to be lessened (but still significant) when Ad vectors are delivered to the CNS, due to the relatively "immune-privileged" status of this tissue. Expression of multiple Ad genes resident in the vector potentially could also obscure the impact of neuroapoptotic gene activity in the targeted cells.[62,63]

These shortcomings can potentially be overcome by the utilization of newer generation Ad vectors, with a significantly decreased potential to replicate, and/or express multiple Ad genes resident in the vector genome. Our laboratory has designed a new Ad vectoring system, as diagrammed in Figure 13.4. This system is based on our isolation of new Ad packaging cell lines that can transcomplement multiple Ad gene functions *in trans*, including the E1, E3, polymerase, and preterminal protein genes normally encoded by the Ad genome.[64,65] With these cell lines we have isolated second-generation Ad vectors that not only have a diminished ability to replicate, but that also have a significantly diminished ability to express multiple Ad genes still resident in the vectors.[60] In addition, the modified vectors have a significantly increased carrying capacity that can theoretically accommodate 11 to 12 kb of transgene sequences.[60] Thus the utilization of these "stealthy" second-generation Ad vectors will be very useful for transgene delivery, both in tissue culture and in *in vivo* investigations of apoptotic gene activities. The full extent of their usefulness will only be realized after extensive investigations in a variety of *in vivo* paradigms.

13.4.2.5 Adeno-Associated Virus Vectors

The adeno-associated virus (AAV) is a small human parvovirus that is dependent on concomitant adenovirus infection for its propagation, hence the name. Wild-type AAV is a single-stranded DNA virus that has a broad tissue tropism (including primary neuronal cell populations) that is not dependent *per se* on cellular replication. AAV infection is not associated with any human

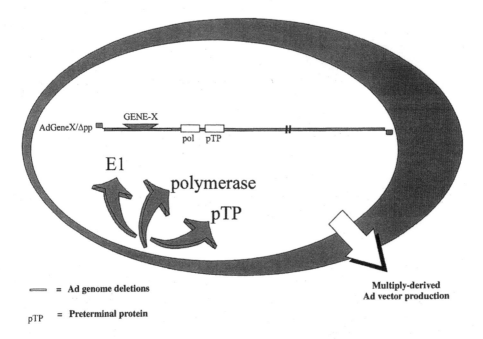

GENE-X

AdGeneX/Δpp

pol   pTP

E1

polymerase

pTP

Multiply-derived
Ad vector production

—— = Ad genome deletions

pTP = Preterminal protein

FIGURE 13.4
Schematic representation of multiple-deleted E1, polymerase, and preterminal protein Ad vector. This highly modified Ad vector can only be grown in cell lines capable of expressing (trans-complementing) each of the respective gene activities. Once the recombinant vector is isolated, it can be propagated by simple reinfection of the transcomplementing cell lines.

disease, and its genome is known to integrate specifically into a region of human chromosome 19. Each of these attributes has driven research exploring the full potential of AAV-mediated gene transfer. Figure 13.5 demonstrates the currently utilized method for production of AAV-based vectors. Multiple elements are required to generate an AAV vector, including cointroduction into human 293 cells of: (1) a plasmid containing the transgene of interest flanked by the AAV-derived 145-bp inverted terminal repeat elements (required for packaging of the genome), (2) a plasmid expressing the AAV-derived *cap* and *rep* genes, and (3) coinfection with an Ad virus that supplies several transcomplementing gene activities.

The Ad virus and the rep-cap plasmid DNA construct supply all the necessary functions required to package the ITR-flanked transgene DNA into AAV capsids. Directly after cell lysis, however, both AAV and Ad viruses are present in the cellular lysate; therefore heat inactivation and a purification step are required to ensure that the AAV vector preparation is not contaminated by the Ad helper virus. In order to amplify the AAV vector, the triple-transfection/infection step must be sequentially repeated in increasing numbers of cells, due to the dependence of the AAV vector genome on transcomplementing functions provided by the rep-cap plasmid and the Ad helper virus. This feature sometimes limits the ease of production of these vectors. Further-

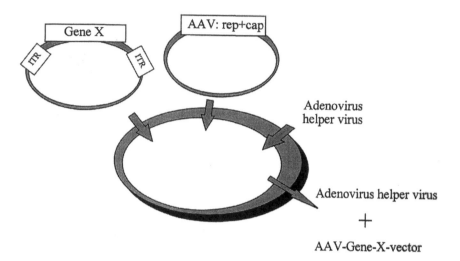

FIGURE 13.5

Schematic representation of AAV vector production. Two plasmid DNA molecules, one encoding the AAV-derived *cap* and *rep* genes, the second containing the AAV-derived ITR elements flanking the transgene, are cotransfected into 293 cells simultaneously infected with an adenovirus helper virus. Note that both Ad helper virus and recombinant AAV vector are produced; therefore purification of the AAV preparation is mandatory. Also, these steps must be repeated for each round of amplification of the AAV vector.

more, the repeated infection/transfection strategy can result in the rapid generation of wild-type AAV, another limitation to this vectoring system.[66]

AAV vectors have in several instances demonstrated the ability to transduce a number of transgenes into a variety of tissues *in vivo*, including muscle-, liver-, and nervous system-derived tissues.[67-69] In addition, after transgene delivery, expression of the transgene is in many instances not accompanied by a host immune response, unlike Ad vector mediated transfer of identical transgenes.[67]

Though technically challenging, the utilization of AAV vectors for transgene expression in neuronal cells *in vivo* has been repeatedly demonstrated. For example, an AAV vector expressing human tyrosine hydroxylase (TH) was injected into the denervated striatum of rats with substantia nigra lesions. Prolonged expression of TH was observed, and this expression was correlated with improved behavioral recovery in the treated rats.[68]

13.4.2.5.1 Limitations of AAV Vectors

There are several limitations of AAV vectors that must also be considered. The carrying capacity of these vectors is physically limited to 4.7 kb of transgene DNA; therefore some transgenes may not be physically accommodated (i.e., those utilizing larger-tissue specific enhancer/promoter elements). A more pressing issue is that many AAV preparations are contaminated by both Ad and wild-type AAV (via recombination between the transgene containing plasmid and the rep-cap plasmid) requiring careful purification steps to ensure purity of the final preparation.[70] However, improved methods of AAV purification have recently been described that include the utilization of a third plasmid encoding only the most important of Ad functions required for AAV production, eliminating the need for an Ad helper virus.[71] In addition, the utilization of plasmids that individually express either the *rep* or the *cap* genes may further reduce the risk of producing wild-type AAV during AAV vector production.[66,71] Clearly, the repeated demonstration of long-term persistence of AAV vector derived transgene expression in brain, liver, and muscle tissues will drive forward the refinement of this vectoring system for *in vivo* use.

13.5 Summary

The rapid accumulation of knowledge concerning the role of apoptosis in the normal and abnormal biology of the human nervous system will have great impact upon the understanding of a multitude of human neuropathological processes. Multiple animal models of these disease states, combined with the utilization of transgenic technology, and/or vectoring systems capable of somatic gene transfer, will contribute greatly to our understanding of, and potential successful intervention into many of these processes.

References

1. Niwa M, Rose SD, Berget SM: *In vitro* polyadenylation is stimulated by the presence of an upstream intron. *Genes Dev* 4:1552–1559, 1990.
2. Hoyle GW, Mercer EH, Palmiter RD, Brinster RL: Cell-specific expression from the human dopamine beta-hydroxylase promoter in transgenic mice is controlled via a combination of positive and negative regulatory elements. *J Neurosci* 14:Pt 1:2455–2463, 1994.
3. Caroni P: Overexpression of growth-associated proteins in the neurons of adult transgenic mice. *J Neurosci Methods* 71:3–9, 1997.
4. Rincon-Limas DE, Geske RS, Xue JJ, Hsu CY, Overbeek PA, Patel PI: 5'-flanking sequences of the human hprt gene direct neuronal expression in the brain of transgenic mice. *J Neurosci Res* 38:259–267, 1994.
5. Whyte DB, Lawson MA, Belsham DD, Eraly SA, Bond CT, Adelman JP, Mellon PL: A neuron-specific enhancer targets expression of the gonadotropin-releasing hormone gene to hypothalamic neurosecretory neurons. *Mol Endocrinol* 9:467–477, 1995.

6. Andra K, Abramowski D, Duke M, Probst A, Wiederhold KH, Burki K, Goedert M, Sommer B, Staufenbiel M: Expression of app in transgenic mice: a comparison of neuron-specific promoters. *Neurobiol Aging* 17:183–190, 1996.

7. Bowman BH, Jansen L, Yang F, Adrian GS, Zhao M, Atherton SS, Buchanan JM, Greene R, Walter C, Herbert DC et al.: Discovery of a brain promoter from the human transferrin gene and its utilization for development of transgenic mice that express human apolipoprotein e alleles. *Proc Natl Acad Sci USA* 92:12115–12119, 1995.

8. Onteniente B, Horellou P, Neveu I, Makeh I, Suzuki F, Bourdet C, Grimber G, Colin P, Brachet P, Mallet J, et al.: Cell-type-specific expression and regulation of a c-fos-ngf fusion gene in neurons and astrocytes of transgenic mice. *Brain Res Mol Brain Res* 21:225–234, 1994.

9. Min N, Joh TH, Kim KS, Peng C, Son JH: 5′ Upstream DNA sequence of the rat tyrosine hydroxylase gene directs high-level and tissue-specific expression to catecholaminergic neurons in the central nervous system of transgenic mice. *Brain Res Mol Brain Res* 27:281–289, 1994.

10. Morita S, Kobayashi K, Mizuguchi T, Yamada K, Nagatsu I, Titani K, Fujita K, Hidaka H, Nagatsu T: The 5′-flanking region of the human dopamine beta-hydroxylase gene promotes neuron subtype-specific gene expression in the central nervous system of transgenic mice. *Brain Res Mol Brain Res* 17:239–244, 1993.

11. Leconte L, Santha M, Fort C, Poujeol C, Portier MM, Simonneau M: Cell type-specific expression of the mouse peripherin gene requires both upstream and intragenic sequences in transgenic mouse embryos. *Brain Res Dev Brain Res* 92:1–9, 1996.

12. Hoesche C, Sauerwald A, Veh RW, Krippl B, Kilimann MW: The 5′-flanking region of the rat synapsin I gene directs neuron-specific and developmentally regulated reporter gene expression in transgenic mice. *J Biol Chem* 268:26494–26502, 1993.

13. Carroll SL, Schweitzer JB, Holtzman DM, Miller ML, Sclar GM, Milbrandt J: Elements in the 5′ flanking sequences of the mouse low-affinity ngf receptor gene direct appropriate CNS, but not PNS, expression in transgenic mice. *J Neurosci* 15:Pt 1:3342–3356, 1995.

14. Deuschle U, Pepperkok R, Wang FB, Giordano TJ, McAllister WT, Ansorge W, Bujard H: Regulated expression of foreign genes in mammalian cells under the control of coliphage T3 RNA polymerase and lac repressor. *Proc Natl Acad Sci USA* 86:5400–5404, 1989.

15. No D, Yao TP, Evans RM: Ecdysone-inducible gene expression in mammalian cells and transgenic mice. *Proc Natl Acad Sci USA* 93:3346–3351, 1996.

16. Christopherson KS, Mark MR, Bajaj V, Godowski PJ: Ecdysteroid-dependent regulation of genes in mammalian cells by a drosophila ecdysone receptor and chimeric transactivators. *Proc Natl Acad Sci USA* 89:6314–6318, 1992.

17. Figge J, Wright C, Collins CJ, Roberts TM, Livingston DM: Stringent regulation of stably integrated chloramphenicol acetyl transferase genes by *E. coli lac* repressor in monkey cells. *Cell* 52:713–722, 1988.

18. Izquierdo RE, Breese K, Jain S, Carestio D, Jung L, Figge J: Stringent regulation of human growth hormone expression in cultured murine C2C12 myoblasts by the *E. coli lac* repressor. *In Vitro Cell Dev Biol Anim* 31:71–76, 1995.

19. Gossen M, Bujard H: Tight control of gene expression in mammalian cells by tetracycline-responsive promoters *Proc Natl Acad Sci USA* 89:5547–5551, 1992.

20. Gossen M, Bonin AL, Bujard H: Control of gene activity in higher eukaryotic cells by prokaryotic regulatory elements. [review]. *Trends Biochem Sci* 18:471–475, 1993.

21. Furth PA, St Onge L, Boger H, Gruss P, Gossen M, Kistner A, Bujard H, Hennighausen L: Temporal control of gene expression in transgenic mice by a tetracycline-responsive promoter. *Proc Natl Acad Sci USA* 91:9302–9306, 1994.

22. Amalfitano A, Chamberlain JS: The mdx-amplification-resistant mutation system assay, a simple and rapid polymerase chain reaction-based detection of the mdx allele. *Muscle Nerve* 19:1549–1553, 1996.

23. Farlie PG, Dringen R, Rees SM, Kannourakis G, Bernard O: Bcl-2 transgene expression can protect neurons against developmental and induced cell death. *Proc Natl Acad Sci USA* 92:4397–4401, 1995.

24. Bernard R, Farlie P, Bernard O: Nse-Bcl-2 transgenic mice, a model system for studying neuronal death and survival. *Dev Neurosci* 19:79–85, 1997.

25. Tanabe H, Eguchi Y, Kamada S, Martinou JC, Tsujimoto Y: Susceptibility of cerebellar granule neurons derived from Bcl-2-deficient and transgenic mice to cell death. *Eur J Neurosci* 9:848–856, 1997.

26. Cenni MC, Bonfanti L, Martinou JC, Ratto GM, Strettoi E, Maffei L: Long-term survival of retinal ganglion cells following optic nerve section in adult Bcl-2 transgenic mice. *Eur J Neurosci* 8:1735–1745, 1996.

27. Greenlund LJ, Korsmeyer SJ, Johnson EM, Jr.: Role of Bcl-2 in the survival and function of developing and mature sympathetic neurons. *Neuron* 15:649–661, 1995.

28. Fukuchi K, Ogburn CE, Smith AC, Kunkel DD, Furlong CE, Deeb SS, Nochlin D, Sumi SM, Martin GM: Transgenic animal models for Alzheimer's disease. *Ann NY Acad Sci* 695:217–223, 1993.

29. Davies SW, Turmaine M, Cozens BA, DiFiglia M, Sharp AH, Ross CA, Scherzinger E, Wanker EE, Mangiarini L, Bates GP: Formation of neuronal intranuclear inclusions underlies the neurological dysfunction in mice transgenic for the HD mutation. *Cell* 90:537–548, 1997.

30. Zhou QY, Palmiter RD: Dopamine-deficient mice are severely hypoactive, adipsic, and aphagic. *Cell* 83:1197–1209, 1995.

31. Cote F, Collard JF, Julien JP: Progressive neuronopathy in transgenic mice expressing the human neurofilament heavy gene: a mouse model of amyotrophic lateral sclerosis. *Cell* 73:35–46, 1993.

32. Moechars D, Lorent K, De Strooper B, Dewachter I, Van Leuven F: Expression in brain of amyloid precursor protein mutated in the alpha-secretase site causes disturbed behavior, neuronal degeneration and premature death in transgenic mice. *EMBO J* 15:1265–1274, 1996.

33. Zhao J, Paganini L, Mucke L, Gordon M, Refolo L, Carman M, Sinha S, Oltersdorf T, Lieberburg I, McConlogue L: Beta-secretase processing of the beta-amyloid precursor protein in transgenic mice is efficient in neurons but inefficient in astrocytes. *J Biol Chem* 271:31407–31411, 1996.

34. LaFerla FM, Hall CK, Ngo L, Jay G: Extracellular deposition of beta-amyloid upon p53-dependent neuronal cell death in transgenic mice. *J Clin Invest* 98:1626–1632, 1996.

35. Czech C, Masters C, Beyreuther K: Alzheimer's disease and transgenic mice. *J Neural Transm Suppl* 44:219–230, 1994.

36. Kennel PF, Finiels F, Revah F, Mallet J: Neuromuscular function impairment is not caused by motor neurone loss in FALS mice: an electromyographic study. *Neuroreport* 7:1427–1431, 1996.

37. Jaarsma D, Holstege JC, Troost D, Davis M, Kennis J, Haasdijk ED, de Jong VJ: Induction of c-jun immunoreactivity in spinal cord and brainstem neurons in a transgenic mouse model for amyotrophic lateral sclerosis. *Neurosci Lett* 219:179–182, 1996.

38. Tu PH, Raju P, Robinson KA, Gurney ME, Trojanowski JQ, Lee VM: Transgenic mice carrying a human mutant superoxide dismutase transgene develop neuronal cytoskeletal pathology resembling human amyotrophic lateral sclerosis lesions. *Proc Natl Acad Sci USA* 93:3155–3160, 1996.

39. Burright EN, Orr HT, Clark HB: Mouse models of human CAG repeat disorders. *Brain Pathol* 7:965–977, 1997.

40. Burright EN, Clark HB, Servadio A, Matilla T, Feddersen RM, Yunis WS, Duvick LA, Zoghbi HY, Orr HT: SCA-1 transgenic mice: A model for neurodegeneration caused by an expanded CAG trinucleotide repeat. *Cell* 82:937–948, 1995.

41. Hennighausen L, Wall RJ, Tillmann U, Li M, Furth PA: Conditional gene expression in secretory tissues and skin of transgenic mice using the MMTV-LTR and the tetracycline responsive system. *J Cell Biochem* 59:463–472, 1995.

42. Amalfitano A, Chamberlain JS: The utilization of dystrophin-inducible mdx mice as a tool to assess the potential of gene therapy for Duchenne muscular dystrophy and other muscle diseases. *Am J Hum Genet* 59 (Suppl):A195, 1996.

43. Mayford M, Bach ME, Huang YY, Wang L, Hawkins RD, Kandel ER: Control of memory formation through regulated expression of a CaMKII transgene. *Science* 274:1678–1683, 1996.

44. Jiao S, Acsadi G, Jani A, Felgner PL, Wolff JA: Persistence of plasmid DNA and expression in rat brain cells *in vivo*. *Exp Neurol* 115:400–413, 1992.

45. Ono T, Fujino Y, Tsuchiya T, Tsuda M: Plasmid DNAs directly injected into mouse brain with lipofectin can be incorporated and expressed by brain cells. *Neurosci Lett* 117:259–263, 1990.

46. Cheng L, Ziegelhoffer PR, Yang NS: *In vivo* promoter activity and transgene expression in mammalian somatic tissues evaluated by using particle bombardment. *Proc Natl Acad Sci USA* 90:4455–4459, 1993.

47. Krisky DM, Marconi PC, Oligino T, Rouse RJD, Fink DJ, Glorioso JC: Rapid method for construction of recombinant HSV gene transfer vectors. *Gene Therapy* 4:1120–1125, 1997.

48. Linnik MD, Zahos P, Geschwind MD, Federoff HJ: Expression of Bcl-2 from a defective herpes simplex virus-1 vector limits neuronal death in focal cerebral ischemia. *Stroke* 26:1670–1674, 1995.

49. Fink DJ, Poliani PL, Oligino T, Krisky DM, Goins WF, Glorioso JC: Development of an HSV-based vector for the treatment of Parkinson's disease. *Exp Neurol* 144:103–121, 1997.

50. Ramakrishnan R, Poliani PL, Levine M, Glorioso JC, Fink DJ: Detection of herpes simplex virus type 1 latency-associated transcript expression in trigeminal ganglia by *in situ* reverse transcriptase PCR. *J Virol* 70:6519–6523, 1996.

51. During MJ, Naegele JR, O'Malley KL, Geller AI: Long-term behavioral recovery in parkinsonian rats by an HSV vector expressing tyrosine hydroxylase. *Science* 266:1399–1403, 1994.

52. Geller AI, Yu L, Wang Y, Fraefel C: Helper virus-free herpes simplex virus-1 plasmid vectors for gene therapy of Parkinson's disease and other neurological disorders. *Exp Neurol* 144:98–102, 1997.

53. Soares K, Hwang DY, Ramakrishnan R, Schmidt MC, Fink DJ, Glorioso JC: Cis-acting elements involved in transcriptional regulation of the herpes simplex virus type 1 latency-associated promoter 1 (lap1) *in vitro* and *in vivo*. *J Virol* 70:5384–5394, 1996.

54. Rasty S, Poliani PL, Fink DJ, Glorioso JC: Deletion of the S component inverted repeat sequence C′ and the nonessential genes U(S)1 through U(S)5 from the herpes simplex virus type 1 genome substantially impairs productive viral infection in cell culture and pathogenesis in the rat central nervous system. *J Neurovirol* 3:247–264, 1997.

55. Le Gal La Salle G, Robert JJ, Berrard S, Ridoux V, Stratford-Perricaudet LD, Perricaudet M, Mallet J: An adenovirus vector for gene transfer into neurons and glia in the brain. *Science* 259:988–990, 1993.

56. Davidson BL, Allen ED, Kozarsky KF, Wilson JM, Roessler BJ: A model system for *in vivo* gene transfer into the central nervous system using an adenoviral vector. *Nat Genet* 3:219–223, 1993.

57. Bajocchi G, Feldman SH, Crystal RG, Mastrangeli A: Direct *in vivo* gene transfer to ependymal cells in the central nervous system using recombinant adenovirus vectors. *Nat Genet* 3:229–234, 1993.

58. Choilundberg DL, Lin Q, Chang YN, Chiang YL, Hay CM, Mohajeri H, Davidson BL, Bohn MC: Dopaminergic neurons protected from degeneration by GDNF gene therapy. *Science* 275:838–841, 1997.

59. Lisovoski F, Akli S, Peltekian E, Vigne E, Haase G, Perricaudet M, Dreyfus PA, Kahn A, Peschanski M: Phenotypic alteration of astrocytes induced by ciliary neurotrophic factor in the intact adult brain, as revealed by adenovirus-mediated gene transfer. *J Neurosci* 17:7228–7236, 1997.

60. Amalfitano A, Hauser MA, Hu H, Serra D, Begy C., and Chamberlain JS,: Production and characterization of improved Adenovirus vectors with the E1, E2b, and E3 genes deleted. *J Virol* 72:2: p926-933, 1998.

61. Yang Y, Li Q, Ertl HC, Wilson JM: Cellular and humoral immune responses to viral antigens create barriers to lung-directed gene therapy with recombinant adenoviruses. *J Virol* 69:2004–2015, 1995.

62. Hardwick JM: Virus-induced apoptosis. *Adv Pharmacol* 41:295–336: 2, 1997.

63. Moore M, Horikoshi N, Shenk T: Oncogenic potential of the adenovirus E4 ORF-6 protein. *Proc Natl Acad Sci USA* 93:11295–11301, 1996.

64. Amalfitano A, Begy CR, Chamberlain JS: Improved adenovirus packaging cell lines to support the growth of replication-defective gene-delivery vectors. *Proc Natl Acad Sci USA* 93:3352–3356, 1996.

65. Amalfitano A, Chamberiain JS: Isolation and characterization of packaging cell lines that coexpress the adenovirus E1, DNA polymerase, and preterminal proteins — implications for gene therapy. *Gene Therapy* 4:258–263, 1997.

66. Allen JM, Debelak DJ, Reynolds TC, Miller AD: Identification and elimination of replication-competent adeno- associated virus (AAV) that can arise by non-homologous recombination during AAV vector production. *J Virol* 71:6816–6822, 1997.

67. Xiao XA, Li JA, Samulski RJ: Efficient long-term gene transfer into muscle tissue of immunocompetent mice by adeno-associated virus vector. *J Virol* 70:8098–8108, 1996.
68. Kaplitt MG, Leone P, Samulski RJ, Xiao X, Pfaff DW, O'Malley KL, During MJ: Long-term gene expression and phenotypic correction using adeno-associated virus vectors in the mammalian brain. *Nat Genet* 8:148–154, 1994.
69. Snyder RO, Miao CH, Patijn GA, Spratt SK, Danos O, Nagy D, Gown AM, Winther B, Meuse L, Cohen LK, Thompson AR, Kay MA: Persistent and therapeutic concentrations of human factor IX in mice after hepatic gene transfer of recombinant AAV vectors. *Nat Genet* 16:270–276, 1997.
70. Rolling F, Samulski RJ: AAV as a viral vector for human gene therapy. Generation of recombinant virus. *Mol Biotechnol* 3:9–15, 1995.
71. Xiao X, Li J, McCown TJ, Samulski RJ: Gene transfer by adeno-associated virus vectors into the central nervous system. *Exp Neurol* 144:113–124, 1997.

Index

A

Acetylcholine receptor, 18
Ac-Leu-leu-methioninal, 84, 216
Ac-Leu-leu-norleucinal, 84, 216
Acridine orange, 130, 132, 143–144
Acrolein, 157
Ac-YVAD-aldehyde, 85
Ac-YVAD-chloromethylketone, 88
Ac-YVAD-nitrile, 217
Adeno-associated virus vectors, 260–263
Adenovirus vectors, 258–260
Adrenalectomy, 26
Aging-associated motoneuron loss, 37
Alzheimer's disease, 4, 6, 38, 89
   evidence for apoptosis, 34
   genes, 34–35
   modeling, 34
7-Aminoactinomycin D exclusion, 193
7-Aminomethylcoumarin (AMC), 214
7-Amino-4-trifluoromethylcoumarin (AFC), 213, 214
Amyloid beta-peptide (AβP), 89
   NF-κB activation and, 118
Amyloidogenic protein, 6
Amyloid precursor polypeptide (APP), 34–35
Amyotrophic lateral sclerosis (ALS), 4, 6, 19, 27, 36–37, 38, 89
Antibody crosslinking, 107
Antioxidants
   mitochondrial ceramide-induced H₂O₂ production and, 114
   motor neuron disease and, 36, 37
Antipain, 219
AP-1, 116
Apaf-1, 5, 6, 74, 77, 79, 82, 83
Apolipoprotein E, 35
Apoptosis, 74, 116
   antineoplastic strategy, 53–55, *See also* Cancer
   background levels, 25
   Caenorhabditis elegans model, 6, 14–15, 24, 74, 154, 206, *See also* specific "ced-" genes

CNS development, *See* Developmental cell death
detection methods, 129–148, *See* Viability assays; specific histochemical methods
flow cytometry and, 186, *See* Flow cytometry
genes, *See* specific genes
interpretation of negative results, 199
macromolecular synthesis, 84–85
mechanisms in neurological disorders, 6–8, *See* Neurodegenerative disorders
methodologies, 8, *See* Flow cytometry; Ultrastructural analysis; Viability assays; specific methods
molecular mechanisms, 4–6
morphological changes, 130
necrosis-apoptosis continuum, 27–28, 156, 165
necrosis morphology/biochemistry versus, 130
neuronal development, *See* Developmental cell death
polypeptide cleavage, 208, 210–211
proteolysis as evidence of, 208
terminological distinction, 13
types of developmental cell death, 155
Arabinofuranosylcytosine, 106
Astrocytes, HIV infection, 62, 65
Astrocytomas, 50
Ataxin-1, 4
Ataxin-3, 4
ATP, radiolabeled, 236–237, 241, 243
Atrophin-1, 4
Autophagic cell death, 155
Avidin-peroxidase, 135

B

Bacillus cereus, 110
Bacterial sphingomyelinase, 110
Baculovirus, 85
Bad, 67
Basal ganglia, 66–67